Lecture Notes on Coastal and Estuarine Studies

Lecture Notes on Coastal and Estuarine Studies

Managing Editors:
Malcolm J. Bowman   Richard T. Barber
Christopher N.K. Mooers   John A. Raven

13

Seawater-Sediment Interactions in Coastal Waters

An Interdisciplinary Approach

Edited by Jan Rumohr, Eckart Walger and Bernt Zeitzschel

Springer-Verlag Berlin Heidelberg GmbH

ISBN 978-3-540-17571-1   ISBN 978-3-662-02531-4 (eBook)
DOI 10.1007/978-3-662-02531-4

© Springer-Verlag Berlin Heidelberg 1987

Originally published by Springer-Verlag Berlin Heidelberg New York in 1987.

2131/3140-543210

Principal Contributors

BRIGITTE BABENERD, Institut für Meereskunde an der Universität
    Kiel, Düsternbrooker Weg 20, D-2300 Kiel, FRG

WOLFGANG BALZER, Institut für Meereskunde an der Universität
    Kiel, Düsternbrooker Weg 20, D-2300 Kiel, FRG

LUTZ-A. MEYER-REIL, Institut für Meereskunde an der Universität
    Kiel, Düsternbrooker Weg 20, D-2300 Kiel, FRG

VICTOR SMETACEK, Institut für Meereskunde an der Universität
    Kiel, Düsternbrooker Weg 20, D-2300 Kiel, FRG

GEROLD WEFER, Geologisch-Paläontologisches Institut der Universität Kiel,
    Olshausenstrasse 40, D-2300 Kiel, FRG

FRIEDRICH WERNER, Geologisch-Paläontologisches Institut an der Universität
    Kiel, Olshausenstr. 40, D-2300 Kiel, FRG

Acknowledgment

The editors wish to acknowledge the help and assistance of all who shared the long and arduous trail starting from a 750 page first draft collected within a few weeks several years ago to this final report:

Sebastian A. Gerlach and Jörn Thiede who read the entire first draft and reviewed the final manuscript contributed with helpful criticism.

Gabriele Kredel together with Dagmar Barthel and Sally Allendorff suffered nerve-racking months and raised limitless endurance in overcoming linguistic and factual obscurities. To them we extend our most sincere gratitude.

Heide Schomann and Alison Walker carried out the typewriting and numerous corrections.

Tom Noji and Petra Stegmann assisted in reading the proofs.

Finally the editors thank the authors for their patience when progress was difficult.

Contents

INTRODUCTION

During the late sixties, the marine scientific community was becoming increasingly aware of the necessity of conducting process-oriented research on specific "problem areas". It was assumed that the results of such detailed analyses would provide an explanatory framework for the descriptive data accumulating from the extensive surveys of the oceans at large that had dominated marine science up to that period. The physical, chemical and biological interaction between the ocean and the sediments was identified as one of the most important interdisciplinary problems at the 1969 meeting of the Intergovernmental Oceanographic Commission. In the same year, a group of scientists from Kiel University - representing the five disciplines: physical, chemical, geological and biological oceanography as well as applied physics - combined forces and, in 1970, submitted a comprehensive proposal to the German Research Foundation (DFG: Deutsche Forschungsgemeinschaft) under the title "Interaction Sea-Seabottom" ("Wechselwirkung Wasser-Meeresboden"). The professors G. Dietrich, G. Einsele, G. Hempel and E. Seibold were the chief initiators of this project. It addressed two themes:
- the relationship between water movement and sediment structure and,
- the interaction between the chemical regime and the organisms at the sediment surface.

An important incentive for this interdisciplinary undertaking was provided by the Special Research Programme (SFB: Sonderforschungsbereich) launched by the DFG with the expressed aim of encouraging interdisciplinary research in German universities. The Special Research Project (SFB 95), with an initial contingent of 31 scientists, was launched in 1971.

Long-term, large-scale interdisciplinary research must be organized within a formal structure that can have considerable bearing on the quality and direction of work. We therefore present a brief description of the organization of the SFB here. In accordance with the recommendations of the DFG, the SFB was democratically organized and divided into project areas and subprojects; the latter were temporary groupings addressed at specific problems. Proposals for funding were written up at the subproject level which thus represented self-contained units. The thrust of research of each subproject was, however, discussed within the framework of the SFB at two higher levels of hierarchy: a) the planning commission comprising one representative of each subproject and several elected members and b) the general assembly of all members of the SFB. The interests of each discipline were represented in all the subprojects by double or even triple membership of those persons representing research fields that were most in demand. The major decisions were made by majority vote in the general assembly that comprised all involved scientists; the general assembly also elected, for two year terms, a speaker from its ranks who represented the SFB in various external and university boards and was also responsible for running the organisation.

The speaker was assisted in his task by a scientific secretary. Meetings of the assembly were prepared by the planning commission whose main function was to coordinate the subprojects and prepare the proposals for funding. The latter had to be sanctioned by the general assembly. Proposals were written up for three year periods and it might be mentioned that approximately two-thirds of the total funds were spent towards salaries of scientists and technicians. The project areas encompassed the fields of sedimentology/hydrography, biology/chemistry and oceanographic metrology. These rable regrouping of the subprojects took place in accordance with shifts in research emphasis.

Research conducted by the SFB 95 can be divided broadly into 2 successive six year phases. During the first phase, the various subprojects launched their research programmes somewhat independently, with the hope that coalescence of the different approaches would occur on the basis of suitable mathematical models that were being developed concomitantly. The desire to combine the results of these individual studies into a unified model of sea/seabottom interaction was very strong as evidenced by the frequent and broad interdisciplinary discussions. Considerable effort was spent on developing a coherent programme and growth was planned with the aim of filling in gaps rather than strengthening areas already represented. By the second three year period, the ranks of SFB members had swelled to 58 scientists of which 25 were employees of the SFB. The pioneering, "grand-design" mood of the early years is clearly discernible in an introductory account of the SFB 95 given by HEMPEL (1975). However, by the mid-seventies, it became apparent that the goal of developing quantitative models linking together all the aspects of sea/seabottom interaction was a remote and increasingly vague proposition. In the second phase, attention was thus shifted to the interdisciplinary study of individual key processes whose importance was realised in the course of the early research.

This reorganisation of research strategy did not, of course, involve the various groups to the same extent. Those groups still grappling with methodological problems were naturally more reluctant to move on to new fields than others who were fortunate enough to have at their disposal an adequate suite of methods commensurate with the demands of their respective research areas. The framework but not the intensity of interdisciplinary communication changed with development of the SFB; this was simply due to the fact that in the second phase, more time was spent on discussing specific results of ongoing investigations than on debating the more general aspects of interdisciplinary strategy. The forum for scientific discussions was shifted from the general assembly to weekly seminars which were well attended but, because the various presentations were not of equal interest to all groups, the entire SFB gathered together less frequently than in the early years. The unified front of the first phase, where most members were aware of all ongoing projects within the SFB, broke up somewhat in the second phase, with interdisciplinary, albeit smaller groups forging ahead in different directions. Another important impetus for the phase change came from the younger scientists who started their scientific careers within the SFB and were hence well acquainted with the problems and potentials of interdisciplinary research but were now more interested in focussing on specific questions than on grand but elusive syntheses.

We consider it worthwhile to recount the development of the SFB in somewhat greater detail than the above outline because our own history reflects, to a considerable extent, that of the marine scientific community in general. Progress here has also been very unequal; many of the individual disciplines are still grappling with both methods and concepts and frequently have problems in achieving a harmonious inter-action between these two legs on which science progresses. Many scientists shy away from interdisciplinary research because they feel that it amounts to tying one of one's own wobbly legs to another possibly even more erratic one. Two legs can certainly better than three, but we have found that in the case of inherent weakness, which is more widespread than admitted, tying legs together can result in mutual support even if locomotion is slow initially. At the risk of trying the reader's patience, we stretch the analogy of the legs by characterising the SFB phases in the following manner: the first phase was a search for the race track and for partners among those assembled there, with similar predilections or even idiosyncrasies (the personal aspects involved here cannot be ignored); the <u>common</u> goal was quantification of the dominant processes and the development of comprehensive models. Thereafter, groups and individuals who found they shared <u>particular</u> goals and could coordinate their locomotion patterns effectively, joined together and broke away in different directions and at different speeds. The "locomotion patterns" were determined by the compatibility of methods and the "direction" by specific problems within the general field of sea-seabottom interaction.

The first phase

Although a sweeping theoretical framework for coordinating all the work of the SFB was already established at its inception, implementation of the plans along a broad front proved difficult. It was decided early that field research was to be given first priority, the only large land-based experimental set-up being an 18 m flume. Simply because of proximity, Kiel Bight was selected as the natural location for field work, although research in other, preferably contrasting areas, was envisaged from the start. The presence of a well-trained, enthusiastic team of scuba-divers, most of whom were also scientists, proved to be an extremely important asset because it provided the prerequisite for conducting in situ experiments. Such experiments were planned by several groups and all necessitated continuous deployment of enclosures, instruments or both. Because of the heavy shipping traffic and trawl fisheries in Kiel Bight, such research could only be carried out in an undisturbed locality. To this end, an area of 400 m x 900 m with depths ranging from 7 - 20 m was selected, marked by surface buoys and officially declared a restricted zone open only to research. This experimental site was dubbed the "Hausgarten". Its usefulness has resulted in its continued maintenance and experiments are still carried out there.

Project Area A: Sedimentology and near-bottom water movement

Interaction between water movement and the sediment surface was the basic theme of this project area and membership comprised primarily hydrographers, geologists and physicists with participation of some zoologists. Field studies were carried out in selected areas - the mud-flat regions of the North Sea and in the Channel System of the Belt Sea (Western Baltic) - characterized by strong currents. The latter are of a tidal nature in the North Sea whereas in the Channel System, they are mainly wind-driven and tend to fluctuate at longer time-scales.

This project area combined complementary strategies: one was to predict texture and structure of the sediment and the morphogenesis of its surface from the hydrodynamics of the overlying water and the other was to draw conclusions on the hydrodynamics from detailed mapping of the sediment surface. The former strategy was based on flume experiments carried out with sediments of specific grain size ranges under well-defined conditions and also on the results of short-term, high-resolution field measurements of near-bottom currents and sediment transport. The intended feedback between field observation and laboratory experiment was achieved to a reasonable approximation in an investigation on the origin of current parallel "comet marks". How-ever, with regard to sediment entrainment conditions, a wide gap between the results of flume experiments and their use in evaluating field observations had to be recognized. It was not until the last year of the SFB that an experiment studying the conditions of the formation of residual sediments provided results that indicated the direction along which this gap could possibly be narrowed.

The stumbling block in these investigations proved to be the characterization of the turbulence field at the water-sediment interface. Field measurements of near-bottom turbulence patterns were carried out at first with hot-wire velocimeters and later with a micropropeller system. Both these instrument types, however, proved too trouble-prone for long-term in situ deployment. In the flume experiments Laser-Doppler velocimetry proved to be an invaluable tool.

The various complexities resulting from organism colonization of the sediments were purposely excluded from the flume experiments. The aim was to focus entirely on the physics of sediment entrainment which by itself is highly complex.

Mapping of the Channel System was carried out with sidescan sonar. These observations revealed the presence of ripples of various sizes whose distribution could be related to past current fields. An important finding was the critical role of short-term events in shaping large-scale features of the sediment surface. Recolonization of the environment following such events occurs in characteristic patterns that, particularly in the case of molluscs, provide a useful indicator for reconstruction of the hydro-dynamic history. The growth rate of macrobenthos and the production of carbonate on virgin substrates, such as would appear after large-scale sediment upheaval, was monitored in floating trays positioned in the "Hausgarten". This work was carried out in close collaboration with zoologists within the framework of the biological/chemical project area dealt with below.

Another aspect of research within this project area dealt with the geochemistry of the sediment and its pore waters in relation to redox state. The Channel System was an ideal site for such studies because of periodic anoxia in some of the more isolated basins. Heavy metal and radioisotope distributions provided interesting information on benthic boundary layer processes as well as on the history of sedimentation in the Channel System of Kiel Bight and the open Baltic.

Project Area B: Shallow-water ecosystems

The biological/chemical project area was by far the largest and hence the most hetero-geneous of the three groupings. This diversity was partly due to prevailing differences in the state of the art of the various branches of marine biology and chemistry involved but was also a result of different backgrounds of the respective scientists. An attempt was made to achieve agreement and collaboration between these branches within the con-ceptual framework of energy flow and cycling of matter. A detailed compartmental model of a shallow-water ecosystem was used as a basis to coordinate the investigations of the various sub-groups. The eventual aim was to convert the data generated by these sub-groups into deterministic mathematical models of the Kiel Bight ecosystems. The ODUM (1972) approach was agreed upon as being the most suitable for this purpose. With this end in mind, it was decided to conduct as much research as possible within the "Hausgarten" area so as to achieve regional compatibility between the results of the different approaches represented within the SFB. Further, the data were to be provided in carbon or caloric units. An underlying assumption of this approach was that the structure of the ecosystem, i.e. the dominant functional compartments as represented by trophic levels in the various systems, and its driving forces were known and all that was required for development of predictive models was quantification of the main flows of energy and matter over an annual cycle.

The research carried out by this heterogeneous grouping during the first six years of study can be loosely summarized under the following headings; some of these results have been presented in the following chapters.

a) Factors affecting macrophytobenthic structure, particularly the influence of depth and type of substrate, were deduced from observations of specific areas by scuba divers and underwater TV. Attempts were made to estimate production and loss of biomass from algal beds in the "Hausgarten" area; seasonality in input of organic matter to the ecosystem was also studied. It proved difficult to typify stable phytobenthic communities in relation to depth and substrate in this region. Species distribution appeared to be random and was described as a mosaic strucure. An additional important finding was that an errant vegetation component consisting of loose algal mats drifting on the bottom contributed significantly to total algal biomass of the region. The structural complexity of macroalgal distribution in time and space rendered simple quantification as demanded by the model extremely difficult. Somewhat in contrast, work carried out on the fairly homogeneous and well defined Zostera meadows yielded data on production, grazing, remineralisation and input to the rest of the system that were more commensurate with the demands of the ecosystem model. Research effort in this field was reduced during the following six year period.

b) The role of benthic algae in functional interrelationships at the sediment surface was studied in in situ enclosed flow-through systems (the "tunnel" experiments) by means of short-term monitoring of oxygen and nutrient fluxes. The effect of macro-algal exudation on microbial growth was also followed. Intense interdisciplinary investigations with these tunnels were carried out on several occasions and consider-able problems were encountered with simply maintaining the experimental set-up under natural conditions. Further difficulties arose in data interpretation: temporal variations in nutrient concentrations, for instance, were surprisingly large and proved difficult to relate to the individual processes comprising the complex interrelationships within the enclosed communites. The correlation between the light regime and oxygen production, however, provided reasonably reliable estimates of net production.

c) Annual cycles of the pelagic system in terms of nutrient concentrations and biomass, production and species composition of phytoplankton and zooplankton in relation to the environment were recorded at weekly intervals in the 20 m water column of the "Hausgarten". The emphasis in these studies was laid on the role of sedimentary nutrient input to the water column on the one hand and the sedimentation of particulate matter on the other. The pelagic system is spatially more homogeneous than coastal benthic systems and is therefore more amenable to modelling exercises; however, budgetary analyses and attempts at quantitative modelling were not very successful. The short-term field observations provided many unexpected results which clearly indicated that existing knowledge of the processes governing structure and function of pelagic systems was by no means sufficient to warrant construction of predictive models. The temporal dynamics of the pelagic community and the importance of event-scale processes, such as mass sedimentation of the spring and autumn blooms, formed the basis for this critical re-evaluation.

d) In the "plankton tower" experiments, budgetary analyses of the cycling of matter between pelagic and benthic systems were conducted on the basis of daily measurements over several weeks on 10 m enclosed water columns. Planktologists, microbiologists and applied physicists collaborated in the experiments; the experimental set-up did not permit simultaneous monitoring of the benthos. The aim of the plankton tower experiments was to study processes within a captive water column where horizontal advection could be definitely ruled out. The plastic enclosures, extending from the surface to the sediment and including both these interfaces, were suspended within a 16 m tall steel framework - the "plankton tower" - standing at 10 m depth in the "Hausgarten". Maintenance of the plastic enclosure under the given hydrographical conditions proved a difficult task and it took much effort before a robust, leak-proof, experimental set-up could be developed. However, a leaky enclosure of one experiment revealed the presence of a previously unsuspected process of considerable bearing to the dynamics of shallow-water systems: flushing of pore water by density displacement and the resulting significant input of nutrients to the water column and oxygen to the infaunal benthos.

e) "Bell jar" experiments were carried out in the "Hausgarten" to study sedimentary nutrient release in relation to redox conditions of the bottom water. The bell jar used in the experiment covered 3.1 m² sediment surface and contained 2 m³ of water. Short-term monitoring of the enclosed water body was rendered possible by in situ sensors and an automatic sampler that collected water samples for analysis at preset intervals. The instrumentation was developed by applied physicists of the SFB.

During simulated transitions from oxic to anoxic conditions, the consumption of successively utilized oxidants was quantitatively related to the release of nutrients from the sea bed. After the onset of anoxic conditions, the sediment releases enormous amounts of nutrients to the bottom water. A close relationship between Eh, pH and phosphate and ammonia release was found; silicate release on the other hand was not Eh dependent. The various factors influencing nutrient concentrations in bottom water layers could be traced by comparing the results of bell jar deployments in different seasons and on different sediment types.

f) Microbiological studies in Kiel Bight addressed a variety of topics. Because of the central role of bacteria in ecosystems, close collaboration with other groups was necessary and microbiologists participated in the "tunnel", the "plankton tower" and the "bell jar" experiments. Bacterial production and remineralistion, with particular emphasis on the interaction between heterotrophic microorganisms and various biotic and abiotic factors at the sediment-water interface, were studied in these experiments. Considerable effort was simultaneously spent on the improvement of methodology. Bacterial breakdown of various naturally occurring substrates: macroalgal exudates and tissue, sedimenting material of planktonic origin, were studied as well as colonization of sediment particles. Another topic addressed was nitrification which was shown to occur primarily at the sediment surface. The microbiological group was substantially strengthened in the second phase of the SFB.

g) Carbonate production experiments on suspended platforms filled with sediments of differing grain size were carried out at different water depths in the "Hausgarten". Some results have been presented in Chapter 6 and will not be repeated here. This experiment was also used to study colonization and biomass production of meiobenthos and macrobenthos and proved to be a fruitful avenue of research in this difficult field. However, the experimental set-up was rather cumbersome to handle and further experiments were carried out on platforms standing on the sea bed. The time of deployment was of critical importance in determining colonization because of seasonality in the availability of meroplanktonic larvae. Another drawback was that the platforms concentrated organic material from the surrounding water and were hence not representative of an equivalent area of natural sediment. Thus, macrobenthic biomass production in suspended substrates was much higher per unit sediment surface than primary production in the overlying water column.

h) Foraminiferal carbonate production was studied over two annual cycles at selected sites in the "Hausgarten" area. These studies, carried out in close cooperation with other biological groups, revealed that reproduction of benthic foraminifera is dependent on a variety of factors of which the abiotic ones appear to play the major role. The growth period extends from 3 - 5 months and only a small portion of the carbonate produced is eventually preserved, the fate of most shells being

mechanical breakdown and dissolution. These processes determining the contribution
of foraminifera shells to sediments of Kiel Bight are controlled by several factors of which intensity of water movement, oxygenation, sedimentation rate and
grain size range are the most important.

i) Zoobenthos/fish relationships were studied in the field and under in situ experimental conditions. In these studies, accent was laid on seasonal dynamics of community structure and biomass of various depth zones; the experiments involved
colonization of virgin sediments and also caging of natural communities to study
the effects of predator exclusion, in this case demersal fish. Some of the results
of these studies have been presented in Chapter 3 and we shall only touch on some
of the problems encountered with the experimental set-ups here. The rationale for
the experiments resulted from the problems associated with estimating production
rates from the population dynamics of natural communites. The latter is also a very painstaking undertaking. One of the experimental problems encountered has been
mentioned above - the extraordinarily high production on virgin substrates exposed
above the seabottom. However, distinct successional patterns in dominant organisms - from crustaceans to polychaetes and finally molluscs - were observed over
a three year period. The cage experiments did not function as conceived but provided valuable insights, nevertheless. The biggest problem was the rapidity of fouling, particularly due to wandering algal clumps (errant flora) that smothered the
cages and resulted in anoxic conditions at the sediment surface. This provided an
opportunity to study macrobenthic succession in transition from oxic to anoxic
conditions. The cages also served as a refuge for smaller fish (gobiids) whose increased predation pressure also effected succession patterns of the benthos. The
structure of the demersal fish populations and their various feeding habits were
also studied intensively.

j) The structure of zoobenthic communities was studied in differing environments: a
Zostera meadow, artificial hard substrates and stagnant mud sediments. Grazing of
Zostera is insignificant in Kiel Bight and this biomass is utilized by heterotrophs
only after leaf-shedding in autumn. A succession of Balanus to Mytilus was observed on the artificial hard substrates and extraordinarily high growth rates were
recorded. One m² of Mytilus is capable of utilizing the primary production of
75 m^3 of water column. Benthic community structure in mud sediments was controlled
by the frequency and duration of anoxia in the overlying water. Interannual variation in community structure, particularly in the case of macrofauna and ciliates,
was hence directly a function of the hydrography. Functional aspects of zoobenthic
ecology were studied in the "tunnel" experiments. The lack of adequate methods for
measuring community activity under fluctuating oxygen concentrations proved a
serious hindrance for these studies.
An assessment of the seasonal and regional distribution of meiobenthos in Kiel

Bight showed that the type of sediment was the major determinant of meiobenthic composition and biomass.

k) The physiological ecology of zoobenthos was studied on selected species under laboratory conditions. The influence of temperature, salinity and oxygen on various metabolic levels was addressed. Metabolic rate and its temperature dependence decreased successively in the molluscs Abra alba, Macoma calcarea and M. baltica. The physiological characteristics of these organisms provide some insight on their distribution patterns in Kiel Bight. Growth and chemical composition of Nereis spp. in relation to the quality and quantity of food supply was also studied.

Project Area C: Metrology, instrumentation and data processing

This project area was exclusively engaged in developing instruments, measuring systems and data processing facilities for the physical, geological, chemical and biological disciplines of the SFB. Therefore, collaboration with the other two project areas necessarily had to be intense. Of the instruments developed, the multisample sediment trap (planktology) and the bell jar stirrer and multisampler (chemistry) deserve special mention. Major effort in this area was, however, focussed on development of continuously profiling sensors and their combination into compact data acquisition systems tailored according to the needs of the various experiments. An automatic profiler system recording pressure, temperature, conductivity, oxygen, pH, irradiation and horizontal attenuation of red light was built for the "plankton tower" experiments and a similar system measuring temperature, conductivity, oxygen, pH and Eh was deployed in the bell jar. The in situ systems were in direct radio contact with the institute; thus, these systems could be operated by remote control and data were available instantaneously for inspection. The "plankton tower" proved an ideal platform for this purpose. The highly sophisticated measuring systems operated successfully above water but immense problems arose with maintenance of the sensor packages positioned underwater. Leakage, fouling and corrosion, particularly of the more delicate sensors during continuous in situ deployment, resulted in serious interruptions of the data stream. For logistical reasons, retrieval and repair of the instruments took several days, and in the earlier experiments, particularly those in the "plankton tower", only limited data sets could be acquired. Another problem that arose was associated with the time and space scales of the relevant processes. It became increasingly apparent that the major processes affecting change in the systems were driven on an event scale and that more frequent monitoring was desirable during the comparatively brief periods when significant changes occurred. However, gearing the data acquisition systems to such events proved a difficult task.

Members of this project area were also involved with the hydrodynamics of the sedi-

ment/water interface and the development of instruments for measuring particle and
water transport, both in the flume and in field experiments.

The second phase

The shift in emphasis that occurred during the transition from the first to the second six-year phase, explained on a general level above, was a result of intense strategy discussions carried out within the three project areas and also at the entire SFB
level. The most fundamental regrouping occurred in the biological/chemical project
area where four new subprojects were formed:
a) turnover of material in tropical and sub-tropical shallow waters,
b) structural dynamics of benthic communities,
c) function of benthic organisms in exchange processes at the benthic boundary layer
   and
d) biogeochemical exchange between water column and sediments.
The difficulties encountered with synthesizing the data obtained from the many and
disparate approaches of the first phase into a general model of the Kiel Bight ecosystem were an important incentive for the regrouping. Attention was accordingly focussed on those subsystems which appeared to be the most important in cycling of
matter within the total ecosystem. As the biggest problems were those associated with
methodology, the new groups reflected coalitions of scientists with mutual interests
and compatible data. The most serious gaps in data coverage were identified in the
fields of organic chemistry and bacteriology; these fields were accordingly strengthened during the second phase. It also became increasingly clear that in situ expe-
rimentation in the marine environment was a formidable undertaking. In some cases,
the artifacts introduced by the experimental set-up necessitated reformulation of the
questions posed at the outset; however, unexpected developments sometimes led to new
insight into the nature of the processes under study. In all cases, interpretation of
the data within the framework of energy flow through trophic levels of the Kiel Bight
ecosystem proved to be a much more complex task than anticipated. The shift in research emphasis also occurred in the metrology project area where a new group dealing
entirely with marine optics was established.

The nature of the Kiel Bight system as a transition area between two vastly differing
water bodies - the North and Baltic Seas - rendered the general applicability of the
results to other shallow water systems questionable. For instance, the low salinity
and pH of Kiel Bight depress carbonate accumulation significantly, thus fundamentally
affecting the geochemical processes at the sediment surface. It was therefore decided
to study the same processes addressed in the Kiel Bight system in other, more typically marine systems with carbonate sediments. Harrington Sound in Bermuda and Hilutan-

gan Channel near Cebu City, Philippines, were selected for conducting this type of comparative research. Other processes, particularly those pertaining to production, breakdown and sedimentation of organic matter were studied in Antarctic waters. All three project areas were involved in the investigations conducted in areas outside Kiel Bight.

It is impossible to summarize briefly and also impartially the results obtained during the second phase of the SFB. Many of the new results were generated by significant improvement of older methods but also by introduction of entirely new methodology. This applied particularly to the fields of near-bottom sediment transport, microbiology, benthic metabolism and analytical organic chemistry. Because of the overriding importance of methodology in marine science, it was our intention to devote an entire chapter solely to this topic. However, the constraints imposed by space led us to abandon, albeit reluctantly, this idea. Many, but by no means all of the results of the second phase of the SFB have been presented in the following chapters; therefore, we shall merely provide brief outlines of the contents of these chapters here.

Chapter 1 sets the stage by providing a detailed description of the topography and sediment distribution of Kiel Bight.

Chapter 2 deals with processes in the pelagic system of Kiel Bight and is divided into two sections. In the first section, discrete phases of the annual cycle of the pelagic system are identified on the basis of nutrient input to the water column and its effect on primary production, zooplankton grazing and sedimentation of biogenic matter. An important finding of far-reaching consequence for the other disciplines was that major sedimentation of organic matter occurred following phytoplankton blooms as a result of mass sinking of cells and fresh phytodetritus. In contrast, copepod grazing tended to retard loss of particles from the pelagic system. Thus, the annual cycle of sedimentation comprised brief periods of heavy sedimentation during late spring and autumn separated by a long summer phase of low and fairly constant sedimentation rates.

The second section of this chapter deals with the relationship between organic substrates and heterotrophic microbiological activity. As the methodology on which these results are based was developed after completion of the work described in the first section, it was not possible to combine the data.

Chapter 3 consists of several sections devoted to functional and structural aspects of benthic biology in Kiel Bight. The first section deals primarily with abundance, biomass and production of the macrofauna but also includes the meiofauna. In a second section, the benthic response to annual patterns of sedimentation are dealt with. It is shown that mass sedimentation of the spring and autumn blooms elicit an immediate response in metabolic activity of the benthos in terms of heat production. Thus the annual cycle of benthic metabolism closely matches that of sedimentation. A final section addresses the role of bacteria in the Kiel Bight sediments. This section deals with qualitative and quantitative aspects of the colonization of sediment particles by bacteria and goes on to discuss seasonal and diurnal fluctuations of bacterial populations. Some estimates of bacterial production are also included.

Chapter 4 examines the geochemistry of the sediments and the overlying water with special reference to diagenesis and exchange processes. This chapter deals with the physico-chemical aspects of input and composition of organic matter via sedimentation, its breakdown on the sediment surface and the release of nutrients to the water column. The results of the bell jar experiments and in vitro investigations of manipulated sediments are presented in this connection. Early diagenesis of organic matter in deeper sediment strata is also dealt with. These and other data are combined and an attempt is made to balance the cycling of organic matter through the benthic system. The dependence of trace element chemistry, such as iron and manganese, on these primarily biological processes is pointed out and the factors leading to dissolution and formation of heavy metal concretions are discussed.

Chapter 5 is a detailed account of the work carried out by the project area: water movement and sediments. This is a long chapter embracing a very wide range of processes that are dealt with both on a theoretical as well as a locality-specific footing. The bulk of this chapter is concerned with sediment transport: one part deals with the hydrodynamics of the boundary layer and the other draws conclusions on hydrodynamics from sedimentological findings. Estimates of the range of bottom shear-stress in relation to wind fields were obtained from a combination of long-term with medium-term, multisensoral measurements of currents in a site of the Channel System of Kiel Bight. These results are compared with others obtained from short-term, high-frequency, three-dimensional flow measurements conducted in the boundary layer of a more accessible site. The effect of surface waves on sediment structure was also studied. These field measurements are discussed in the light of critical erosion velocities determined in the flume experiments. In other sections of this chapter, sediment distribution patterns in Kiel Bight, both on abrasion platforms as well as in the Channel System, were mapped and morphological features related to long-term and event-scale hydrodynamic processes. The role of bioturbation in disturbing sediment lamination in Kiel Bight is also discussed.

Chapter 6 is devoted to production and accumulation of biogenic carbonates in the widely differing environments of Kiel Bight and Harrington Sound, Bermuda. Carbonate production in Kiel Bight is in the order of grams per m² and year, but it is two orders of magnitude higher in Harrington Sound. Similar vertical sections of the saturation

state of the water with respect to carbonate minerals were found in both areas although the degree of saturation was much lower in Kiel Bight; hence, accumulation rates of biogenic carbonate in Kiel Bight are low whereas almost all the carbonate
produced in Harrington Sound is preserved. The stable isotopes of the dominant groups
of calcareous organisms showed that the shells contain information on seasonal temperature ranges and life-history stages. In spite of the differences in shell preservation rates in the two environments, living and dead assemblages generally corresponded well with one another; thus, the fossil record should indeed allow a realistic
assessment of the palaeoenvironment.

Concluding remarks

The development of interdisciplinary science within the SFB over the twelve years of
its existence occupied an intermediate position between the ideal and the worst case
scenario: the former would have been represented by an orderly front of individual
scientists, tightly interlocked by forces emanating from theoretical insight and
practical necessity, methodically working the ground of their respective fields, ever-
willing to lend a helping hand whenever requested. This vision may thrill the heart of
many an administrator or even some budding young scientist; however, the majority of
scientists who would also wish to be active researchers need not be told that such a
situation will remain an illusion as long as basic research funding remains thinly
spread and scientists are kept more occupied with administrative and _formal_ teaching
duties than with research and its _active_ teaching. The worst case scenario for the
SFB would have been a chaotic free-for-all of individuals, entrenched in their respective disciplines, but contriving clever interdisciplinary excuses to delve deeply
into the commonly earned pot of funds. We leave it to the reader to judge, on the basis of this book, the performance of the SFB against the scale offered above. We know
that we were not as close to the ideal as we would have liked to have been, but this
was due more to weakness of flesh rather than willingness of the spirit. Throughout
its existence, the SFB was permeated with good faith and enthusiasm for interdisciplinary research, a spirit felt and shared by all. We would like to take this opportunity to thank all our former colleagues, both the authors of the following chapters
and all those who worked diligently, although away from the limelight, and whose names
do not appear here but whose contribution to whatever success we enjoyed will not be
forgotten. We are gratefull to the German Research Foundation (DFG) for financial
support and to the management of the Christian-Albrechts-University, Kiel, who
supported the project in many ways.

Victor Smetacek   Eckart Walger

CHAPTER 1:   BATHYMETRY AND SEDIMENTS OF KIELER BUCHT

B. BABENERD and S.A. GERLACH

1.1   INTRODUCTION

For the purpose of a proper understanding of this synopsis it seemed appropriate to include a brief account of the definition of geographic names and boundaries of the area under investigation. This includes a statistical description of the depth structure and sediment cover of the area.

Kieler Bucht is one part of that transitional area between the Baltic and the North Sea which formerly was named Belt Sea (WATTENBERG 1949, SCHÜTZLER and ALTHOF 1969), but which nowadays more often is called Western Baltic. The term "Western Baltic" as used in weather forecasts had different meaning in different Baltic countries (DHI 1979). Fortunately, since August 1, 1984 this confusion is overcome by the World Meteorological Organization (WMO): An area B 11 "Western Baltic" is defined in the limits up to now used by Denmark and Poland: the region south of the Belts and the Sound, and west of the island of Ruegen (Fig. 1-1).

The Internatinal Council for the Exploration of the Sea (ICES) classifies the Belt Sea (without the Sound) as Fishing Area III c. Most of Kieler Bucht is covered by ICES statistical rectangles 38 G 0 and 37 G 0. For the purpose of Danish fishery statistics rectangles of smaller size had been in use. In 1974 new subdivisions for fishery statistics have been proposed (INTERNATIONAL BALTIC SEA FISHERY COMMISSION 1975) which define Subdivision 22 with an eastern boundary at 12°E, south of Gedser Odde, while the eastern boundary of ICES Fishing Area III c was a line from Gedser Odde to Darsser Ort.

Prior to about 1920 the southern limit of the Little Belt, from the Danish point of view, was the line Poels Huk to Vejsnaes Nakke, and for instance KOLDERUP ROSENVINGE (1909) like other marine researches used this border for considerations on algae distribution. His symbols Lb (Little Belt), Sf (South Fyn Waters), Sb (Great Belt), Bw (Western Baltic) are sometimes used even now. The official German publication (MARINELEITUNG 1922), however, took the line Falshoeft to Vejsnaes Nakke as limit of the Little Belt, and Flensburg Fjord was then part of the Little Belt area. Recent official publications, however, differ the other way: from the Danish side (DANSKE LODS 1983) Falshoeft to Vejsnaes Nakke is the boundary, while the official Federal Republic of Germany publication (DHI 1978) and the official British publication (BALTIC PILOT 1974) now define the southern border of the Little Belt by a line from Poilshuk to Vejsnaes Nakke, thus making Flensburg Fjord an appendix to Kieler Bucht.

Nautical charts

Denmark

DK 185 Oestersoeen vestlige del, Kielerbugt 1:130 000, 8. ed. 1980 (originally prin-
    ted 1925)
DK 152 Lille-Baelt sydlige del (Little Belt, southern part) 1:70 000, 7. ed. 1980
DK 170 Farvandet syd for Fyn (The waters south of Fyn) 1:50 000, 5. ed. 1980
DK 142 Store-Baelt sydlige del (Great Belt, southern part) 1:70 000, 9. ed. 1980

Federal Republic of Germany

D   30 Kieler Bucht 1:100 000, 4. ed. 1980
D   32 Falshöft bis Holtenau 1:50 000, 4. ed. 1981
D   33 Ansteuerung der Kieler Förde 1:25 000, 3. ed. 1980
D   26 Flensburger Förde 1:50 000, 11. ed. 1980
D   43 Gabelsflach bis Heiligenhafen 1:50 000, 2. ed. 1983
D   31 Gewässer um Fehmarn 1:50 000, 4. ed. 1982

Charts D 32, D 26, D 43 and D 31 also available with UTM-grid from Deutscher Mili-
tärgeographischer Dienst (Militärhydrogeographie), Bernhard-Nocht-Str. 78, 2000 Hamburg 4.

1.2   NAMES

By international convention, geographic names should not be translated, but written
the way they are locally used. We then have two names: Kielerbugt and Kieler Bucht
with equal rights. Kieler Bucht is used not only officially by German authorities,
but also on British nautical charts and in The Times Atlas of the World. If one
prefers, for reasons of better scientific understanding, an English language term,
the choice is between Kiel Bay and Kiel Bight. "Bay" is the term most widely used for
geomorphological structures which we call "Bucht" in German. The term Kiel Bay was
used by the British BALTIC PILOT (1974), by the Westermann-Rand McNally International Atlas (1974), by Lloyd's Maritime Atlas (1979) and on US nautical chart no.
44067. But in scientific publications the term Kiel Bight was used about as often as
the term Kiel Bay; Kiel Bight and Gulf of Kiel was used e.g. in the publications of
the Helsinki Commission (MELVASALO et al. 1981). Despite all complications we use
Kieler Bucht as a logical alternative in this introduction but we will refer to Kiel
Bight in the following contributions because the editors believe that this term is
best understood in the scientific community.

18

In the following we avoid the Germain "Umlaut" ä,ö,ü by writing ae, oe, ue and the Danish æ, ø, å by writing ae, oe, aa.

We use the following English language terms:
Gelting Bay (Geltinger Bucht)
Flensburg Fjord (Flensburger Förde, Flensborg-Fjord)
Schlei Fjord (Die Schlei)
Eckernfoerde Bay (Eckernförder Bucht, could be named Eckernfoerde Fjord according to its shape)
Kiel Fjord (Kieler Förde)
Hohwacht Bay (Hohwachter Bucht)
Orth Bay (Orther Bucht)
Fehmarn Belt (Fehmarnbelt, Femer Baelt)
Fehmarn Sound (Fehmarnsund)
Great Belt (Store-Baelt, Großer Belt); Langeland Belt (Langelands Baelt) is the southern part of the Great Belt
Marstal Bay (Marstal-Bugt; this term is used, e.g. by BALTIC PILOT (1974) and in nautical charts DK 185 and D 30; Vejsnaes-Bugt is a synonym)
Little Belt (Lille-Baelt, Kleiner Belt)

Fig. 1-1 Kieler Bucht and its position within the different concepts of the Belt Sea or Western Baltic area.
   a) stippled area: Western Baltic as in Federal Republic of Germany weather reports up to 1984, which at the same time is the area of "Belt Sea".
   b) stippled area: Western Baltic as in German Democratic Republic weather reports up to 1984.
   c) stippled area: Western Baltic as in weather reports of Denmark and Poland and as defined by WMO 1984.
   d) stippled area: Subdivision 22 of the International Baltic Sea Fishery Commission.

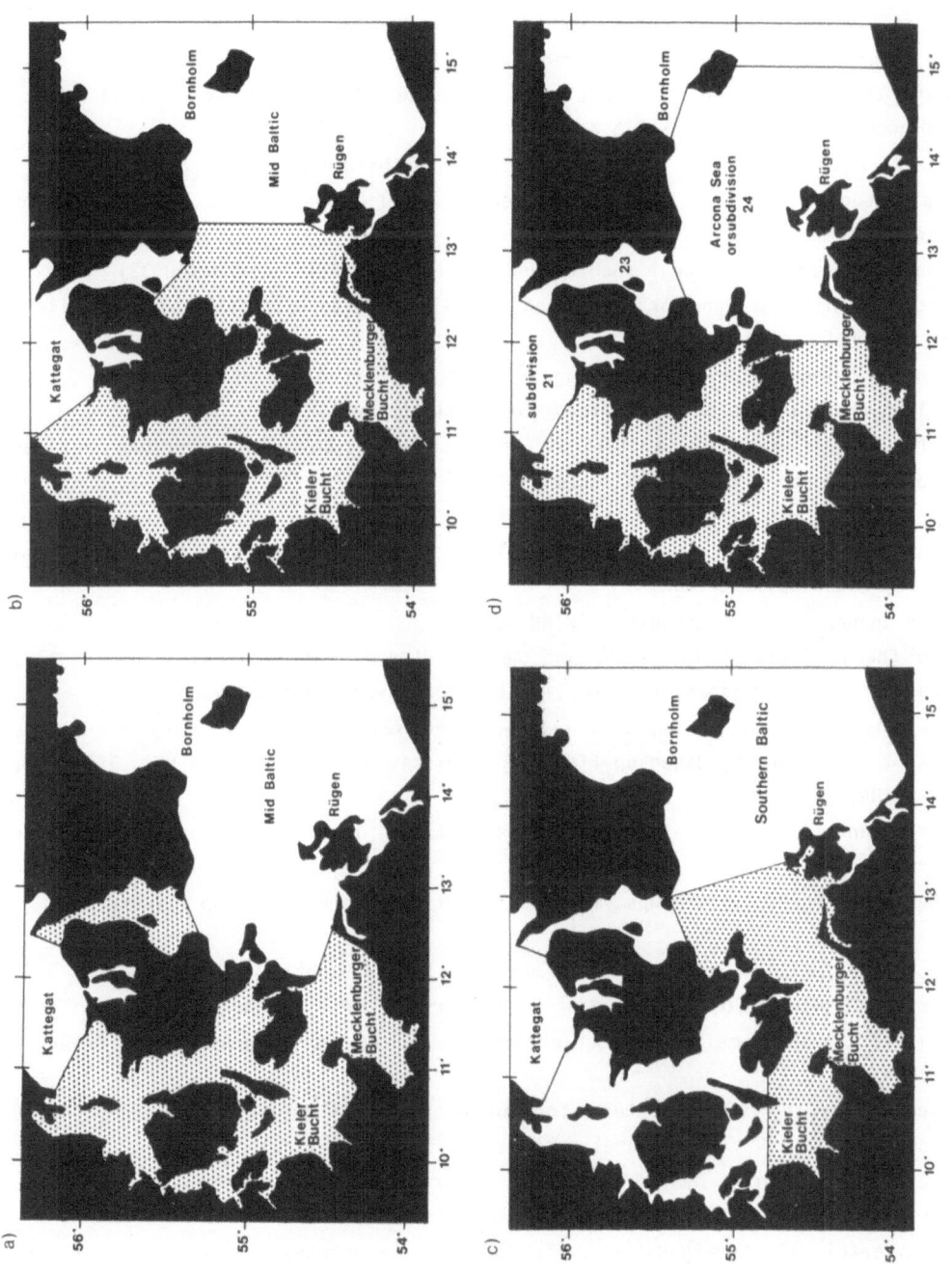

Fig. 1-1

We propose the following terms for channels in the Kieler Bucht region:

Kegnaes Channel between Kegnaes and Bredgrund

Falshoeft Channel between Falshoeft and Bredgrund (Falshoeftrinne; called Breit-grundrinne this volume chapter 5.4.2.3)

Wattenberg Channel 4 nm east of Schleimuende (Wattenbergrinne, in honour of the for-mer director of Institut of Marine Research, Kiel; sometimes called Aaskuhle by German fishermen)

Schleimuende Navigation Channel (Schleimünder Seegat in nautical chart D 32)

Bokniseck Channel between Bokniseck and Mittelgrund (Boknisrinne, called Meilen-fahrt by THUROW 1970)

Stollergrund Channel (Stollergrundrinne in nautical chart D 32)

Gabelsflach Channel east of Gabelsflach (Gabelsflachrinne)

Schlauch Channel extending southwards into Hohwacht Bay (Schlauchrinne, called Schlauch by German fishermen; WEIGELT 1985)

Fluegge Channel south of Fluegge (Flüggerinne)

Vindsgrav Channel northwest of Fehmarn (Vinds Grav in nautical chart D 30; called Graven by Bagenkop fishermen)

Langeland Channel east of Langeland (WINN 1974)

Kloerdyb Channel between Langeland and Aeroe (Kloerdyb in nautical chart DK 170)

Land Channel east of Vejsnaes Flak (called Landrenden by Bagenkop fishermen)

Vejsnaes Channel south of Vejsnaes Flak (called Vejsnaesrinne in many scientific papers, Vesterrenden by Bagenkop fishermen, Gulstav SW Dybe Rende in DHI, 1983, Nr. 14100-14280)

Thuriner Channel north of Vejsnaes Flak (called Thuriner Renden by Bagenkop fishermen)

Rise Channel southwest of Vodrups Flak (called Riserenden by Bagenkop fishermen).

We propose the following terms for some areas in Kieler Bucht:

Dorschmulde (THUROW 1970) for an area 8 nm north of Kiel Lighthouse, deeper than 20 m (also called Rummelloch or Schietloch by German, and Kalhovederne by Bagenkop fishermen)

Platengrund (German fishermen term for an elevation 4 nm east of Damp; called Liselottesbank by BREY 1984, Kugelknast by NELLEN et al. 1985)

Suederfahrt (Süderfahrt, THUROW 1970) for an area southwest of Vindsgrav Channel, deeper than 20 m (called Sydvest by Bagenkop fishermen)

Millionenviertel (KÜHLMORGEN-HILLE 1963) for an area between Vindsgrav Channel and Vejsnaes Channel, deeper than 20 m.

1.3   GEOLOGICAL HISTORY, AND DEFINITION

Kieler Bucht was shaped during the late Pleistocene and Holocene. Glaciers excavated, among other fjords, Kiel Fjord, Eckernfoerde Bay, Flensburg Fjord and the southern Little Belt region. Around 12,000 B.C. the ice withdrew; the area was continental. Around 6,500 B.C., during the Ancylus-Lake freshwater period of the Baltic, a large river through Fehmarn Belt and Great Belt started to transport the freshwater runoff of the Baltic towards the Kattegat region. Around 6,000 B.C. the sea level of the Kattegat was about 23 m below that of present day Kieler Bucht sea level, and flooding of the Kieler Bucht area with marine water started. The Baltic was changed into the brackish Litorina-Sea. Present sea level in Kieler Bucht was achieved around 1,000 B.C.

According to its late glacial and Holocene development and to its present day hydrography Kieler Bucht must be understood as a marginal bay of the Fehmarn Belt - Great Belt section of the estuarine system which connects the brackish Baltic with the marine Skagerrak. Because of the different density of the water masses sharing in the exchange we distinguish two vertically separated, counter-rotating transports: outflow of low salinity (light) Baltic water at the surface and inflow of saline (heavy) Skagerrak, Kattegat or Great Belt water at the bottom. This general picture is very often disturbed by wind effects.

Kieler Bucht hydrography is mainly influenced by the water masses passing through Fehmarn Belt - Great Belt; the Little Belt is narrow and shallow and - compared with the Great Belt - allows for less than 10 % of the water exchange. For details of the hydrography refer to the contributions in chapters 2 and 5.3.1, this volume.

We define Kieler Bucht as separate from the Fehmarn Belt - Great Belt system. This has already been done in the frame of "The Belt Project" by AERTEBJERG NIELSEN et al. (1981). They suggest as boundaries of Fehmarn Belt: in the west a line from Gulstav (the southern end of Langeland) to the northernmost point of Fehmarn, in the east a line from the easternmost point of Fehmarn to Hyllekrog. However, the western part of the deep channel which runs through Fehmarn Belt towards the Great Belt is not entirely covered by the area described as above. So we think it appropriate to draw a line from Gulstav (Dovns Klint) southwards to 54°39' (i.e. to the southern end of Gulstav Flak, the bank south of Langeland) and from there to Westermarkelsdorf Lighhouse on the island of Fehmarn.

We keep the boundary between Kieler Bucht and the Little Belt region as officially defined by the Danish authorities: a line from Falshoeft to Vejsnaes Nakke which touches the southern end of Bredgrund (Breitgrund). Bredgrund is a bank which divides the areas of Little Belt and Flensburg Fjord, respectively. As Flensburg Fjord has so

many peculiarities it should be treated separately from Kieler Bucht. In our calculations of Kieler Bucht we do not include the area of Kiel Fjord and Kiel Harbour south
of Friedrichsort Lighthouse, nor the Schlei Fjord, nor the many brackish water lakes
along the coastline (often called "Noor" in German or "nor" in Danish).

1.4 PLANIMETRY

It should generally be known that squares on a nautical chart do not represent squares in nature, this being a consequence of the Mercator projection. The deviation is
negligible, however, when regarding rather small areas like Kieler Bucht.

We intended to characterize each individual square kilometer of the Kieler Bucht area
according to its position, water depth and sediment type. We use the 1 km grid of the
UTM (Universal Transversal Mercator) projection, which is normally applied to terrestrial topographic maps, but is also indicated in Danish nautical charts (for example on chart "DK 185, Kielerbugt"). Kieler Bucht as a whole lies in Grid Zone
Designation 32, 100,000 m Square Identification NF (western part) and PF (eastern
part). The coordinates bear running numbers (0-99), counting from the west to east
and from south to north, and each 1 km square is defined by the coordinates west and
south of the respective square.

According to the UTM system, a 1 km grid has been drawn covering the Kieler Bucht
area on a 1:50 000 projection of nautical chart "D 30, Kieler Bucht". As Kieler Bucht
topography is rather complicated, and water depth differences of 10 m or more do occur within one square kilometer, it was necessary to subdivide each square kilometer
into 16 subsquares of 250x250 m. For each subsquare the mean water depth and the mean
sediment type was determined. These data were used for further computations of the
bathymery and sediment cover.

The total area of Kieler Bucht is 2571 km². The lenght of the coastline is:

South coast of Aeroe (Vejsnaes Nakke to Kloerdyb)	9 km
West coast of Langeland (Kloerdyb to Dovns Klint)	16 km
West coast of Fehmarn (Westermarkelsdorf to Fehmarn Sound)	22 km
Coast of Holstein (Fehmarn Sound to Friedrichsort)	70 km
East coast of Schleswig (Friedrichsort to Eckernförde)	38 km
East coast of Schleswig (Eckernfoerde to Falshoeft)	41 km
Total coastline of Kieler Bucht	196 km

Table 1-1: Areas (km²) of Kieler Bucht with different range of water depth and sediment cover.

Depth range (m)	Lag sediment	Patchy	Sand	Muddy sand	Sandy mud	Mud	Mixed sediment	Total
0- 2	6.3	-	30.3	0.8	-	-	-	37
2- 4	15.0	-	25.3	2.3	-	-	-	43
4- 6	31.7	-	47.6	3.3	-	-	-	83
6- 8	73.3	-	49.1	4.4	0.1	-	-	127
8-10	82.9	-	39.3	4.6	0.1	0.1	-	127
10-12	98.8	-	67.8	5.0	-	0.6	-	172
12-14	116.3	-	101.1	9.5	-	1.4	-	228
14-16	66.8	-	138.4	45.6	8.5	3.8	-	263
16-18	31.6	3.1	151.6	152.3	49.5	11.3	-	399
18-20	7.1	25.0	49.8	218.9	95.0	30.8	0.1	427
20-22	0.3	6.3	6.4	97.0	41.7	52.6	0.8	205
22-24	0.3	0.1	0.5	59.3	62.3	86.2	4.1	213
24-26	-	-	0.3	8.4	33.6	103.4	3.3	149
26-28	-	-	0.1	1.4	15.4	48.0	4.1	69
28-30	-	-	-	0.3	6.1	10.3	1.2	18
30-32	-	-	-	-	3.5	5.6	0.9	10
32-34	-	-	-	-	0.1	0.8	0.1	1
34-36	-	-	-	-	-	0.1	-	0.1
Total	530	34	708	613	316	355	15	2571 km²

According to the signature on nautical chart "D 30, Kieler Bucht", about 60 km of the coastline are formed by cliffs cut of glacial deposits, the rest is mostly coastal sand. In addition there are several lagoonal areas, e.g. at Orth (4 km), Heiligenhafen (11 km), Bottsand (3 km) and Aschau (1 km) which provide about 20 km of additional coastline not included in the figures, mentioned above.

1.5 BATHYMETRY

The basis for the evaluation of the depth structure is a chart 1:50 000 of Kieler Bucht drafted in 1970 by R.S. NEWTON, Geological-Palaeontological Institute, Kiel University, on the basis of nautical charts and additional soundings. This map with 2 m isolines was kindly provided to us by F. WERNER, Geological-Palaeontological Institute. In a simplified form, with 5 m isolines, it has already been published by HINZ et al. (1971) and SEIBOLD et al. (1971). It has been used as reference in many Kieler Bucht publications since then. Using the information from 41,118 squares of 250x250 m each, the bathymetry of Kieler Bucht has been analyzed in Tables 1-1, 1-2 and Fig. 1-2.

Table 1-2: Percentage of areas with different range of water depth and sediment cover in Kieler Bucht.

Depth range (m)	Lag sediment	Patchy	Sand	Muddy sand	Sandy mud	Mud	Mixed sediment	Total
0- 2	0.246	-	1.180	0.029	-	-	-	1.5
2- 4	0.584	-	0.983	0.090	-	-	-	1.7
4- 6	1.233	-	1.853	0.129	-	-	-	3.2
6- 8	2.850	-	1.909	0.170	0.002	-	-	4.9
8-10	3.227	-	1.527	0.180	0.005	0.002	-	4.9
10-12	3.843	-	2.636	0.195	-	0.024	-	6.7
12-14	4.524	-	3.935	0.370	-	0.054	-	8.9
14-16	2.600	-	5.387	1.775	0.331	0.146	-	10.2
16-18	1.231	0.122	5.898	5.927	1.926	0.438	-	15.5
18-20	0.275	0.973	1.938	8.519	3.697	1.199	0.002	16.6
20-22	0.010	0.243	0.248	3.775	1.622	2.045	0.029	8.0
22-24	0.012	0.002	0.019	2.306	2.422	3.354	0.158	8.3
24-26	-	-	0.010	0.326	1.306	4.025	0.126	5.8
26-28	-	-	0.002	0.054	0.598	1.868	0.161	2.7
28-30	-	-	-	0.010	0.236	0.399	0.046	0.7
30-32	-	-	-	-	0.136	0.216	0.034	0.4
32-34	-	-	-	-	0.002	0.029	0.002	0.03
34-36	-	-	-	-	-	0.005	-	0.005
Total	20.6	1.3	27.5	23.9	12.3	13.8	0.6	100.0 %

About 6 % of Kieler Bucht is shallower than 6 m, 16 % is shallower than 10 m, 32 % is shallower than 14 m; the common depth range (42 %) in Kieler Bucht is from 14 to 20 m. 26 % of Kieler Bucht is deeper than 20 m, but only 4 % is deeper than 26 m, and the channels deeper than 30 m make up only 0.4 % of the total area. None of the subsquares has an average depth of more than 36 m. The mean depth is 16.3 m, the median depth is 17 m, i.e. half of the whole area is shallower and half is deeper than 17 m. The modal depth is 19 m, i.e. this is the most frequent depth of Kieler Bucht.

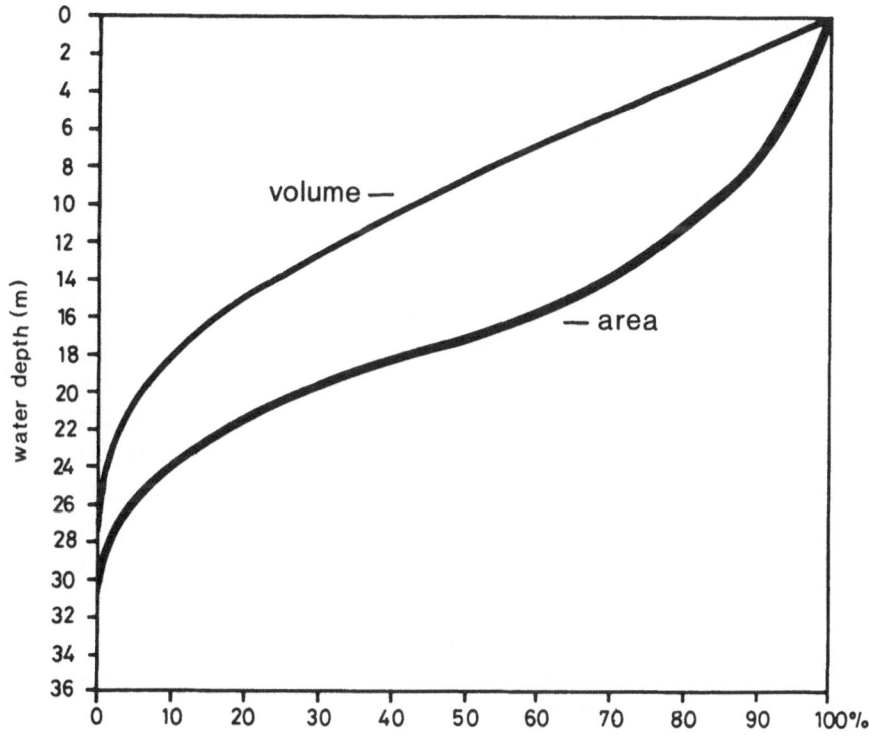

Fig. 1-2: Depth distribution of Kieler Bucht shown by cumulative curve and cumulative curve of volumes of water layers calculated for 2 m depth intervals.

1.6   VOLUMES

From the original data (250x250 m subsquares) volumes have been calculated for each 2 m water layer of Kieler Bucht assuming a gradual decline of water depth within each 2 m interval (Table 1-3, Fig. 1-2). The water above 8 m water depth makes up about half of Kieler Bucht water (47.0 %). Only 5.6 % of Kieler Bucht water belongs to the water mass below 20 m, only 1.2 % is deeper than 24 m, and only 0.1 % deeper than 28 m.

Table 1-3: Volumes of different water layers in Kieler Bucht.

Water layer	Volume km³	%	Below	Volume km³	%	Above	Volume km³	%
0- 2	5.105	12.2	0 m	41.952	100	0 m	-	-
2- 4	5.025	12.0	2 m	36.847	87.8	2 m	5.105	12.2
4- 6	4.899	11.7	4 m	31.822	75.9	4 m	10.130	24.1
6- 8	4.688	11.2	6 m	26.923	64.2	6 m	15.029	35.8
8-10	4.434	10.6	8 m	22.235	53.0	8 m	19.717	47.0
10-12	4.135	9.9	10 m	17.801	42.4	10 m	24.151	57.6
12-14	3.735	8.9	12 m	13.666	32.6	12 m	28.286	67.4
14-16	3.243	7.7	14 m	9.931	23.7	14 m	32.021	76.3
16-18	2.581	6.2	16 m	6.688	15.9	16 m	35.264	84.1
18-20	1.755	4.2	18 m	4.107	9.8	18 m	37.845	90.2
20-22	1.123	2.7	20 m	2.352	5.6	20 m	39.600	94.4
22-24	0.706	1.7	22 m	1.229	2.9	22 m	40.723	97.1
24-26	0.344	0.8	24 m	0.523	1.2	24 m	41.429	98.8
26-28	0.126	0.3	26 m	0.179	0.4	26 m	41.773	99.6
28-30	0.040	0.1	28 m	0.053	0.1	28 m	41.899	99.9
30-32	0.012	-	30 m	0.013	-	30 m	41.939	100.0
32-34	0.001	-	32 m	0.001	-	32 m	41.951	100.0
34-36	0.0001	-	34 m	0.0001	-	34 m	41.952	100.0
			36 m	-	-	36 m	41.952	100.0

Total   41.952   100 %

1.7 SEDIMENT

The basis for the evaluation of the sediment cover is a chart drafted by R.S. NEWTON and F. WERNER in the years 1969 - 1971, which has been published by SEIBOLD et al. (1971). The original drawing 1:50 000 was kindly provided to us by F. WERNER, Geolo-gical-Palaeontological Institute, Kiel University. Sediment types are characterized as follows:

a) Lag sediment or relict sediment, patches or stripes of 10-30 cm thick coarse se-diment over glacial till, and patches of exposed till. There are pebbles, cobbles and boulders covered with fixed algae and sessile animals. Lag sediment indicates areas of erosion.

b) Patchy: areas with patches of lag sediment and muddy sand.

c) Sand: areas of well sorted medium and fine sand, adjacent to areas of erosion, where eroded sand comes to rest, while finer material is transported to deeper areas. With increasing water depth the percentage of fine material ($<$ 40 μm) increases up to 5 %.

d) <u>Muddy sand</u> with 5 - 50 % of fine material (≼ 40 μm). Muddy sand areas roughly commence at water depths where ripple marks disappear.

e) <u>Sandy mud</u> with 50 - 80 % of fine material (≼ 40 μm).

f) <u>Mud</u> with more than 80 % of fine material (≼ 40 μm).

g) <u>Mixed</u> sediment: an unusual mixture of silt and clay with gravel, which occurs either on the slopes between sandy and muddy areas or when gravel is swept into mud areas, or in the deep channels where occasionally bottom currents resuspend fine material and prevent its final desposition.

Tables 1-1, 1-2 and Figs. 1-3, 1-4 present the distribution of different sediment types in different water depths in Kieler Bucht. Almost one third (32.5 %) of the 0 - 6 m depth range is covered by lag sediment, most of the rest by sand. In the 6 - 8 m water depth range lag sediment makes up more than 50 % of the sediment cover; the rest is mostly of sand. Lag sediment disappears at 20 m water depth, but some patchy sediments (lag sediment with muddy sand) occur down to 23 m water depth, which is also the lower limit of sand.

In sheltered regions muddy sediments start at 12 m water depth, but their contribution is significant only at water depths exceeding 18 m.

The area below 28 m water depth makes up little more than 1 % of the total Kieler Bucht area. It comprises small areas of final sedimentation as well as the bottom of the deep channel system where the sediment is influenced by bottom currents. Therefore, the percentage of sandy mud and of mixed sediment below 28 m is higher, and the percentage of mud is lower, as compared with the 24 - 28 m depth range. However, the data are not sufficient for a significant evaluation, therefore these depth ranges have been omitted from Fig. 1-4.

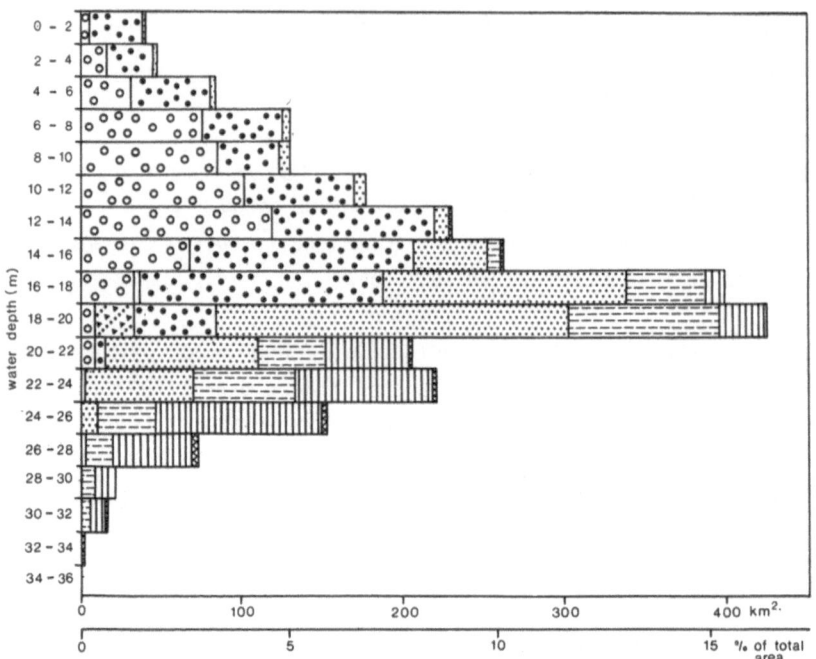

Fig. 1-3: Depth distribution of Kieler Bucht and distribution of sediment types shown by histogram. (For legend see Fig. 1-4).

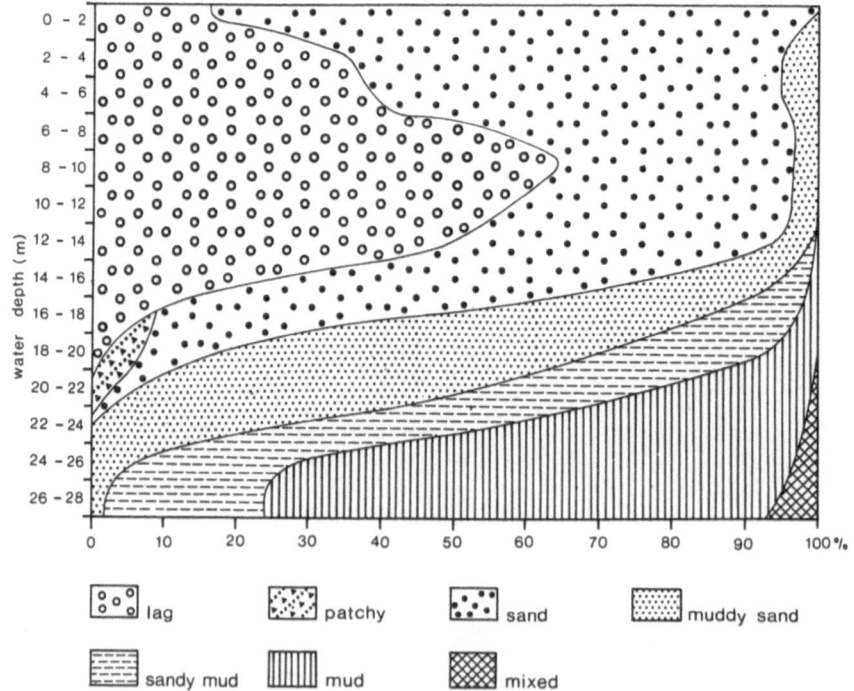

Fig. 1-4: Percentage of the different sediment types within particular ranges of water depth in Kieler Bucht.

29

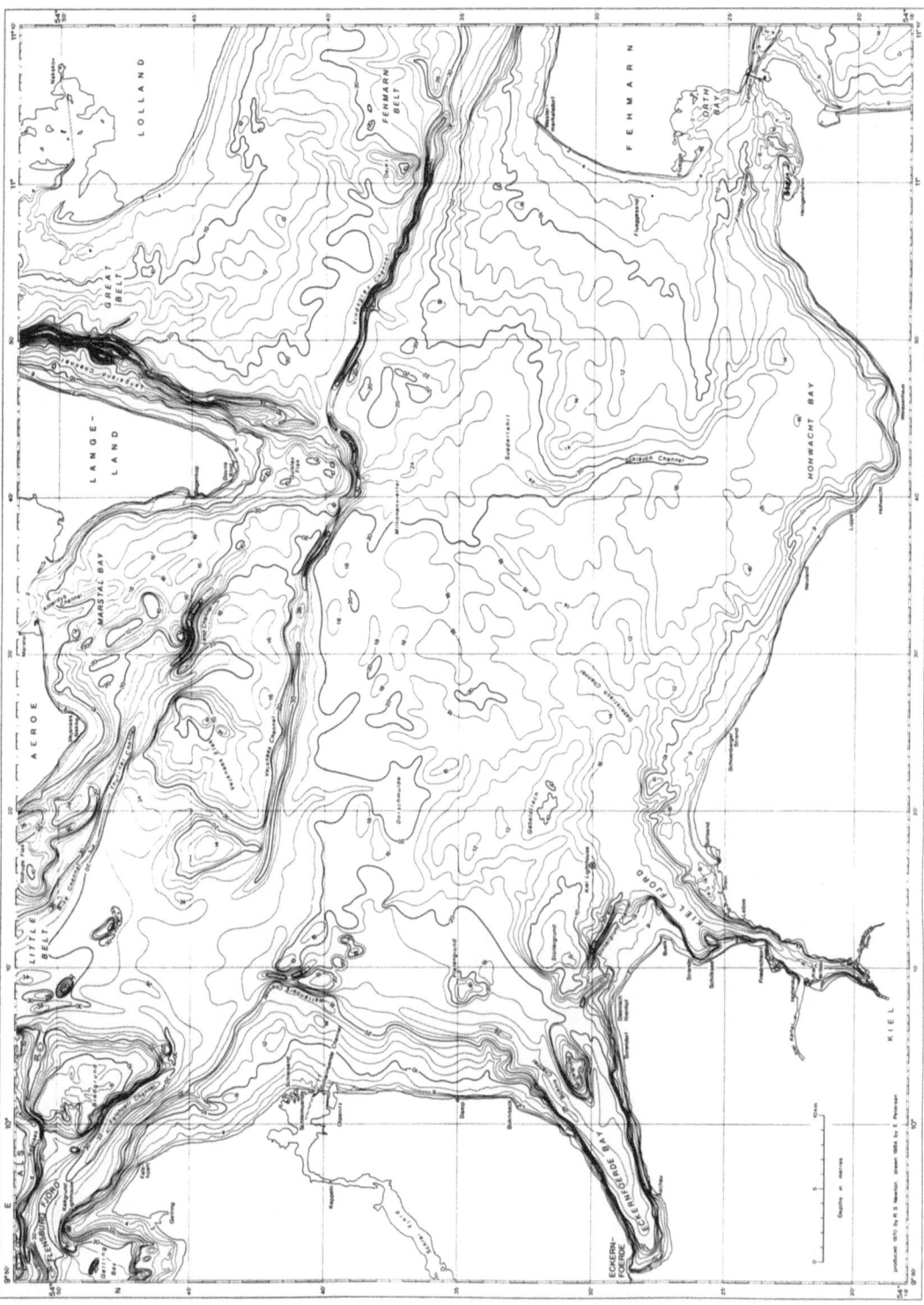

Fig. 1-5: Map of Kieler Bucht with bathymetric isolines (see also attached map).

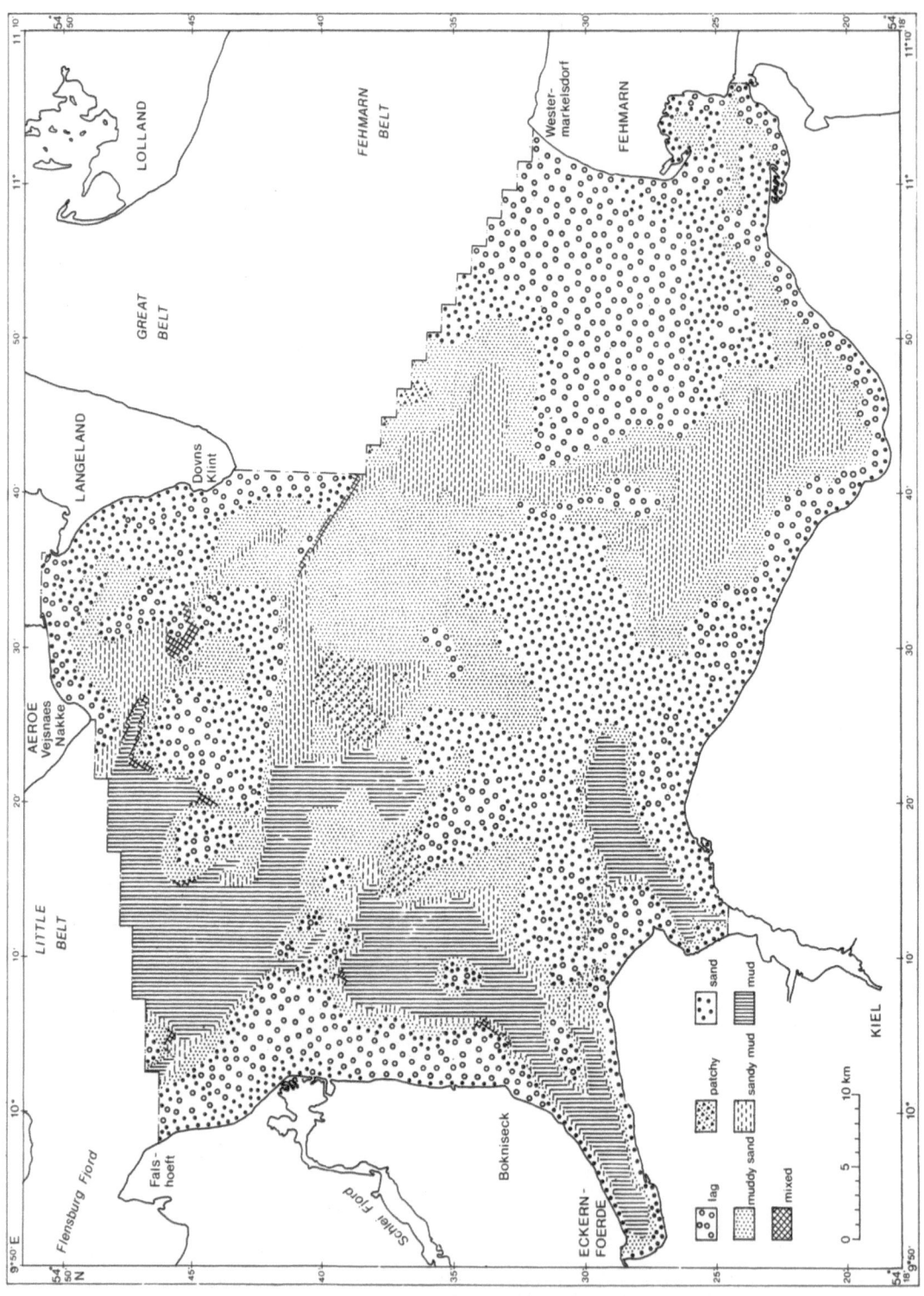

Fig. 1-6: Map of Kieler Bucht showing sediment cover

1.8   CONCLUDING REMARKS

In order to facilitate further research in Kieler Bucht, we present maps of Kieler Bucht depth distribution and sediment cover (Figs. 1-5, 1-6 and attached map). The data for these compilations were collected up to 1971; no new soundings or new sediment analyses have been included. For detailed interpretation of sediment distribution patterns and corresponding processes on geological and recent time scales see chapter 5 this volume.

1.9   ACKNOWLEDGEMENTS

We thank Dr. Torben Jacobsen (Marine Pollution Laboratory, Charlottenlund), Prof. Dr. N. Kingo Jacobsen (Institute of Geography, Copenhagen), Prof. Dr. Gunnar Kullenberg (Institute of Physical Oceanography, Copenhagen) for valuable information about limits of geographical regions, and Dr. Ole Bagge (Danish Institute for Fisheries and Marine Research, Charlottenlund) for information on geographical names used by Bagenkop fishermen. We thank our colleagues and staff of Fischereiamt des Landes Schleswig-Holstein, Kiel, for the names used by German fishermen and Dr. W. Bettac, German Hydrographic Institute, for information about terms. Thanks are due to many colleagues in Kiel who contributed to the manuscript. The drawing of the bathymetric and sediment charts was made possible by a grant from the Umweltbundesamt, Berlin within the project "Eutrophication of the North Sea and the Baltic" and was done by Mrs. M. Petersen. Other figures were drawn by Mrs. H. Kähler.

CHAPTER 2:   THE PELAGIC SYSTEM

V. SMETACEK, B. v. BODUNGEN, M. BÖLTER, K. v. BRÖCKEL,

R. DAWSON, B. KNOPPERS, G. LIEBEZEIT, P. MARTENS, P. PEINERT,

F. POLLEHNE, P. STEGMANN, K. WOLTER and B. ZEITZSCHEL

PREFACE

In the initial phase of the SFB 95, the thrust of research on the pelagic system was directed towards assessing the factors controlling vertical flux of particles settling out of the water column. Zooplankton faecal pellets were considered to be the major vehicle of vertical particle flux and attention was hence centered on the factors determining the production rate of pellets in the water column. These were identified as the standing stock and production of phytoplankton and zooplankton. Annual cycles of various properties of the water column were recorded in the "Hausgarten" from 1972 onwards. The results showed that the processes within the pelagic system leading to sedimentation of particles were more complex than expected. Further, it became apparent that research on the processes retarding the vertical flux, particularly those pertaining to bacterial breakdown of particles in the water column, should be given equal importance to those enhancing sedimentation rates. This chapter is hence divided into two sections. The first presents the results of the annual cycles of the physico-chemical environment in relation to those of the phytoplankton, zooplankton and sedimentation recorded in the "Hausgarten". The second section deals primarily with microbial secondary production in relation to organic substrates in the water column. For technical reasons, much of the latter work could not be carried out in the "Hausgarten" but was conducted instead in the more accessible Kiel Fjord.

2.1   SEASONALITY OF PLANKTON GROWTH AND SEDIMENTATION
       (V. SMETACEK, B. v. BODUNGEN, K. v. BRÖCKEL, B. KNOPPERS, P. MARTENS,
       R. PEINERT, F. POLLEHNE, P. STEGMANN, B. ZEITZSCHEL)

2.1.1   Introduction

The pelagic community is basically self-sustaining as it is composed of both autotrophic (phytoplankton) as well as heterotrophic organisms (bacteria, protozoa and
metazoa). Their patterns of interaction constitute the pelagic energy flow and cycling of matter. The size of the pelagic community, in terms of total biomass per
area, is determined by the amount of radiant energy as well as biogenic elements
available for plant growth in the upper mixed layer. The biomass-carrying capacity of
a pelagic system is determined by the combination of these two factors in relation to
the depth of mixing. The light supply is primarily a function of the season, whereas
the quantity of limiting elements present in the productive surface layer is largely
determined by the balance between input of dissolved biogenic elements (nutrients)
via water transport and loss of these elements via biogenic particles sinking out of
it. In contrast to nutrient input, which is driven by physical transport processes,
loss of biogenic elements due to sedimentation is mediated by biological processes
that govern the production rate, size and sinking speed of the various biogenic particles formed in the water column. Accordingly, a quantitative study of a pelagic system will have to define the boundaries, and monitor the import and export of bioge-
nous material passing through these boundaries.

In a shallow-water system such as Kiel Bight, the close proximity of the
sediment/water interface to the productive surface layer exerts a profound influence
on the structure and functioning of the entire system. Sedimentation of organic
substance from the pelagial is compensated by the release of nutrients from benthic
remineralization. The latter can be of equal importance to pelagic production as the
nutrient supply from pelagic heterotrophs (ROWE et al. 1975). Benthic suspension
feeders living within the mixed layer gather food from and release nutrients directly
to the plankton community of the productive zone. These benthic organisms (mussels,
barnacles etc.) are closely geared to the phytoplankton, in contrast to heterotrophs
living below the mixed layer, whether pelagic or benthic. In this chapter we use the
terms "new" and "regenerated" production, first introduced by DUGDALE and GOERING
(1967), to denote imports to the productive surface layer and turnover within it
respectively. This is irrespective of whether nitrogenous nutrients are present as
nitrate or ammonia or whether released by benthic or pelagic heterotrophs (SMETACEK
1984). The important point here is to distinguish between a diffuse balance between
photosynthesis and remineralization within the mixed layer (regenerated production)
on the one hand and sudden, event-type nutrient input via vertical water transport
(new production) on the other.

The main objective of the research carried out by the plankton group of the SFB 95 was directed towards understanding the processes involved between the input of new nutrients to the pelagic system and the subsequent loss via sedimenting particles out of it. For this purpose, comprehensive field studies and experiments with natural plankton populations enclosed in plastic bags and tanks were carried out.

2.1.2   Data Base

From 1972 to mid 1974 and again from 1980 onwards, data pertaining to the plankton and its environment were collected at intervals of one to two weeks from the routine station situated at 20 m depth in the "Hausgarten".

The following parameters were recorded routinely from a 20 m water column: Temperature, salinity, oxygen, pH, nutrients, particulate organic carbon and nitrogen, species composition and biomass of phytoplankton, protozooplankton and metazooplankton. In some years, dissolved organic carbon and caloric content of seston were also recorded. Primary production with in situ incubation was measured from 1973 onwards. Throughout this period, sedimentation rates were monitored, at first with simple inverted bottles and later with open funnels. A hydrodynamically improved version of the multi-sample trap, described in its final state by ZEITZSCHEL et al. (1978), was developed later and deployed continuously from 1976 through 1978.

From 1974 to 1978, experiments with natural plankton populations enclosed in the 30 m^3 plastic bags of the "plankton tower", which was located in the "Hausgarten", were carried out. The experiments were run for 4 - 5 weeks and the same parameters as in the field studies were also recorded from the enclosed plankton communities as well as from the outside water column at daily or 2-day intervals. An overview of the results of these experiments has been presented by SMETACEK et al. (1982).

Much of the data base has already been published and in the following, we shall draw from these accounts to present a generalized picture of the seasonal development of the plankton community and its interaction with the benthic system. We have not cited the relevant papers in the description of the seasonal cycles presented below but shall do so here in the form of a short list: nutrients, hydrography, primary production (v. BODUNGEN 1975); caloric content of seston and dissolved organic carbon distribution (v. BRÖCKEL 1975); phytoplankton and protozooplankton biomass and species composition (SMETACEK 1975 and 1981, STEGMANN 1981, STEGMANN and PEINERT 1984); metazooplankton biomass and composition (MARTENS 1975 and 1976, STEGMANN 1981); organic seston composition, suspended faecal pellets (SMETACEK and HENDRIKSON 1979, SMETACEK 1980a, KNOPPERS 1981); sedimentary nutrient input and sedimentation of organic matter (POLLEHNE 1980, SMETACEK 1980b); annual budget of energy flow (v. BRÖCKEL

1978); production, sedimentation and structure of the pelagic system February-June 1980 (PEINERT 1981, STEGMANN 1981, PEINERT et al. 1982, STEGMANN and PEINERT 1984); short-term observations of enclosed natural plankton communities (KNOPPERS 1976, POLLEHNE 1977, v. BODUNGEN et al. 1976, SMETACEK et al. 1982); long-term analysis of the Kiel Bight pelagic system (SMETACEK 1985a). SMETACEK et al. (1984) have distinguished four stages of the pelagic growth season based on the relationship between the imports/exports and structure and functioning of the pelagic system. In the following we shall examine these four stages in greater detail and also consider the winter period as well.

2.1.3   Hydrography and Nutrients

The hydrography of Kiel Bight is driven by the exchange of water masses between the North Sea (S ~ 33 °/··) and the Baltic Sea (S ~ 7 °/··). The surface outflow of low salinity water from the Baltic Sea is compensated by the bottom inflow of high salinity water from the North Sea via the Kattegat. The two water masses are divided by pronounced density gradients. The higher salinity bottom water originates from mixing of North Sea and Baltic water in the surface layers of the 30 meter deep Kattegat, whereas the lower salinity water entering Kiel Bight results from mixing of Kattegat and Baltic water to the east of the Bight. Wind induced current components and meteorological conditions have a major impact on the actual patterns of water exchange (DIETRICH 1951). Thus, Kiel Bight is subject to meteorologically controlled influx of both water masses and exhibits considerable variability in salinity and water column structure not only from day to day but also year to year. The range of these variations in salinity is generally between 10 to 16 °/·· and 20 to 30 °/·· at the surface and the bottom respectively.

The main channels of water exchange - the Great Belt and Fehmarn Belt - lie to the northeast and east of Kiel Bight (Fig. 1-5 and attached map this volume). Thus, the western part of Kiel Bight, where most of the measurements referred to in this chapter have been carried out, is not directly affected by the short-term variation characteristic of eastern Kiel Bight. Salinity distribution· in the western Bight exhibits a seasonal pattern: a generally well mixed water column during winter with occasional stratification and a permanent halocline with high density gradients during summer months (Fig. 2-1, from LENZ 1981). The long-term average salinity of the water column at the "Hausgarten" station of 18.7 °/·· (BABENERD 1980) indicates that approximately equal portions of Baltic Sea and Kattegat water contribute to water in Kiel Bight. The influx of 'pure' water of either origin is exceptional.

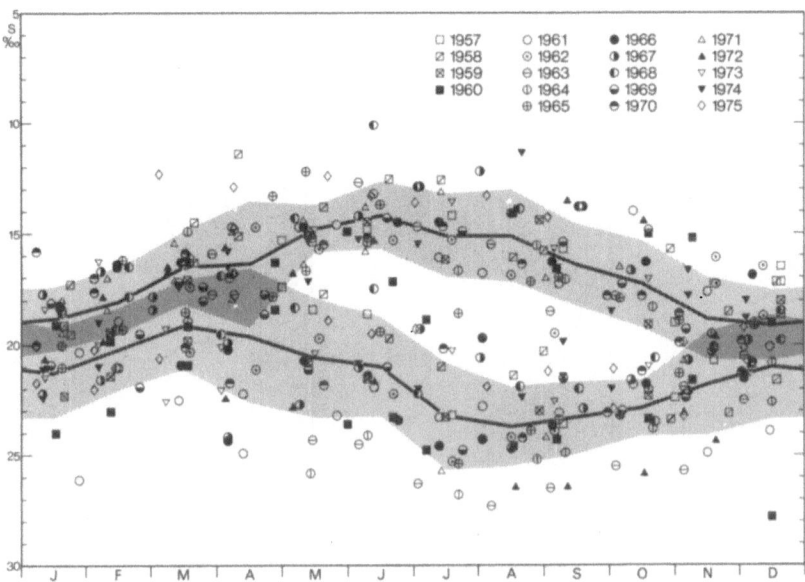

Fig. 2-1: Mean annual cycle of salinity at 0.5 m (above) and 26 m (below) near the
"Hausgarten" station. Data are based on monthly observations from 19 years
(KREY et al. 1978); shaded areas indicate the standard deviation for the
monthly means (from LENZ 1981).

In spite of the great variability in salinity, annual cycles of temperature show a
recurring pattern each year. Temperature and salinity are correlated only between
late May and late September, during the rest of the year, vertical temperature distribution barely or not at all reflects the saline structure of the water column. The
minimum temperature recorded in each year ranges between -0.5° to 4°C around
February/March and the maximum between 18° to 21°C around July/August. In all years,
thermal stratification starts after warming of the entire water column has reached
6° to 9°C around May/June, autumnal breakdown occurring in the temperature range of
14° to 12°C.

The residence time of bottom water in summer is considerably longer than that of surface water, as indicated by the more frequent short-term changes in salinity of the
latter. This is evident from the gradual temperature increase of less than 2°C in the
bottom water from May to September, considering that this water originates from the
Kattegat surface. This is also substantiated by declining oxygen and increasing nutrient levels over periods of weeks to months. In Fig. 2-2 salinity, oxygen and phosphate are depicted from three stations along the major pathway of bottom water. On
its way from the point of entry in the eastern Bight (Vejsnaes-Rinne) towards the
"Hausgarten" area, the nutrient load of bottom water increases and its oxygen content
decreases.

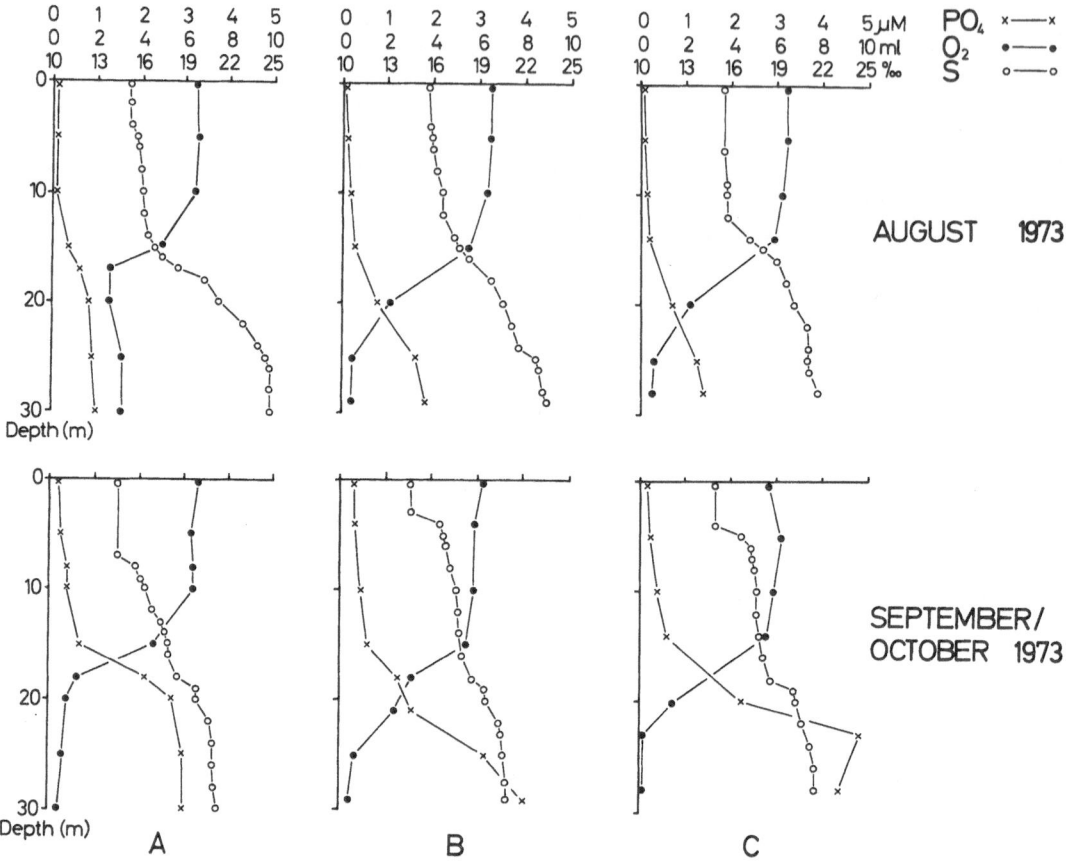

Fig. 2-2: Typical salinity, nutrient and oxygen distributions in late summer in the channel system of Kiel Bight. A: Vejsnaes-Rinne, near the point of entry of the bottom water; B: off Schleimünde; C: direct vicinity of the "Hausgarten" station.

To a certain extent, vertical mixing does occur in summer as indicated by intermittent downward transport of heat. This can happen without break-down of stratification. Apparently, this vertical transport is largely due to current shear and the resultant mixing between layers moving in different directions with different speeds. The frequency and intensity of this type of vertical mixing also varies considerably from summer to summer. Temperature homogeneity is achieved when gradual downward transport of heat is matched by surface cooling, whereas homogeneity of salinity is only brought about by strong storms in fall and winter.

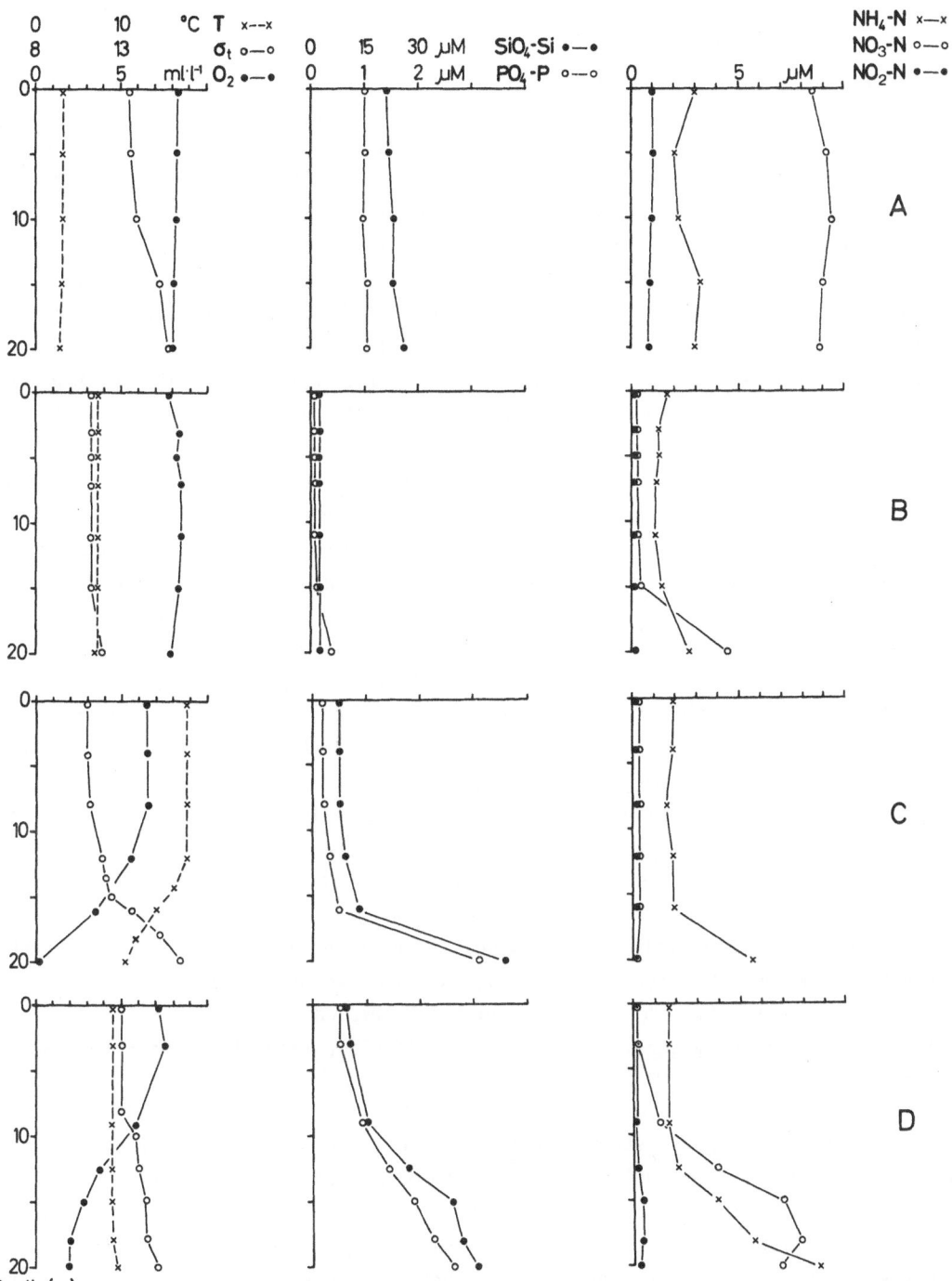

Fig. 2-3: Physical and chemical properties of the water column at the "Hausgarten" station for the different seasons. A: December to February; B: Late April to late May; C: August to early October; D: Mid-November.

As in the case of temperature, the annual course of oxygen and of the nutrient salts phosphate, silicate and TIN (total inorganic nitrogen comprising ammonia, nitrite and nitrate) shows a recurring pattern rather independent of that of salinity. In Fig. 2-3 typical seasonal oxygen and nutrient profiles are depicted. Winter accumulation from a five years average (1973 - 1977) results in surprisingly constant nutrient levels of 1.05 ± 0.04 μMol l^{-1} for phosphate, 24.67 ± 3.02 μMol l^{-1} for silicate, and 12.08 ± 0.67 μMol l^{-1} for TIN; the composition of the nitrogen pool shifts from ammonia to nitrate domination during the winter. The winter nutrient levels are attained by early December by phosphate and silicate whereas TIN reaches its winter levels some weeks later. Nutrient depletion commences with the beginning of the phytoplankton growth season in late Febraury (Fig. 2-3A). Only sudden influxes of almost 'pure' Baltic Sea surface water, which carries a lower nutrient load, may alter the winter pattern. However, such intrusions are exceptional short-term events, as the normal winter nutrient levels are restored following vertical mixing. The presence of stable nutrient levels in the well mixed water column indicates that nutrient concentrations are regulated by interaction with oxic upper layers of the sediments. Such geochemical 'equilibrium'-reactions between water column and sediments have been reported for phosphate in shallow oxic environments. In the case of silicate and TIN little is known about such mechanisms.

Depending on the onset and duration of the spring phytoplankton bloom, all nutrients are depleted by end of March to middle of April in the entire water column, which by then is supersaturated in oxygen. This indicates frequent mixing, as significant production occurs only in the upper 10 m of the water column. Recycling of nutrients within the pelagic environment is negligible at this time, as the bulk of organic material produced by the spring bloom settles on the bottom (see section 3.3.1, this volume). Benthic regeneration will be the major nutrient source during and following this period. Bottom waters and the sediment surface, the main site of nitrification (SZWERINSKI 1981), are still well aerated, which is evident from the high nitrate content of the bottom water (Fig. 2-3B). Vertical salinity gradients do not yet result in stable stratification, thus, nutrients removed by the spring bloom can be reintroduced into the water column, where an efficient recycling by the developing pelagic food web commences.

With the onset of stable stratification, nutrient accumulation in deeper waters and low levels in the surface layer characterize the summer period, i.e. low temperatures correspond with high nutrients and low oxygen (Fig. 2-3C). Deep water nutrient accumulation and oxygen impoverishment appears to be a gradual process, as renewal of bottom water from the Kattegat surface is unlikely to occur (Fig. 2-2). With progressing oxygen depletion in sediments and bottom waters, the inorganic N:P ratio decreases to less than 6:1, due to the excess release of phosphate by the dissolution of particulate inorganic phosphorus compounds. Silicate shows essential similarity to phosphate during summer, although redox-dependency of silica dissolution could not be demonstrated in enclosure-experiments (see also chapter 4, this volume).

The pelagic recycling of nutrients can be augmented by nutrient input events due to vertical mixing as described above, as well as by density displacement of nutrient-rich pore waters caused by oscillation of the halocline which flushes vast areas of sandy sediments (SMETACEK et al. 1976). Nutrients derived from these sources are not measurable in the water column as they are taken up rapidly by the summer phytoplankton (v. BODUNGEN et al. 1976). The processes governing nutrient input to the water column during summer are dependent on weather conditions and are thus highly variable from year to year. During long periods of stable high pressure conditions, which favours steep density gradients in the water column, nutrient input of this kind is largely reduced as was the case in 1972. Consequently fewer numbers of summer phytoplankton blooms were recorded than in other years.

Successive erosion of the thermocline from late summer onwards results in upward mixing of nutrients as well as in the gradual oxygenation of the bottom water by October. In November, another short phase of oxygen depletion and nutrient accumulation in the bottom water occurs, favoured by the presence of haline stratification (Fig. 2-3D) before the uniform distributions of oxygen and nutrients typical for the winter months are established.

There is little indication of nutrient limitation of phytoplankton growth in the physiological sense, but nitrogen appears to be the nutrient in shortest supply during summer and autumn which thus regulates the magnitude of total production through these periods. This is different in spring, when nitrogen and phosphorus are taken up by the spring bloom at the same ratio of 12 - 13:1 in which they occur in the water column.

2.1.4   Development of the Pelagic System

Winter or Non-Growth Phase

This period, which extends from December to February, is characterized by low plankton biomass and low rates of change in the various system components. The water column is as a rule homogeneous with respect to its dissolved and particulate components. The contribution of detritus to the total organic particulate pool is at its maximum for the year (> 90%). The phytoplankton biomass decreases steadily till its annual minimum is reached in February, before the onset of the spring bloom. In spite of its low biomass, the winter phytoplankton community is fairly diverse. Measurable primary production is restricted to the upper 5 m and the production/biomass ratios of 0.3 compare well with values from other seasons. There do not appear to be any specific adaptations to winter conditions in Kiel Bight phytoplankton, as also indicated by the absence of a winter community dominated by species attaining their biomass peak only at that time.

The same is true for the protozoan and metazoan plankton. It has been shown that most of the holozooplankton species are present in winter and the copepods also reproduce (FAHLTEICH 1981). Breeding success is low, however, as indicated by the low levels of the population. LENZ (1977) suggested that the winter zooplankton of Kiel Bight maintains its apparent activity by utilizing detritus to a greater extent than phytoplankton. Winter detritus levels are fairly high (150-300 mg C m^{-3}), but this material is of poor food value, as demonstrated by the low heterotrophic carrying capacity of the system.

Renewal of the winter detritus pool is due to sediment resuspension brought about by storms. During stormy weather, very high seston values have been encountered in the water column (9 g C m^{-3}). However, sedimentation of this material is rapid following storm cessation. Winter sedimentation patterns are very variable as they are weather dependent. The sedimenting material in winter has a higher carbon content and the proportion of humic acids to total organics is lower, as compared to bottom sediments. This indicates that particle selection occurs during resuspension by the winnowing effect of turbulence; the lighter, more organically rich particles are maintained longer in suspension by lower turbulence levels than the heavier, inorganic particles.

Diatom Spring Bloom Phase

The spring bloom phase, which marks the advent of the plankton growth season, begins in late February and ends by mid-April. Its onset is signalled by increasing phytoplankton biomass and changing species composition. Under calm conditions, the bloom can culminate by mid-March and in stormy springs, the comparatively short, characteristic exponential growth phase can be shifted to as late as early April. The bloom is terminated following nutrient depletion by senescence of the dominating species which results in their wholesale sedimentation. As a rule, phytoplankton biomass levels immediately following bloom termination are at their minimum for the growth season. Thus, 3 developmental periods - initial, exponential and sedimentation - can be distinguished, their time scales being a function of the weather. The blooms studied in 1972, 1974 and 1980 exhibited the typical pattern described above in contrast to the anomalous bloom of 1973 (Fig. 2-4). In the former blooms, distinct

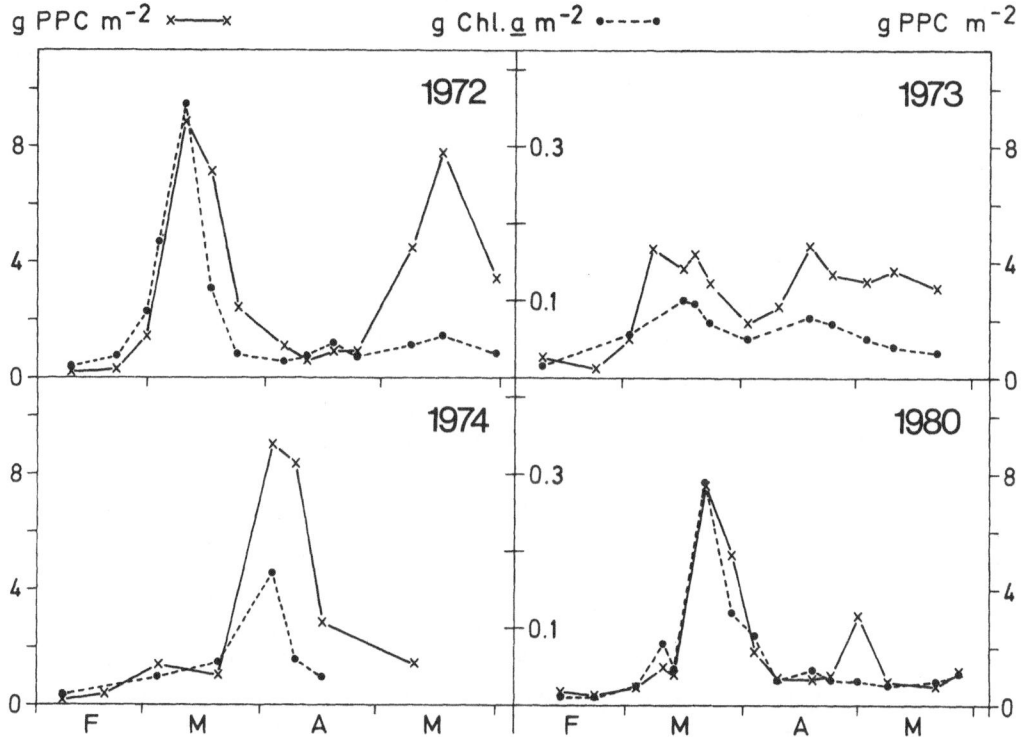

Fig. 2-4: Developmental patterns of 4 spring blooms as reflected in chlorophyll content and phytoplankton biomass (PPC) (calculated from cell counts) of the
of the 20 m water column (from SMETACEK 1985 a).

differences in the duration and timing of the developmental periods can be
ascertained. The 1972 and 1980 biomass peaks in March were dominated to over 60 % by
a single species - Detonula confervacea - the 1974 peak in April was composed of
several Chaetoceros species, which together contributed over 70 % of biomass.
Selection of the dominant species probably occurs immediately prior to the exponential
period. During the initial period, several diatom species, and the autotrophic
ciliate Mesodinium rubrum can be of similar importance. There is considerable
year-to-year variation in the composition of the diatom population during this phase
(SMETACEK 1985a) particularly in the case of the accompanying species, i.e. those
contributing 1-10 % of total biomass. It therefore appears likely that selection of
the dominant species or genus occurs twice - once during the initial period and then
again just prior to the exponential period. This would explain why species dominant
in one year can be virtually absent in the next. Possible mechanisms of selection
have been speculated upon by SMETACEK (1985b).

The spring bloom of 1973 was quite unlike the other blooms, not only in its developmental pattern - i.e. blurring of the 3 periods - but also in its species composition.
Three distinctly different populations of similar size - dominated by Thalassiosira,
nanoflagellates and Chaetoceros respectively - were observed, and as nutrient
depletion was not achieved before mid-April, this phase merged with the next without
undergoing the biomass minimum characteristic of the 3 'normal' blooms. This atypical
pattern was due to several massive advective events clearly evident in salinity,
that not only brought in allochthonous populations but also, by rapidly disrupting
the structure of the water column, led to radical short-term changes in the physical
environment of the phytoplankton. In normal years, salinity did not change
appreciably during this phase.

In all cases, termination of the bloom is signalled by almost total nutrient depletion
in the entire water column. The carrying capacity of plant biomass is fixed by the
winter nutrient levels if only new production is considered. Assuming C/N and C/P
ratios (by atoms) of phytoplankton to be 6/1 and 106/1 respectively (REDFIELD et al.
1963), 20 g C m^{-2} based on nitrogen and 24 g C m^{-2} based on phosphorous should be
produced by the spring bloom each year. These figures compare well with actual
measurements of primary production; in 1973 and in 1980   21 and 18 g C m^{-2} were
recorded respectively during this phase. However, the maximum standing stocks
encountered so far have always been around 10 g C m^{-2}. Significant primary production
in spring is restricted to the upper 5 m. Vertical biomass distribution tends to be
uniform, however, indicating frequent vertical mixing during this phase.

The steep decline in biomass following bloom culmination each year, also observed in
the POC in the water column, was due to sedimentation of the phytoplankton population
in the form of living cells, resting spores and "fresh" phytodetritus. Sedimentation
of spring blooms was monitored over a period of 8 years and in each case the bulk of
the organic carbon produced by the bloom, equivalent to the maximal standing stock
recorded, was transferred to the benthal.

The composition of sedimenting matter during this phase changes abruptly from
resuspended sediment prior to and during the bloom to a massive input of fresh organic
matter of high nutritive quality following its decline. This bloom input triggers an
immediate response from the benthos (see chapter 3, this volume). The total amount
of primary organic material reaching the sediments is in the range of 10 g C m^{-2}.

In Kiel Bight, zooplankton plays only a minor role during the spring bloom. Differences
between zooplankton standing stocks of 1973 and 1980 are much less than phytoplankton
(Fig. 2-5). In both years protozooplankton responded more rapidly to phytoplankton
biomass build-up than the metazooplankton, due to faster growth rates of the former,

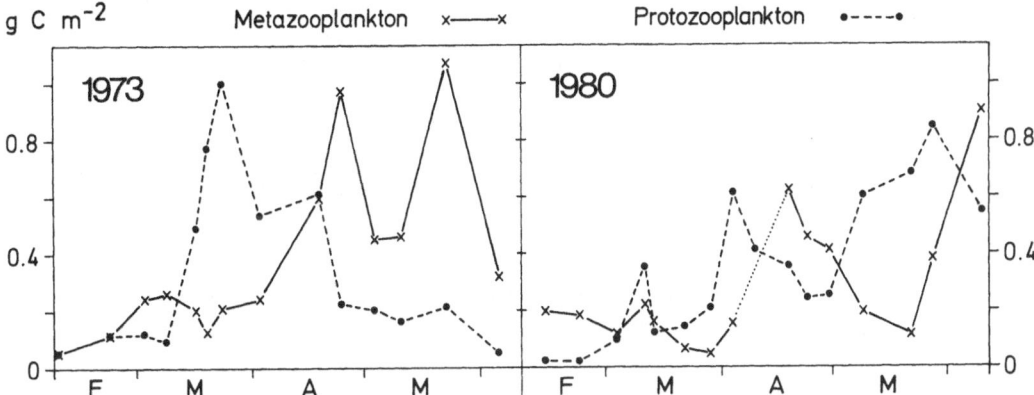

Fig. 2-5: Protozooplankton and metazooplankton biomass in the 20 m water column during early and late spring of 1973 and 1980 (1973 redrawn from SMETACEK 1981; 1980 from PEINERT et al. 1982).

However, there is apparently a reciprocal relationship between these 2 planktonic components, mediated by predation of protozoans by the metazoans.

Species composition of protozooplankton, also recorded in 1974, showed considerable yearly variation and the peaks were dominated by different phagotrophic species each year. Both ciliates and naked dinoflagellates attained comparable biomass levels. Non-loricated ciliates dominated biomass in 1973 and 1974 whereas tintinnids were more important in 1980. The developmental pattern of metazooplankton, including species composition, is much less variable than that of the protistan components. The copepods Pseudocalanus elongatus, Paracalanus parvus, Centropages hamatus, Acartia spp. as well as Oithona similis are all present in similar quantities during the initial period of the bloom, a slight increase in biomass of most components occurring during this period relative to the winter levels. A sharp increase in biomass of Pseudocalanus and Paracalanus occurs in April during the terminal period of the bloom. Oithona also increases in biomass but not to the same extent as the suspension-feeding species. Meroplanktonic larvae, rotifers and cladocerans are also present but their biomass is insignificant when compared to that of the copepods.

    Post-Spring Bloom Phase

This phase commences after sedimentation of the spring bloom in April and is terminated by stable thermal stratification of the water column in early June. At the onset of this phase, nutrients and phytoplankton biomass are at their lowest level for the growth season. However, a fairly large population of protozooplankton is invariably present, as is also a rapidly growing cohort of herbivorous, i.e.

suspension-feeding copepods (Pseudocalanus and Paracalanus), apparently spawned by over-wintering adults feeding on the spring bloom.

Small dinoflagellates (10-20 µm), in many years Prorocentrum balticum, are the characteristic phytoplankton of this period, although a silicoflagellate of similar size, Distephanus speculum, dominated in 1983 (NÖTHIG 1984). They apparently grow by utilizing nutrients "left behind" in the water column in DOC and POC pools, although sedimentary nutrient input also plays an important role in biomass build-up. In this phase, production maxima are shifted from the surface to 5 m and even 10 m depths. In 1973, the deepest water column maximum of the year (50 mg C m^{-3} d^{-1} at 12 m) was found in a population of P. balticum (Fig. 2-6). Year-to-year variation in phytoplankton biomass is most marked during this phase. In one year (1972), the dinoflagellate population increased steadily till mid-May and attained a biomass peak rivalling that of the spring bloom. In another year (1980), biomass remained at low levels with occasional small peaks. Conditions in 1973 and 1974 were intermediate between these 2 extremes. This yearly variation in phytoplankton biomass is apparently caused by variation in sedimentary nutrient input. Zooplankton grazing must have a much greater impact on the phytoplankton in this period as compared to the spring bloom. However, yearly variation in phytoplankton biomass has apparently little effect on the meta-zooplankton population that exhibits remarkably similar developmental patterns each year. A reciprocal relationship between protozooplankton and metazooplankton biomass is evident from Fig. 2-2. Maximal zooplankton biomass is in the range of 0.8 - 1.2 g C m^{-2} (SMETACEK 1985a).

Sedimentation rates during this period are the lowest of the year and there is an increase in the total biomass of the plankton. The copepod peak declines fairly rapidly by mid-June each year. The concomitant massive increase in Aurelia biomass and hence grazing pressure (MÜLLER 1979) could well be an important cause of the co-pepod decline. However, it is also possible that this first cohort simply dies off after spawning, in which case, their carcasses would be transferred to the benthic system. Although the sediment traps frequently collected large numbers of dead copepods, it was difficult to determine whether the animals had actively swum into the trap and been killed there by the preservative or whether they were dead or moribund before falling into the trap.

47

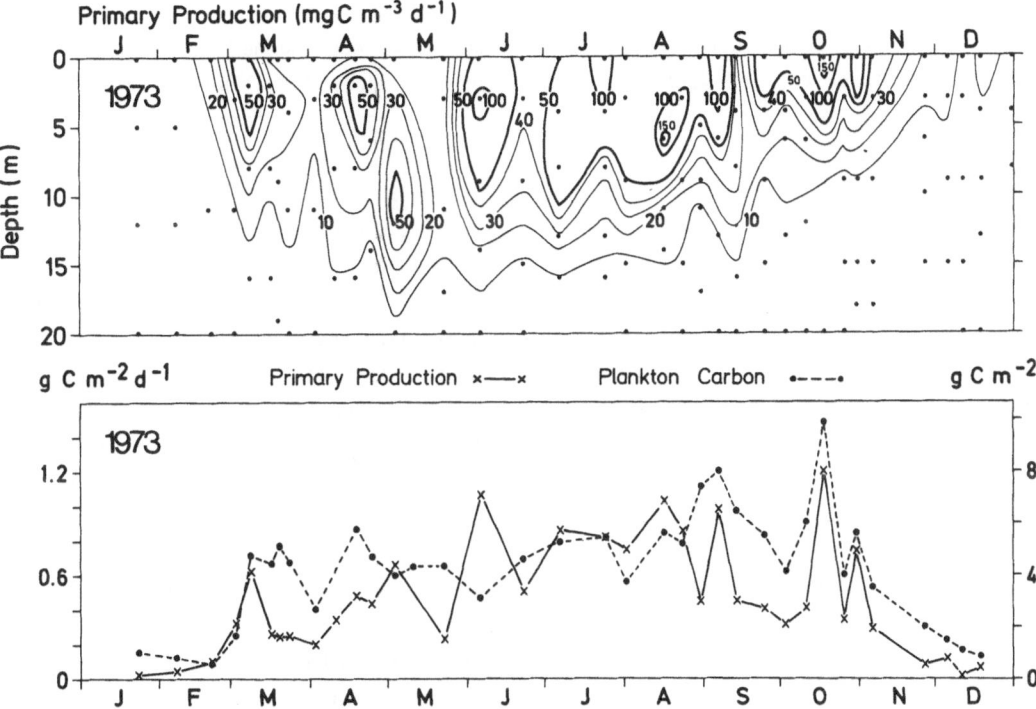

Fig. 2-6: Isopleths of primary production (in situ incubation) (from v. BODUNGEN
1975) and integrated values of primary production and total plankton biomass (Plankton Carbon) for the 20 m water column (redrawn from v. BODUNGEN
1975 and SMETACEK 1981).

Summer Recycling Phase

This phase is the period of thermal stratification extending from early June to mid-
September. Primary production during this period accounts for about half the annual
total whereas sedimentation is generally low, although in some years higher rates
occur sporadically. The composition of summer sedimenting material shows the greatest
degree of variation for the whole year. When sedimentation rates are low, the C/N,
C/P and C/chlorophyll ratios tend to be very high, as does also the proportion of
organic carbon of the material. These high ratios - highest annual values were
recorded in summer - indicate this material to be refractory in nature; these particles represent, therefore, wastes of a recycling system. Particles settling out
during sporadic bursts of higher sedimentation (> 0.2 g C m^{-2} d^{-1}), most common in
August, invariably have low C/N, C/P and C/Chl. ratios, indicating that they repre-

sent comparatively "fresh" phytoplankton material that has undergone a lesser degree of modification than the particles comprising the background summer vertical flux. In spite of the large copepod stocks present in summer, their faecal pellets contribute only a minor portion to the material collected in traps. Their fate is break-down in the water column rather than sinking out of it.

There is a gradual build-up in pelagic biomass through the summer, evident in the long-term mean (BABENERD 1980), which indicates that nutrient input to the system is greater than output via sedimenting matter. As the summer progresses, nutrients accumulate in oxygen depleted bottom water. These nutrients originate to a smaller extent from particles settling out of the euphotic zone that are remineralized in deeper water. The bulk of these nutrients emanate from the sediments as shown by the high mobilization rates of phosphate (see chapter 4, this volume). A part of these nutrients is gradually or sporadically introduced into the euphotic zone during the summer, resulting in the increase in plankton biomass from June to August. The sediments within the illuminated zone are also an important contributor of nutrients. It has been shown that oscillations of the halocline lead to flushing of pore water from porous sediments by density displacement (SMETACEK et al. 1976). As organic particles including plankton are also flushed into the sediments, their remineralization proceeds more rapidly here than in the water column. Nutrient input to the pelagial through flushing, although quite frequent, is difficult to quantify. It has been shown in one "plankton tower" experiment that sporadic nutrient input driven by sediment-flushing events was responsible for the day-to-day fluctuations in primary production of the overlying water column. Nitrogen was shown to be the controlling element in this period (v. BODUNGEN et al. 1976).

Nutrient input from stagnant deeper water and from sediment pore water can be regarded as new nutrients as they are not part of the pool being recycled within the pelagial. Most of the summer primary production is, however, based on regenerated nutrients. New nutrient input will lead to an increase in phytoplankton biomass resulting, in turn, in an imbalance of the regenerating system. A part of these new nutrients is retained within the system by heterotrophic incorporation and increase in the size of this component. However, a remainder sediments out, the amount of which is determined by the phytoplankton biomass increment in relation to the response rate of pelagic heterotrophs. Although this process has not yet been directly observed, we assume that the sporadic higher sedimentation rates described above are the result of such input of new nutrients to the regenerating system. An indirect indication is provided by the data presented in Fig. 2-4. The rapid decline of the diatom population was most probably due to sedimentation as there was no corresponding increase in heterotrophs. Nutrients contributing to the Ceratium biomass build-up were evidently new nutrients, as indicated by the correlation between flushing events and days with higher production levels (v. BODUNGEN et al.

1976). Lateral advection can be ruled out as there was no corresponding change in salinity during the critical period of species change. The major advective event, reflected in low surface salinity from day 19 onwards, had only a minor effect on the species composition of the phytoplankton.

Such rapid changes in species composition of phytoplankton, as in the example above, are not typical for the Kiel Bight summer. Rather, successional patterns vary from year to year. A large diatom population (Rhizosolenia fragilissima) maintained itself for over 2 months (July/August) in 1972 whereas in 1973 diatom stocks were much smaller and more species were involved. Dinoflagellates, particularly Prorocentrum micans, and various ceratia dominated biomass in August 1973. Nanoflagellates and cyanophytes were of lesser importance. The depth distribution of summer phytoplankton varies with the species composition. Diatoms tend to be fairly uniformly distributed above the pycnocline whereas dinoflagellates congregate in specific layers at variable depth. Differences between vertical distribution patterns of diatoms and dinoflagellates can be seen from Fig. 2-7. Weather conditions and their effect on vertical density distribution are important factors in determining whether diatoms or dinoflagellates dominate the plankton. Both groups are equally likely to contribute to sedimenting matter and large populations of apparently moribund dinoflagellates (Ceratium tripos and P. micans) have been observed in close proximity to the sediment surface, reminiscent of the sedimenting diatom spring bloom.

The high summer production rates, apart from sudden increases in biomass, are largely balanced by pelagic remineralization. All 3 major pelagic heterotrophic compartments are involved, although the relative roles are as yet unclear. Metazooplankton appears to be particularly important even though phytoplankton/zooplankton ratios, at least during 1973, were always well above 1:1 and generally as high as 4:1. However, this ratio declines considerably if the unpalatable ceratia are disregarded and the omni-vorous Aurelia aurita included. This medusa can contribute up to 50% of total zooplankton biomass (MÖLLER 1979). It is, however, difficult to estimate the biomass of Aurelia because of its highly patchy vertical and horizontal distribution. If grazing by Aurelia is indeed the decisive factor in reducing the early summer Pseudo-calanus stock, then either Aurelia grazing pressure declines or the copepod growth rate increases during the summer months. The Aurelia abundance remains fairly steady through the summer, its decline occurring as late as September (MÖLLER 1979). The pro-tozoan stock of summer is not only smaller in its biomass but also in the size of its individuals as compared to the spring and autumn. However, occasional outbursts of large ciliates occur. It has been suggested that the large metazoan stocks rather than the availability of food are responsible for the low biomass of large pelagic ciliates during the summer.

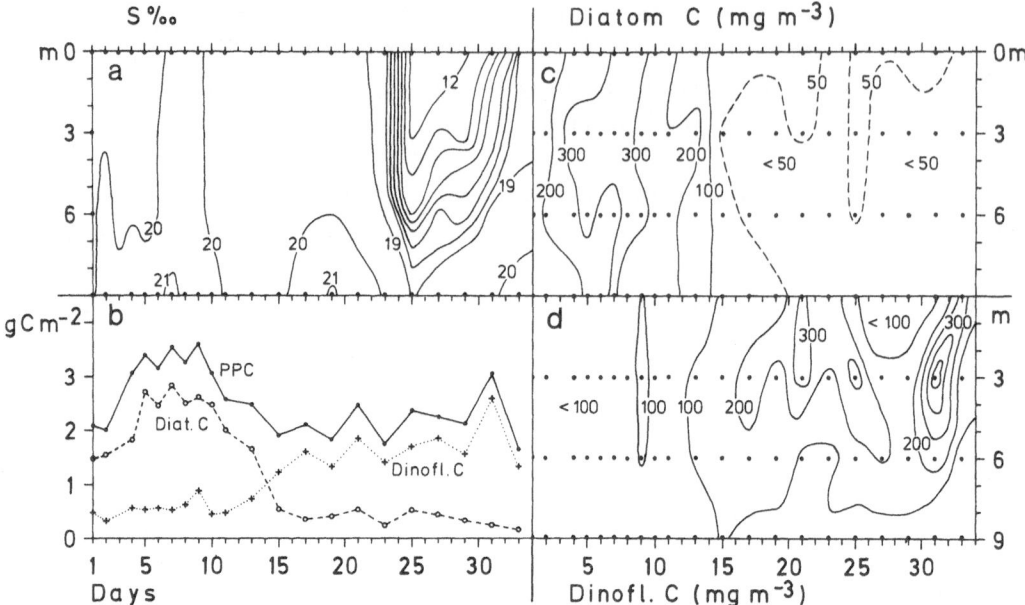

Fig. 2-7: An example of phytoplankton succession from the summer (26 July - 27 Aug. 1974). A: Salinity distribution; B: Integrated values (10 m water column) for total phytoplankton biomass (PPC), diatom biomass (Diat. C) and dinoflagellate biomass (Dinofl. C); C: Vertical distribution of diatom biomass (mostly Rhizosolenia alata); D: Vertical distribution of dinoflagellate biomass (mostly Ceratium tripos). The transition from diatom to dinoflagellate dominance of phytoplankton biomass took place before the intrusion of low salinity surface water between days 23 and 25.

The metazoan stock increases through the summer, reaching a peak in August/September (Fig. 2-5). Copepods contribute the bulk of the biomass and suspension-feeding types (Pseudocalanus, Paracalanus), raptorial types (Oithona) as well as genera of intermediate feeding types (Centropages, Acartia) are all important biomass contributors. The appendicularian Oikopleura, which feeds on nano- and picoplankton, attains considerable importance regularly in August and early September. Together with Aurelia, the summer metazooplankton is certainly the most diverse of the year. Bacterial standing stock and activity are also at their highest level and, in combination with the highly diverse zooplankton community, these heterotrophs ensure a high retentive capacity of essential elements by the system.

The pycnocline also serves as a barrier to smaller sedimenting particles and frequent occurrence of high particle concentration immediately above it has been shown by LENZ (1965). Intense heterotrophic activity occurs in this zone of particle accumulation and KREY (1974) stressed the importance of this "false bottom" in maintaining cycling of matter within the surface layer.

An additional aspect of the summer system was noticed in the "plankton tower" experiments and was studied closely in 1978: inorganic nutrients alone can dramatically enhance the growth of heterotrophic microbial populations, i.e. both bacteria and heterotrophic protozoa. When sufficient light is available, phytoplankton outcompetes the bacteria in taking up inorganic nutrients because of their much larger standing stock. Under continuous darkness, however, heterotrophic microbes can absorb considerable amounts of inorganic nutrients ($>$ 1 μMol P and $\geqslant$ 10 μMol N 1^{-1}) within a week, indicating that the available organic substrate is deficient in these elements. We conclude that it is the animals, including protozoa, rather than the bacteria that are responsible for releasing nutrients for plant growth in regenerating systems. This observation, if it is a universal feature, puts the controversy regarding the relative roles of microbial vs. zooplanktonic remineralization in regenerating systems in a new light (WILLIAMS 1981). Input of new nutrients not only stimulates phytoplankton growth but also makes those organic substrates poor in essential elements available to bacterial utilization. In the surface layer, major accumulation of such "refractory" material occurs presumably in the dissolved pools, as refractory particles are liable to sink out of the system.

Termination of the summer phase generally coincides with break-down in thermal stratification, although this is not necessarily a cause and effect relationship. Characteristic summer phytoplankton and zooplankton decline at about the same time and there is a concomitant increase in the autumn flora and fauna. Nutrients do not appear to be recycled within the water column and there is strong indication, though no conclusive evidence yet, that the bulk of the late summer biomass is diverted to the sediments.

The Autumn Bloom Phase

Break-down in thermal stratification during September initiates the autumn phase. This is characterized by a massive build-up of the Ceratium fusus population that grows steadily through the summer and culminates in a biomass maximum in October rivalling that of the spring bloom. The rise in the C. fusus biomass is accompanied by an increase in nutrient input from the large pools accumulated in anoxic bottom waters during the summer. Vertical mixing does not necessarily have to proceed to the bottom in order to tap these pools as they can also be displaced upward by influx of higher salinity water and concomitant removal of surface water. The oxygen-depleted zone rich in nutrients steadily rises upward during August. Nutrient uptake by the autumn bloom follows a completely different pattern to that of the spring bloom and culmination of this bloom is apparently not determined by nutrient depletion. However, this bloom is similar to the spring bloom in other respects: it utilizes accumulated nutrients rather than regenerated nutrients as it develops under high phytoplankton/zooplankton ratios; in spite of its using ammonia as a nitrogen source, it essentially represents new production. The depth distribution of biomass is more uniform than that of the summer ceratia which might be due to more intense mixing or to C. fusus being a feebler swimmer than the summer ceratia. Another common feature between spring and autumn blooms is the fate of their biomass. The C. fusus population abruptly vanishes from the water column within a week after culmination (Fig. 2-5). This population sediments out en masse, although the mechanisms triggering the collapse of the population are not as well known as in the case of the spring blooms.

In most years, a population of mixed diatoms follows the culmination of C. fusus in October. The diatoms attain a fairly large biomass peak in early November (Fig. 2-8). Nutrients are not limiting throughout; competition between the dinoflagellates and the diatoms could only be for light. Diatom growth is aided by weak stratification and the population is then restricted almost entirely to the upper layer.

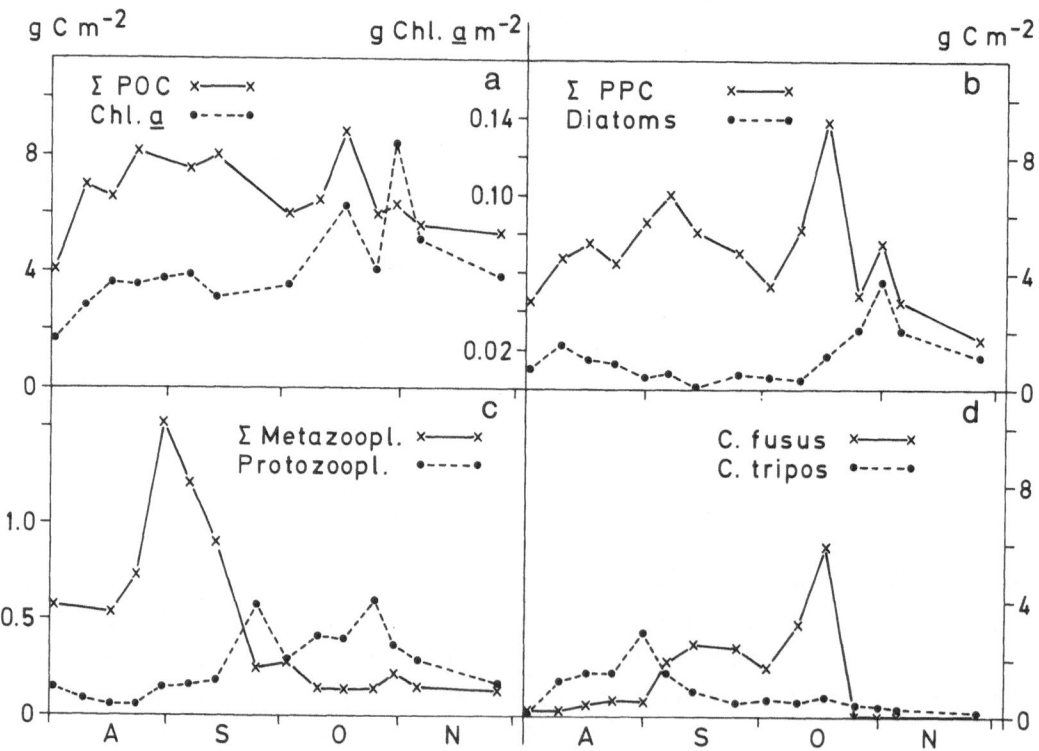

Fig. 2-8: Plankton parameters from late summer and autumn of 1973. Particulate organic carbon (POC) and chlorophyll a concentration (a); total phytoplankton biomass (PPC) and diatom contribution (b); Ceratium fusus
and C. tripos biomass (d), metazooplankton and protozooplankton
biomass (c). All values integrated for the 20 m water column.

This bloom can also be considered as new production but further growth is terminated by a combination of increasing turbulence and decreasing light levels rather than nutrient depletion. The fate of biomass produced by this bloom is not as clear as in the case of the spring and ceratia blooms. Recent studies (GRAF et al. 1983, CZYTRICH et al., 1986, NOJI et al. 1986) indicate sedimentation to be of importance as well. Zooplankton apparently play a minor role in the autumn phase, protozooplankton being more important than the metazoans (Fig. 2-5). The autumn protozoans are a mixture of ciliates and large phagotrophic dinoflagellates, and although one of the common autumn ciliates is reputed to feed on ceratia (ELBRÄCHTER 1971), grazing is apparently not of importance. The diatom bloom of 1973 had no apparent effect on the zooplankton. A short period of oxygen depletion and nutrient accumulation in bottom water has frequently been observed in November; the input of fresh organic matter from the autumn blooms must contribute significantly to this increase in benthic activity (GRAF et al. 1983).

2.1.5 Discussion

The seasonal cycle of plankton in Kiel Bight, in terms of biomass and species succession, is similar to those found in other coastal, temperate waters. The spring diatom bloom followed by a herbivorous copepod maximum is characteristic of the temperate Atlantic. The succeeding sparse flagellate population leading to the summer regenerating community is also the rule. However, the summer phytoplankton biomass in Kiel Bight is higher than in most other regions, a feature apparently related to the hydrography and shallowness of the area in general. Autumn blooms also appear to be a seasonal feature characteristic of many regions (RAYMONT 1981). This pattern fits well in the classical schematic seasonal cycle of phytoplankton and zooplankton biomass of the North Atlantic set out by HEINRICH (1962). The structural changes recorded by us in Kiel Bight are therefore not exceptional.

The five phases of the seasonal cycle dealt with above obviously differ from each other in both structural and functional aspects. On a general level, it is also possible to identify the main driving forces bringing about change in the system. Whereas physical factors of the environment directly shape the processes occurring at the beginning and end of the growth season, the role of biological feed back loops in maintaining system structure is of greatest importance in the middle phases. The maximum biomass that can be attained by the spring bloom is determined by the winter nutrient concentrations, i.e. the carrying capacity of the system, once sufficient radiant energy is available, is controlled by geochemical mechanisms located at the sediment interface (see chapter 4, this volume). The intensity of vertical mixing together with radiant energy supply rather than the advent of stratification determines the onset of the bloom.

The spring bloom triggers spawning and growth of the overwintering copepod adults and young stages respectively; further, by transferring nutrients from the water column to the sediments, it fundamentally changes the pelagic environment, that is, it limits the potential for further development of the system. However, the timing and species composition of the bloom has little impact on the ecosystem as it is largely utilized by non-specialized opportunists (bacteria) (GRAF et al. 1982).

Sedimentation of the spring phytoplankton bloom is not a phenomenon peculiar to Kiel Bight, although it has not been studied with the same degree of detail elsewhere (SMETACEK 1984). High sedimentation rates of plankton were reported from the 70 m deep Bornholm Sea (Southern Baltic) by SMETACEK et al. (1978). COACHMAN and WALSH (1981) from data of the Bering Sea suggest that much of the material produced by the spring bloom in shallow shelf regions is transferred to the benthal in contrast to deep areas. Spring bloom sedimentation similar to that in Kiel Bight has been reported from Norwegian polls (SKJOLDAL and LÄNNERGREN 1978, WASSMANN 1983) and the Baltic coastal waters (JANSSON 1978). After spring bloom sedimentation, the nutrient-poor, light-rich environment is regularly colonized by small flagellates. Prorocentrum balticum dominates the May phytoplankton with a regularity remarkable for such an ubiquitous species. LOHMANN (1908) states that its occurrence in Kiel Bight is independent of salinity and temperature; the reasons for its appearance apparently lie in the structure of the system at that time, i.e. a water column not yet stabilized but practically devoid of nutrients. Such a situation can only be found in shallow regions such as Kiel Bight, where the spring bloom development is not triggered by seasonal water column stratification.

P. balticum is known to carry out vertical migration (WHEELER 1966), i.e. it is an active swimmer and this feature, coupled with its small size which increases efficiency of nutrient uptake, might explain its success after spring bloom sedimentation. The importance of this stage for the further development of the pelagic system as compared to that of the spring bloom is demonstrated by the fact that establishment of the zooplankton population and differentiation of the pelagic food web is initiated by the population of small dinoflagellates. As they grow in nutrient-poor water, they utilize freshly regenerated nutrients that originate from both pelagic and benthic remineralizers. This population is eventually grazed down by zooplankton and, as faecal pellets are utilized within the water column and sedimentation rates are at their lowest for the year, essential elements are effectively passed on to the summer populations. During summer, the high production, low sedimentation and also high remineralization rates within the water column, indicative of rapid turnover, point to biological balance within the system; i.e. competition within trophic levels is of greater importance, as is also the grazing pressure, than at other times and biological rather than environmental selection increasingly influences the course of succession. HENDRIKSON (1976) found a turnover time of particulate organic matter of approximately one week during summer. The phytoplankton is represented by all the different "ecological types" occurring in Kiel Bight: large, medium and small, motile as well as non-motile forms; amongst zooplankton, fine, medium and coarse suspension feeders as well as raptorial feeders are all represented. This greatest "ecological diversity" of the year is an expression of the increasing complexity of the system. This is evidenced not only by the extreme vertical inhomogeneity of the physical and chemical properties of the water column but also by the greater degree of vertical segregation of the phytoplankton population than at other times. Sedimentary input of nutrients from the sides, locally intense vertical mixing or even restricted upwelling and the patchy distribution of large organisms such as jellyfish all also contribute to horizontal inhomogeneity in the pelagic environment.

The autumn bloom phase bears some similarity to the spring bloom in terms of its import/export behaviour; however, growth patterns of the two phases differ radically. The spring bloom is characterized by an outburst of fast-growing diatoms, in contrast to the first autumn bloom which consists of slow-growing, long-lived ceratia. They build up their population over the entire summer period to reach dominance in autumn. The major biological changes characterizing transition from the summer to the autumn stage are the generally rapid decline in metazooplankton biomass and the continued increase in the biomass of ceratia, during which Ceratium tripos is replaced by C. fusus in dominance. To what extent the zooplankton decline is related to the rise of the ceratia is uncertain. Most of the species decline at the same time; the decline might well be of a self-regulatory nature, i.e. the adults die of old age after completing reproduction in late summer. However, it is also possible that the rapidly increasing phyto-/zooplankton ratios during September are related to the unpalatability of the ceratia for the metazoans. That ceratia are indeed eaten by crustaceans has been reported occasionally but it remains a fact that they are not relished by zooplankton as the majority of the smaller dinoflagellates are. Whether this is due to morphological characteristics or to more subtle factors such as the presence of herbivore deterrants requires clarification. Whatever the reason, longevity as a success strategy in a system subject to heavy grazing must be coupled to some other features that reduce the risk of being eaten. Many terrestrial plants have adopted this strategy and there is no reason to believe that defence against grazing should not have evolved in pelagic plants as well. This would explain the long residence time and comparatively slight fluctuation of the ceratia population each year.

The autumnal diatom bloom also differs from the spring diatom bloom in that it apparently grows comparatively slowly through the ceratia phase and reaches its maximum not before sedimentation of the dinoflagellates. Its growth is not terminated by nutrient depletion but rather by declining light and increasing turbulence. Heavy sedimentation following this bloom was recorded in November 1981 (GRAF et al. 1983).

The autumn phytoplankton populations utilize light with much greater efficiency than the spring phytoplankton bloom. This is not due primarily to intrinsic properties of the species involved. Apparently, the size of the initial population is decisive. In spring, the bloom builds up from a very small seeding population within 2 to 3 weeks, whereas the autumn blooms represent culmination of populations growing slowly over a much more extended period. Light utilization is more efficient simply because of the larger phytoplankton standing stock. However, the effect of temperature, which is much higher in autumn than in spring (10°C and 0 to 4°C respectively), cannot be ruled out as a factor in increasing efficiency.

In this presentation of the temporal progression of the entire pelagic system we have shown that the developmental continuum comprises discrete phases that differ fundamentally in various aspects of their structure and function. Treatment of a region on the basis of annual averages will not further our understanding of marine ecosystems. For example, the annual primary production figure of a given area tells us little about the availability of food in that system, unless it is differentiated into "new" and "regenerated" production. Similarly, as discussed in chapter 3, this volume, the pattern of sedimentation is of fundamental importance in determining structure and functioning of the benthic system.

The characteristic quantitative features of the four phases of the growth season have been listed in Table 2-1. These generalized figures have been derived from measurements or from indirect estimates. They summarize the relationship between import, patterns of utilization and export of the four phases. As pointed out by SMETACEK et al. (1984), differentation of annual cycles into phases based on quantitative relationships of the dominant components will provide the necessary context for interregional comparisons. It is our belief that such ecosystem comparisons will greatly aid in furthering our understanding of the dynamics of these systems.

Table 2-1: Quantitative relationships expressed as approximations between inputs, outputs, and production within the system as well as range in biomass ratios of some of the major planktonic compartments characteristic of the four stages of the growth season in Kiel Bight (from SMETACEK et al. 1984).

Stage	Duration (weeks)	Total primary production ($gCm^{-2}stage^{-1}$)	Average primary production ($gCm^{-2}d^{-1}$)	New/total production (%)	Sedimentation of total production (%)	Phytoplankton/total zooplankton biomass	Metazooplankton/protozooplankton biomass
.....	6-8	~20	0.4	>75	>50	>10:1	0.3-0.1:1
2.....	6-8	~20	0.4	~50	<25	3-1:1	5-0.3:1
3.....	12-14	~70	0.8	<25	<25	6-3:1	2-5 :1
4.....	6-8	~30	0.6	>75	>75	>10:1	1-0.3:1

2.2   PELAGIC MICROBIAL PRODUCTION
       (M. BÖLTER, R. DAWSON, G. LIEBEZEIT, K. WOLTER)

2.2.1   Introduction

The bulk of primary production in the pelagic environment is mainly due to phytoplankton. This material is present either as particulate matter or as exudates in dissolved form and can be metabolized at different trophic levels. In the open ocean the
particulate material is mainly grazed by the different members of the zooplankton,
i.e. the secondary and tertiary producers. Microheterotrophic organisms are mainly
responsible for the use of the dissolved organic material and for its conversion into
particulate matter. Within the framework of the detritus food chain (FENCHEL and JÜR-
GENSEN 1977, FENCHEL 1977, HARGRAVE and PHILLIPS 1977) the main role of the microheterotrophs lies in remineralization. SIBERT and NAIMAN (1980), however, regard
bacteria as comparable to the original primary producers since they utilize dissolved
organic material emanating from all trophic levels for their own biomass production
and thus make this material available as a food source for phagotrophs.

Fig. 2-9 depicts microorganisms in their function as decomposers and producers in an
aquatic system and illustrates their role as a mediator amongst the various organic
compounds. In this section, we will consider dissolved organic material as a food
source for the microbial population, whereby special emphasis is laid on its activity
and to seasonal and spatial variations.

Of the various compounds of dissolved and particulate organic material, analysis of
those substances metabolized by the microheterotrophs is particularly interesting.
Autoradiography and other radio tracer methods have shown that monomeric substances
are taken up rapidly by natural bacterial populations (e.g. HOPPE 1977, GOCKE 1977,
BÖLTER et al. 1982).

Chemical analyses of the dissolved organic material indicate that monomeric compounds
are present in significant amounts in the natural environment (LIEBEZEIT 1980, PALM-
GREN 1981). However, they represent only a small proportion of the total amount of
dissolved organic matter. Different analytical approaches lead to various classifica-
tions of the quantifiable parts of the organic material, e.g. refractory carbon,
structural components, polymeric and monomeric substances (DAWSON and LIEBEZEIT
1981).

Although individual compounds may be defined by chemical techniques, this does not
imply availability to bacteria (GOCKE et al. 1981, LIEBEZEIT and DAWSON 1982).

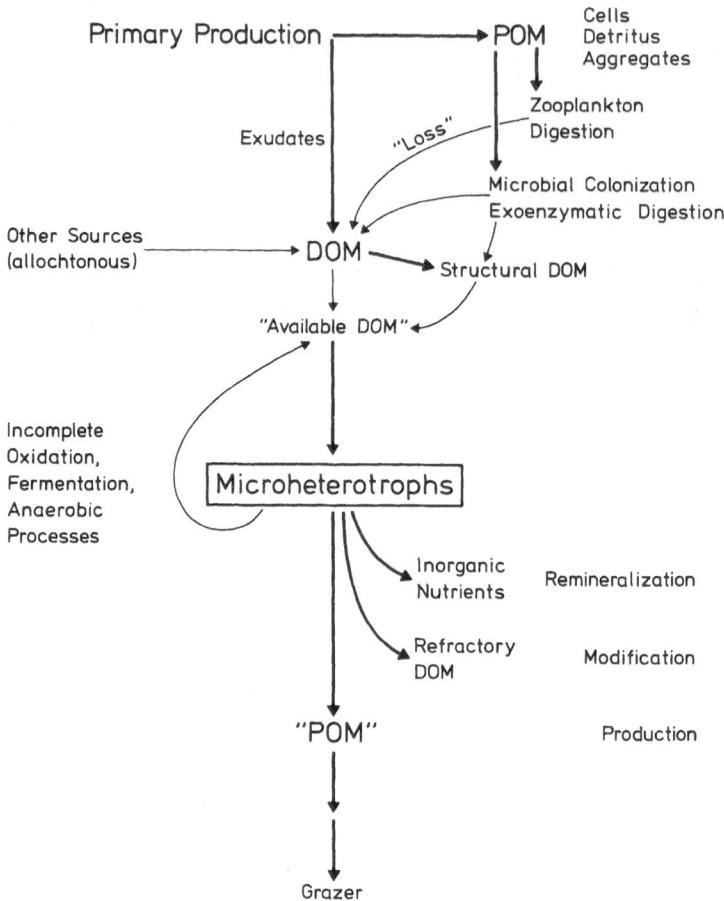

Fig. 2-9: Diagram of the role of microheterotrophs in the pelagic environment.
        DOM: Dissolved organic material,
        POM: Particulate organic material.

Thus, the microbial activity is often described by indirect methods, e.g. measurements of the heterotrophic potential. As only a few of these methods may be regarded as measurements of actual processes, there are few definite results expressed as production or remineralization rates.

Large amounts of dissolved organic material are produced by phytoplankton via exudation. Concentrations and availability of these substances are an important aspect in the analysis of relationships between phytoplankton and bacteria (LARSSON and HAGSTRÖM 1979, WOLTER 1982). However, the utilization of these products by bacteria (e.g. respiration, remineralization, metabolism) seems to be mainly influenced by environmental conditions.

2.2.2   Food Sources

As it is impossible to characterize all organic substances in the marine environment, it has been our aim to analyse mainly those components which are known to play a significant role in the food web.

Carbohydrates, for instance, are produced in large amounts by marine organisms, e.g. as storage products, structural compounds and exudates. Together with amino acids and lipids they account for a substantial part of primary production and represent a significant part of the autochthonous material. The remainder is often described as refractory material and "Gelbstoff" (KALLE 1966) and is of high molecular weight. GAGOSIAN and LEE (1981) point out that this material is also of marine origin, although in near shore environments we have to consider a certain contribution of terrestrial material.

Approximately 40% of the dissolved organic material (DOM) can be positively identified (LIEBEZEIT and DAWSON 1982). However, from the investigations of LIEBEZEIT (1980) and WOLTER (1980) it can be inferred that only 2 - 15% of the DOM is available to the microheterotrophic population. These values correspond to those of YURKOWSKI (1971) who states that 3 - 10% of the DOM in the Baltic is labile and hence may be also available to osmotrophic microorganisms.

Components which can be analysed definitely (without drastic manipulations), such as monomeric carbohydrates, amino acids and fatty acids, are particularly interesting because the mild analytical conditions employed do not seriously change the nature of the original material (LINDROTH and MOPPER 1979, LIEBEZEIT 1980, DAWSON and LIEBEZEIT 1981, DAWSON et al. 1983).

The determination of classes of substances, such as free dissolved, extractable or refractory compounds (JOHNSON and SIEBURTH 1977, DAWSON et al. 1983) can considerably help in describing the state of the original material or even the condition of the phytoplankton population (BÜLTER and DAWSON 1982).

The multitude of possible chemical, biochemical and biological reactions which can occur in the ecosystem is presented in Fig. 2-10. Glucose is used as an example to demonstrate some of these reactions:
Free dissolved glucose is found as a result of functional exudation, stress exudation, cell lysis or enzymatic degradation of polymers. It can be used directly by microorganisms. Another function is the formation of complexes (MOPPER et al. 1980), of primary condensates (GOCKE et al. 1981) or of Schiff's bases with compounds containing primary amine groups. Furthermore, it can adsorb on particulate material or colloids.

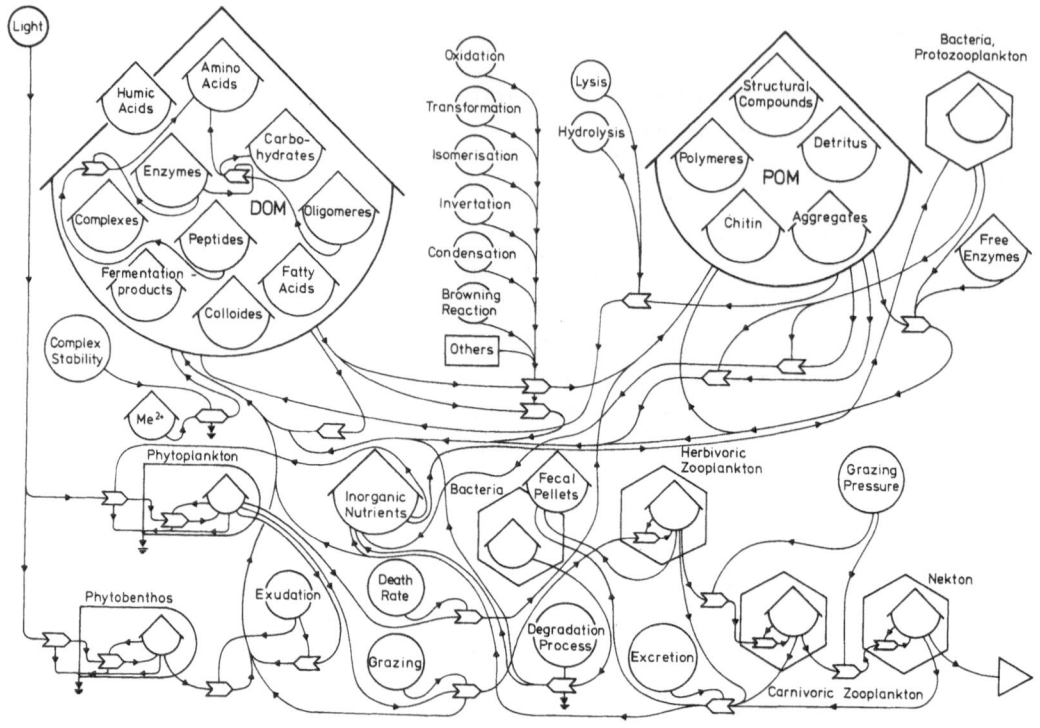

Fig. 2-10: Proposal of biological and chemical interactions between dissolved (DOM) and particulate (POM) material.

The relatively constant ratio between glucose and fructose in seawater (MOPPER et al. 1980) can be explained by the Lobry de Bruyn-Alberda van Erkenstein rearrangement. However, this process is very slow at _in situ_ temperatures and enzymatic transformation has been discussed as a major mechanism in maintaining this balance.

Comparable phenomena are discussed by BÖLTER et al. (1980) for the abundance of ribose in seawater. This carbohydrate must be regarded in close connection with the nucleic acids.

2.2.3   Spatial and Temporal Distribution of Dissolved Organic Matter

Thermoclines and haloclines may be regarded as barriers for the flux of particulate and dissolved organic matter. These boundary layers are of special interest in the analysis of pelagic environments and it is possible to use chemical and biological parameters for their characterization.

LENZ (1974) showed for the Kiel Bight that particulate matter can be trapped within density gradients. In an oceanic environment, dissolved organic compounds such as amino acids and carbohydrates also showed maxima in pycnoclines which were accompanied by an increase in microbiological activity (LIEBEZEIT et al. 1980).

Beyond the seasonal and spatial variations in production and decomposition, one has to consider short term variations in these processes. Diurnal and hourly changes of the concentrations of individual substances as well as in the bulk of the DOM have been monitored. These changes are often in the range of the seasonal fluctuations or even higher. They may be due to oceanographic phenomena (currents, turbulence) and internal rhythms of the organisms. GOCKE (1975a) and MEYER-REIL et al. (1979) showed variations in microbial activity and the amount of organic and inorganic substances to occur on a large scale.

2.2.4   Seasonal Influences on Food Sources and Microbial Production

For the investigation of seasonality, a comprehensive study was carried out 1978/79 at a fixed station in Kiel Fjord, which provided data used in a number of papers focussing on chemical, microbiological and planktological parameters (LIEBEZEIT 1980, WOLTER 1980, 1982, PALMGREN 1981, BÖLTER 1981, 1982a, 1982b, BÖLTER et al. 1982). The total data file comprises about 150 parameters for which measurements were taken at least every second week. For this analysis the data set was reduced to 80 parameters.

Because of the great variation in the individual parameters, their rapid changes and interactions, the use of an adequate mathematical procedure for analysis was imperative. As the data set concerns time series, parameter-free methods were necessary. The programme block developed by BÖLTER et al. (1980) contains suitable procedures for use on such ecological data sets.

By using a combination of three different cluster analyses (complete linkage, average linkage and median; LANGE and WILLIAMS 1966, BOCK 1974) the stations, i.e. sampling data, are sorted into 4 groups. They correspond generally to the seasons spring, summer, autumn and winter. The cluster "autumn", however, also contains some data of late spring. It should be mentioned here that Kiel Fjord differs in many respects, both in biomass levels and dominant species, from the open Kiel Bight described in the previous section.

As it is not possible to present all courses of the measured parameters here, presentation is concentrated on those parameters which are important for describing main items of the system (see Fig. 2-11 and Table 2-2). In the following, the 4 seasonal situations are described with special regard to the microheterotrophic organisms. All data are given in terms of the median (X(m)) of the corresponding data set, i.e. time period.

Winter phase

This season is characterized by a high variation in the individual parameters which, however, act on low levels. Water temperature has a median value of 3°C, the energy input by solar radiation a value of 320 J cm^{-2}. Corresponding to the homogeneous water column in winter (c.f. section 2.1, this chapter), the inactive phytoplankton (primary production, X(m) = 1.6 µg C l^{-1} h^{-1}), the low standing stock of particulate organic carbon and phytoplankton carbon, the dissolved organic carbon also shows low values (dissolved carbohydrates, X(m) = 19.2 µg C l^{-1}; exudates, X(m) = 0.86 µg C l^{-1}). However, the standing stock of bacteria does not show such a drastic decrease in comparison to other seasons. Microbial biomass levels are approximately at 12 µg C l^{-1}. Similar values were recorded by ZIMMERMANN (1977) at a comparable station in Kiel Fjord.

In spite of the availability of sufficient dissolved organic carbon, microbial activity is significantly lower than in other seasons. The bacterial carbon production shows the lowest value (X(m) = 0.16 µg C l^{-1} h^{-1}, using glucose as substrate). This is also shown by the low ratio of production/biomass. The reason for this low overall microbial metabolic activity may be that during autumn and winter only 9% of the total bacteria have active metabolism (HOPPE 1977). Hence, it is proposed that either the available substrate has fallen below a threshold, or some essential substances are absent, or other parameters are limiting the activity of microorganisms.

As such, temperature is often discussed as a controlling factor for microbial activity (e.g. HOPPE 1977, GOCKE 1977). However, detailed consideration of this parameter shows that during the period November to March, with water temperatures between 0 and 5°C, highest fluctuations in microbial activity are measurable (BÖLTER 1982a). Thus, discussions on these fluctuations generally centre around the assumption that the activity of the microorganisms depends primarily upon changing phytoplankton production and concomitant changes in stocks of organic carbon. This conclusion may also serve as an explanation for the drastic increase in microbial activity during the spring bloom of phytoplankton at low water temperatures.

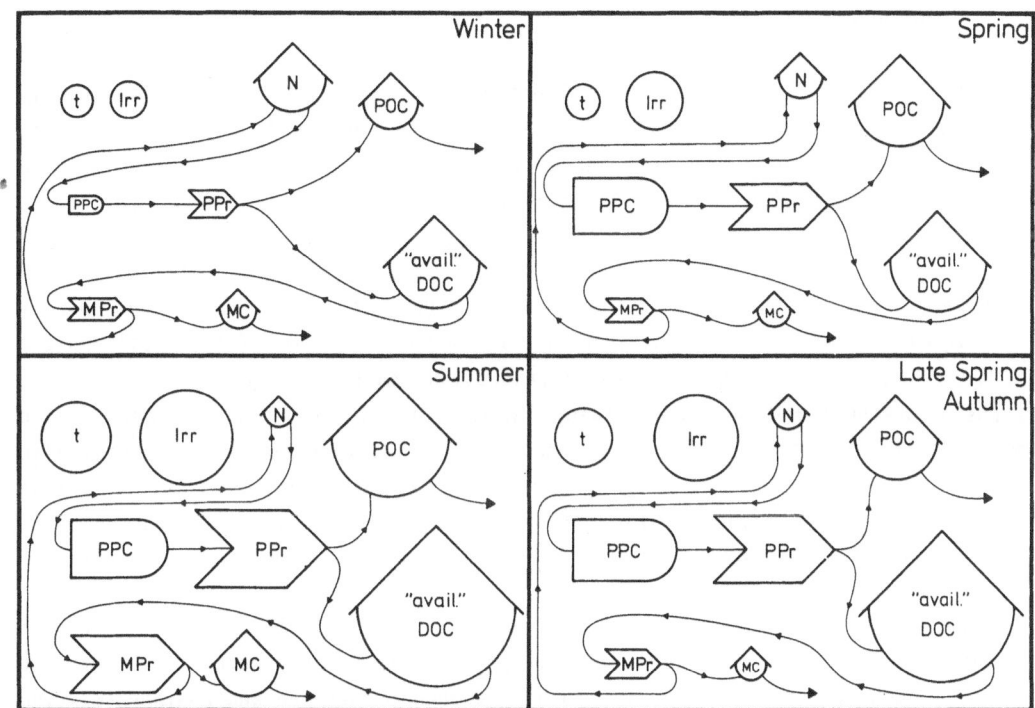

Fig. 2-11: Fluxes of organic and inorganic matter during the seasons as evaluated by the cluster analysis (c.f. Table 2-2). The individual compartments give a relative figure of the amount of material by data given in Table 2-2.

Table 2-2: Median values of parameters measured during the seasons as evaluated by cluster analysis.

	winter	spring	summer	late spring autumn
t (temperature) °C	4.4	3	15.3	9.8
Irr (global irradiation) (J cm^{-2})	324	763	1671	978
N (inorganic nitrogen) (µg at 1^{-1})	55.2	49.9	5.4	19
PPC (phytoplankton carbon) (µg C 1^{-1})	31.5	467	406	320
PPr (primary production) (µg C 1^{-1} 6h^{-1})	10.6	200	362	226
POC (particulate organic matter) (mg C 1^{-1})	0.3	0.7	1.3	0.9
"available DOC" (dissolved free amino acids and dissolved free carbohydrates) µg C 1^{-1}	31	34	68	37
MPr (microbial production based on glucose) µg C 1^{-1}	0.9	3.8	24.3	4.1
MC (microbial carbon) µg C 1^{-1}	10.8	21	52.4	18.4

Spring Phase

Although only few data from our set fall under this season they differ significantly from those of the other periods. As indicated by the results of earlier investigations (c.f. RHEINHEIMER 1977) this season can be regarded as a relatively isolated event in the annual cycle of the pelagic system as also suggested by SMETACEK et al. (section 2.1, this chapter).

The temperature rises only up to 3.8°C during spring (X(m)), whereas energy input by global irradiation increases significantly and reaches 714 J cm^{-2} (X(m)). Diatoms are the dominant primary producers; primary production reaches 26 µg C l^{-1} h^{-1} (X(m)), and phytoplankton standing stock attains 501 mg C l^{-1} (X(m)) (WOLTER 1980). At the same time, the concentrations of the components of dissolved organic material also increase: exudates, X(m) = 13.3 µg C l^{-1}; dissolved carbohydrates, X(m) = 38.4 µg C l^{-1}; and free amino acids, X(m) = 241 nMol l^{-1}. This is in contrast to the total dissolved organic matter (DOM) which increases only slightly (BÖLTER et al. 1982).

The standing stock of bacteria, however, remains nearly uneffected and increases negligibly to 12.3 µg C l^{-1} (X(m)), with respect to the winter months. A significant increase is observed in its activity, both in carbon production and remineralization and in the ratio of production/biomass.

The values of heterotrophic potentials during spring show drastic increases, as found by GOCKE (1977) some years earlier. The increase in parameters of microbial activity is only explainable by an increase on the part of the actively metabolizing bacteria as shown by HOPPE (1977). He found an enhancement for the fraction of active bacteria from 4 to 11% (in 1974) and from 25 to 52% (in 1975) in Kiel Fjord for the period February to March. Even more drastic are the values for Kiel Bight: 0.2 to 17% (March - April 1974) and 8.6 to 40.7% (February - March 1975).

Concomitantly, the turnover rates for dissolved organic compounds decrease (GOCKE 1977, BÖLTER et al. 1982), and we find a considerable remineralization, which, however, does not seem to be important for the pelagic environment since the amount of inorganic nutrients is still rather high (sum of inorganic nitrogen, X(m) = 62.3 µMol l^{-1}).

The main phytoplankton spring bloom ends with nutrient depletion and sedimentation and changes to a post-spring bloom which is characterized by small dinoflagellates and protozooplankton (see section 2.1.4). The analysis of our data file results in a special group of data which characterizes this period of the annual cycle. On the basis of the cluster analysis, this group cannot be defined precisely and it is closely connected to the autumn data group.

Neither the particulate organic material nor the DOM show significant higher values during the original spring bloom, regarding the free monosaccharides and free amino acids. A significant increase, however, is observed with the exudates; concentrations rise to 60 µg C l^{-1} (X(m)) and hence point to a higher carrying capacity of the microbial system.

The above mentioned changes in the phytoplankton population towards flagellates and the increase in protozooplankton may serve as an explanation for the low increase in microbial biomass. The protozooplankton obviously exerts a grazing pressure on the bacteria (GAST 1983), though only few results of field research on this topic are available. In addition, osmotrophic flagellates may compete with the bacteria for dissolved organic nutrients. The combination of these factors seems to be responsible for the special behaviour of the system after the spring bloom, before it is possible to define the situation in summer.

Summer phase

During this season all activity parameters show their highest values. The only exception is the relation between microbial production and biomass, which decreases slightly; this may be due to the high standing stock of bacterial biomass (X(m) = 63 µg C l^{-1}).

As opposed to primary production, the microbial production clearly decreases, only reaching maximum values of 1 % of the autotrophic production. Hence, a significant contribution of the microbial production to the total heterotrophic production cannot be observed.

An important aspect of the microbial activity during this season seems to be remineralization. The proportion of the respired material (measured as ^{14}C-carbon dioxide) generally amounts to more than 30% (BÖLTER 1982b). Due to the high primary production (X(m) = 46 µg C l^{-1} h^{-1}), both particulate organic matter (X(m) = 1.1 mg C l^{-1}) and DOM (e.g. exudates, X(m) = 51 µg C l^{-1}; dissolved free carbohydrates, X(m) = 38 µg C l^{-1}) are sufficiently abundant for the microheterotrophic organisms. In particular, the freshly produced material accounts for high turnover rates.

The high ratios of C/N, C/P and C/Chl a of the sedimenting material and the low sedimentation rates during summer (see section 2.1.4) indicate that most of the particulate material is remineralized within the water column by various zooplankters. The decomposition of particulate material by bacteria seems to be significant only near the bottom where high numbers of bacteria attached to detritus (ZIMMERMANN 1977) and considerably higher activity on particulate material (MEYER-REIL 1983) have been found.

As there is still sufficient nitrogen in the form of amino acids ($X(m)$ = 243 nMol 1^{-1}), the limitation through inorganic nitrogen and competition with the autotrophs does not seem to be evident. Though individual amino acids show high fluctuations, their values do not decrease to zero; moreover, during July the highest values of concentrations in amino acids were recorded (LIEBEZEIT 1980, PALMGREN 1981).

The fluctuations of the different size fractions of bacterial populations point to the high dynamics of this food web. PALMGREN (1981) found variations that were 4 times higher than the minimum value of the total number of bacteria during this season. However, this minimum value, compared to those of various other parameters, is significantly above a threshold value which characterizes the summer season and marks the limits of the other seasons.

Autumn phase

Only under certain biological conditions can this season be regarded as a separate event of the annual cycle. The cluster analysis of this season shows similarity to the post-spring bloom phase. The still high rate of primary production ($X(m)$ = 44.3 µg C 1^{-1} h^{-1}) serves to provide a high surplus of DOM, especially in the form of exudates ($X(m)$ = 68 µg C 1^{-1}) which are, however, not represented within the total carbohydrates ($X(m)$ = 18 µg C 1^{-1}) or amino acids ($x(m)$ = 190 nMol 1^{-1}). Hence, the bacterial production based on glucose is below the spring value ($X(m)$ = 0.3 µg C 1^{-1} h^{-1}). The corresponding value for the exudates as substrate was found to be still 1 µg C 1^{-1} h^{-1} ($X(m)$). This may point to a special adaptation of the bacterial population to this natural substrate. The dominating phytoplankton alga in this season was Rhizosolenia (WOLTER 1980). Although the system is characterized by the marginal conditions of the regenerated nutrients, this situation may be generally compared to the post-spring bloom, which is also evidenced by the associated zooplankton.

2.2.5 Discussion

Comparing the results of the study of plankton dynamics (see section 2.1) from Kiel Bight with this analysis of the data set from Kiel Fjord, concerning mainly the microheterotrophs, it becomes evident that the seasonal cycle, including its subunits, can be regarded as a recurring pattern typical for the area of investigation. Differences between the seasons are mainly influenced by the external energy input, e.g. the global irradiation.

In addition to this, the hydrographic parameters characterizing the spatial variations deserve mention. These factors determine activity at the level of primary production and thus the input of dissolved organic matter. The input by exudation, excretion or cell lysis produces different qualities and quantities of dissolved organic material. WOLTER (1980) has monitored values between 0 and 44.5 % of the primary production as exudates during an annual cycle.

Investigations of BAUERFEIND (1982) on the exudates of Chlorella sp. showed that their molecular weight is between 200 and 5000 D and that they are mainly identifiable as carbohydrates and proteins. Other authors found the main part of the dissolved organic material for exudates in the range below 1500 D (NALEWAJKO and SCHINDLER 1976). Hence, it can be argued that this low molecular weight fraction of the DOM can be directly utilized by the heterotrophic bacteria.

HOPPE (1981) assumes that phytoplankton exudates may provide 2/3 of the total bacterial production. However, a comparison with investigations of BÖLTER et al. (1982) in this area showed that exudates and ^{14}C-labelled carbohydrates behave quite differently, obviously due to differences in composition. Especially in spring and autumn the exudates are used much less than ^{14}C-labelled carbohydrates. During summer, when we find a mixed phytoplankton population and thus a mixture of various exudates, values for the uptake of exudates and ^{14}C-labelled hexoses are quite similar, so that glucose can be regarded as a model substrate for bacterial uptake or remineralization rates.

The low molecular weight substances, however, represent only a small portion of the total DOM. LIEBEZEIT (1980) found up to 3% of DOM in the pool of carbohydrates and amino acids. However, as determined by GOCKE (1977) and GOCKE et al. (1981) using the heterotrophic potential approach, the natural substrate concentrations of various monomeric substances are generally significantly lower.

It seems to be of importance that in spite of the low percentage of available compounds from the total DOM, the microbial standing stock and its activity exhibit wide fluctuations and that the microheterotrophic population is not able to make complete use of this stock of DOM.

Even in winter we find high concentrations of total DOM, about 3 mg l^{-1}, as well as of free dissolved carbohydrates. It is not possible to explain the decline in microbial activity by the low temperatures alone. Thus we must also consider the likelihood that the microbial activity is primarily controlled by other factors or a combination of them. The planktonic standing stock and the exudates, however, are closely related to microbial activity, but not to individual carbohydrates and amino acids as shown also in a graph theoretical approach of BÖLTER and MEYER (unpubl.).

This points to the fact that the exudates are of special importance and seem to have a trigger function for microbial activity. Obviously, this is not the case when a single species of alga dominates the phytoplankton as decribed for the seasons spring and autumn.

In summer the diverse phytoplankton populations serve to provide the more plentiful substrate, the heterogeneous microbial population acts as a buffer and therefore maintains the high trophic level of the system. From this analysis we may deduce that inorganic nutrients have no direct influence on the remineralization of the bacteria. On the one hand, the parameters describing the actual remineralization rates are closely connected to the overall activity parameters, while on the other the inorganic nutrients form a separate group of parameters as found in analyzing the data set.

Further, the different time scales of the interacting levels (e.g. DOM, bacteria, plankton) should be mentioned. As such, the microbial biomass can be characterized by turnover times ranging from some hours in summer to several weeks in winter. Likewise, the phytoplankton bloom in spring is known to have a faster growth rate than the bloom in autumn.

Special adaptation to substrates and high affinities to various compounds point to the high dynamic of the system. This was shown during several container experiments (e.g. WOLTER 1980, v. BODUNGEN et al. 1984). Thus it seems impossible to describe all boundary conditions of this trophic level in general.

The dominant role of the microorganisms in this pelagic system is their use of freshly produced dissolved organic matter; thus they are mainly secondary producers. The utilization of this part of primary production is important to the food web of the lower organisms because considerable amounts of organic material are transported via bacteria to filter feeders. In Kiel Fjord this may be up to 15.6% of primary production (WOLTER 1982).

CHAPTER 3:   ASPECTS OF BENTHIC COMMUNITY STRUCTURE AND METABOLISM

L.-A. MEYER-REIL, A. FAUBEL, G. GRAF and H. THIEL

3.1   INTRODUCTION

The biology of benthic communities has achieved increasing interest during the last
decade. This is mainly because the benthos was predominantly understood as an integrated part of the marine ecosystem. Especially in shallow waters such as Kiel Bight,
mutual relationships exist between the water column and the sediment, mediated by the
sediment/water interface, a boundary which can only be defined theoretically.

The supply with organic material turned out to be the dominating factor determining
structure and activity of benthic communities of Kiel Bight (GRAF et al. 1983, MEYER-
REIL 1983). Simplistic earlier approaches that assumed a continuous and steady nutrient flux of particles from the pelagic into the benthic system (STEELE 1974) must
be replaced by a more complete understanding of the coupling between both systems. In
boreal marine environments, the seasonal cycle of sedimentation events (SMETACEK
1980; c.f. chapter 2, this volume) supports the benthos with organic material. Only
benthic communities in shallow waters which allow the penetration of enough light to
the bottom may be regarded as self-sustaining systems.

Sediments exhibit a complex structure. Through the activity of benthic organisms,
micro-environments are created which range from bacterial fibrous webs and diatom mucus to tubes and burrows produced by higher organisms (polychaetes, mussels). By
their intensive extracellular secretion of high molecular weight material, the organisms may even influence the texture of sediments. In a system such as the sediment,
which tends to preserve its stratification, organisms that actively support the exchange of substances are highly important.

With increasing sediment depth and decreasing availability of energy, a range of
electron acceptors is provided which guarantees the final oxidation of organic material even under anaerobic conditions. Through the oxidation of organic material,
inorganic compounds are released which may be oxidized anaerobically within the sediment or aerobically at the sediment water interface. Within the sediment stratum,
elements may circulate repeatedly thus complicating mass balances (c.f. chapter 4,
this volume). The organisms mediating these processes, exihibit highly complex tro-
phodynamic characteristics. By decomposing organic material, the benthic organisms
meet their requirements for energy and cell constituents thus being the driving force
for the cycling of elements.

This chapter illustrates aspects of the community structure and metabolism of benthic
communities. Certainly this is a reflection of the historical development in benthic
research. In pioneering studies, emphasis was laid on the abundance of meio- and macrofauna, their biomass and production. With increasing knowledge of the coupling
between pelagic and benthic systems, research was concentrated on the response of the

benthic community to the annual sedimentation pattern using recently developed methods for measuring community metabolism. A separate section was devoted to the role
of bacteria in benthic communities.

3.2   COMMUNITY STRUCTURE: ABUNDANCE, BIOMASS AND PRODUCTION
         (A. FAUBEL and H. THIEL)

3.2.1   Abundance

In line with the areas of interest in various research groups, work on benthos in
Kiel Bight in recent years has concentrated on macrofauna standing stock and macrofauna production in relation to fish production, on meiofauna and on microbial studies. This section gives a survey of the abundance, biomass and production of benthic
organisms, while the following sections report on community metabolism and its relations to pelagic processes.

Due to the lower salinity values regularly found in Kiel Bight, viz. 10 - 20 °/oo in
surface layers and 15 - 30 °/oo in deeper waters, the number of species of the benthos
distributed in Kiel Bight is lower than for example in the fully marine North Sea
(ARNTZ 1971, 1978b, REMANE and SCHLIEPER 1971). With respect to Mollusca, Polychaeta
and Crustacea, this reduction amounts to about 50 % or more in terms of species numbers (ARNTZ 1971). Nevertheless, more than 150 macrofauna species are abundant and
most of them belong to polychaetes, molluscs, and crustaceans. Predominant species in
shallow areas of Kiel Bight are Cerastoderma edule (syn. Cardium edule), Macoma baltica, Mya arenaria, Montacuta bidentata and Mytilus edulis (WORTHMANN 1975). On muddy
sand bottoms down to approximately 22 m water depth Astarte sp. dominate. In Abra
alba communities the mollusc Arctica islandica (syn. Cyprina islandica) with respect
to biomass, the polychaete Nephtys sp., and the crustaceans Diastylis rathkei and Ga-
strosaccus spinifer in number are the prevailing species, although Pherusa plumosa,
Terebellides stroemii, and Lagis koreni may also become dominant (ARNTZ and BRUNSWIG
1975b, ALHEIT 1978, ARNTZ 1981). Besides Abra alba, altogether about 40 prey species
for demersal fish are observed in Kiel Bight. The dominating benthic predators are
Gadus morhua, Merlangius merlangus, Limanda limanda, Platichthys flesus, Pleuronectes
platessa, and Gobiidae (ARNTZ 1980).

In accordance with the concept of PETERSEN (1914), ARNTZ and BRUNSWIG (1976) distinguish two major associations, the Macoma baltica community (generally above the pycnocline down to a depth of 15 m on sandy sediments) and the Abra alba community

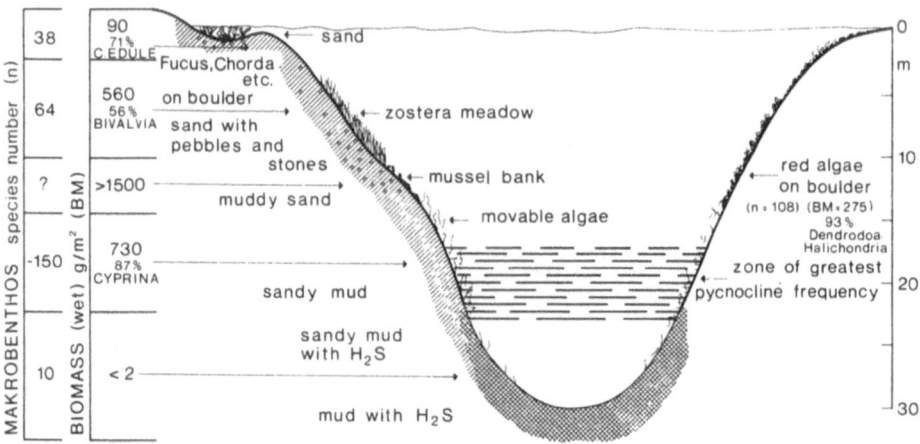

Fig. 3-1: Number of species, biomass and zonation of macrozoobenthos in channel systems of southwesterly Kiel Bight (according to ARNTZ 1981 and THEEDE 1981).

(below 15 m on muddy sand and mud, Fig. 3-1). However, seasonal fluctuations in species standing stock and changes in species distribution due to environmental factors - in Kiel Bight especially the oxygen deficiency - do not allow a clear classification of communities. This fact should be kept in mind when the community concept is still used in describing findings of earlier authors.

In Kiel Bight the water column is normally highly stratified in summer, thus preventing water exchange and oxygen renewal. This factor normally leads to seasonal H_2S stress in deeper channel systems. In these biotops the Arctica-Capitella communities (SCHULZ 1969) occur as an impoverished variant of the Abra alba community. The dominating species which can tolerate low levels of oxygen are Capitella capitata, Polydora ciliata, Harmothoe sarsi, Scoloplos armiger, Pectinaria koreni, Diastylis rathkei, Arctica islandica, Astarte spp., and Corbula gibba (syn. Aloides gibba) (ARNTZ and BRUNSWIG 1976, KÜLMEL 1979). However, annual and year to year variations determine species composition. Pectinaria koreni is partly replaced by Halicryptus spinulosus under low oxygen conditions (ARNTZ and BRUNSWIG 1976). In shallow bottom areas without seaweed, the lowest species diversity of the macrobenthos is met with about 20 species (WORTHMANN 1975), whereas species numbers increase in phytal regions (REMANE 1940, GRÜNDEL 1980, WORTHMANN 1975, LÜTHJE 1978).

Looking at the channel system (Vejsnaes Rinne, Boknis Rinne, and Stoller Grund-Rinne) (Fig. 3-1), a tendency of increasing species diversity and biomass of the macrobenthos is observed with increasing depth, down to depths of 10 - 15 m, just above the zone of greatest pycnocline frequency (Fig. 3-1, 3-2) on fine sand, muddy sand or sandy mud (ARNTZ and BRUNSWIG 1976, ARNTZ 1980, see also Fig. 5-3, this volume). Abundance data of macrobenthos for eight complete years (1968 - 1971 and 1975 - 1978) are pre-

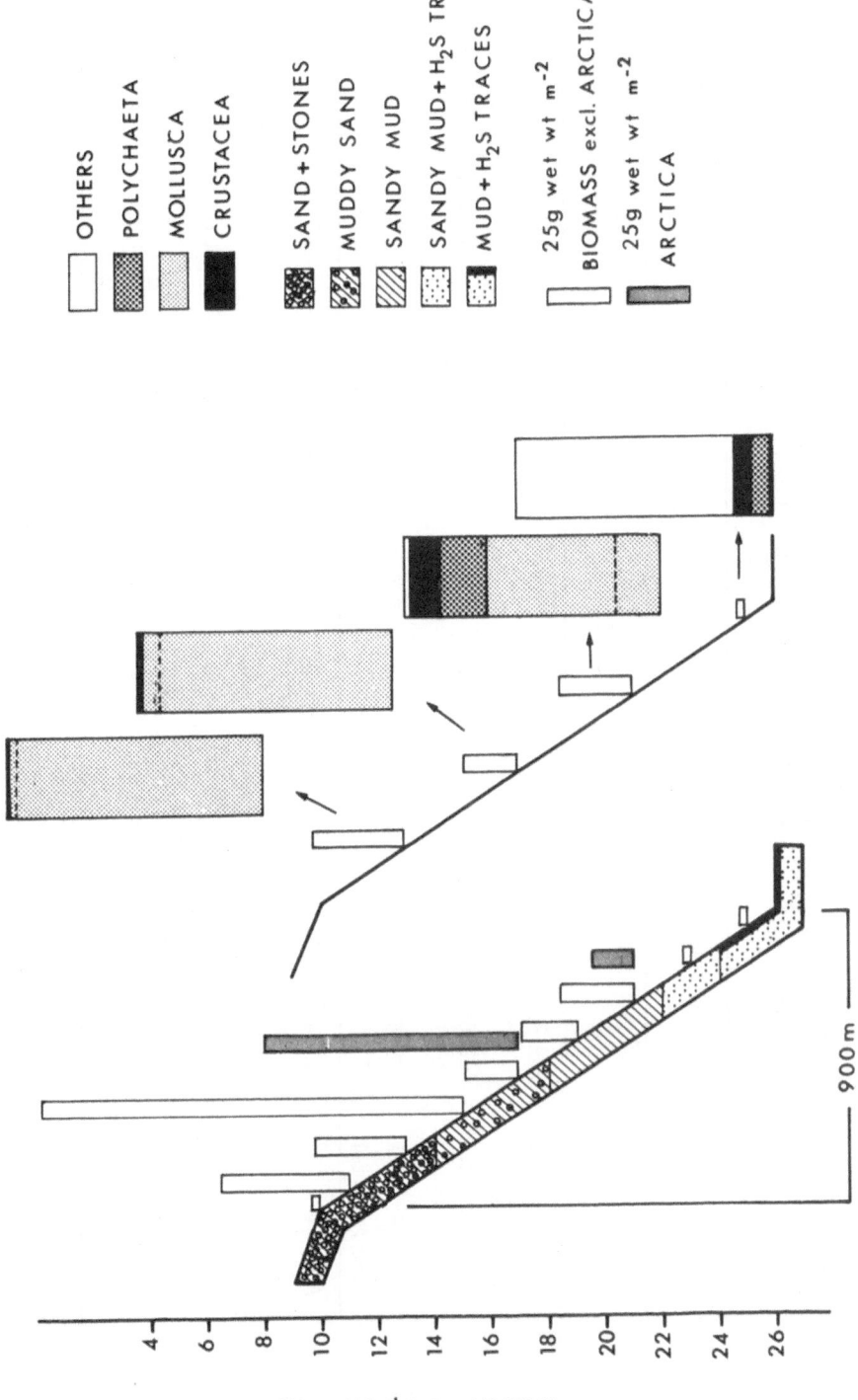

Fig. 3-2: Zonation of macrobenthos in Boknis Rinne, Kiel Bight (ARNTZ and BRUNSWIG 1976 modified).

Table 3-1: Abundance and biomass (AFDW: ash free dry weight) of macrobenthos in the Süderfahrt/Millionenviertel trawling area (cf. attached map). Annual means for years 1968 to 1971 and 1975 to 1978 and mean values for the period of investigation (ARNTZ 1980).

	1968	1969	1970	1971	1975	1976	1977	1978	mean and standard deviation
Number of individuals	955	1223	1695	1715	1640	1801	1558	1395	1498 $\pm$ 288
Wet weight ($g \cdot m^{-2}$)	914.8	1073.5	816.5	1082.7	705.3	942.9	944.3	744.8	906.4 $\pm$ 135.1
(excl. Arctica)	122.8	89.5	105.5	104.7	125.3	144.9	134.3	139.8	120.4 $\pm$ 19.4
(excl. Arctica and Astarte)	49.8	48.5	37.5	44.7	70.3	74.9	54.3	55.8	54.4 $\pm$ 12.6
AFDW ($g\ m^{-2}$)	37.4	44.5	33.8	45.0	30.3	39.4	39.4	32.8	
AFDW (excl. Arctica)	5.8	5.1	5.3	5.8	7.1	7.5	7.0	7.4	
AFDW (excl. Arctica and Astarte)	4.3	4.3	4.0	4.6	6.0	6.1	5.4	5.7	

Abundance (▲ 1979-80; ▽ 1977 BOJE 1977) n·10^4/100 cm^2
Biomass (○) µg·10^4/100 cm^2

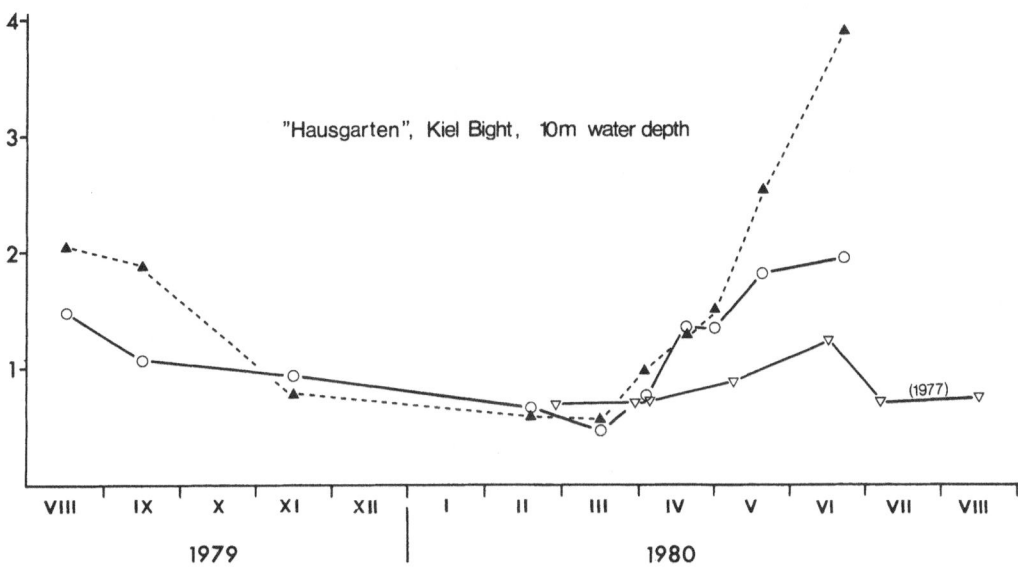

Fig. 3-3: Abundance and biomass of meiobenthic organisms at a 10 m station (Haus-garten, Kieler Bucht). Abundance according to BOJE 1977.

sented in Table 3-1 for the _Abra alba_ community (Süderfahrt/Millionenviertel, 17-19 m water depth). Mean individual numbers range between 955 and 1801 m^{-2} (ARNTZ 1980). In deeper mud areas below the pycnocline, the species number rather abruptly decreases where O_2 deficiency and H_2S are the controlling factors (ARNTZ and BRUNSWIG 1976, ARNTZ 1981, THEEDE 1981). Like the macrofauna, the meiofauna also shows greatest abundances at medium depths. Lower values are found with increasing depths below 18 m water depth, due to a decrease in species diversity. The abundance of Foraminifera increases and the density of Nematoda is relatively constant with increasing depth (KÜLMEL 1977, SCHEIBEL 1976, WEFER and LUTZE 1976). SCHEIBEL and NOODT (1975) state that sediment texture determines the composition of the meiofauna. In comparing the fauna living in medium fine sand with that found in coarse sand, a clear dependence becomes obvious: decreasing grain size results in an increase in nematode abundance and a decrease in individual numbers of harpacticoids (SCHEIBEL 1976). As Table 3-2 shows, species diversity and abundance increase with increasing water depth and in-creasing grain size down to zones of medium depths (10 to 12 m) with substrates of coarse sand. In these regions species diversity attains maximal values. In deeper areas with mud bottoms, particularly in the channels with seasonal H_2S stress, nema-todes, ciliates, and foraminifers predominate (KÜLMEL 1979, REIMERS 1976, WEFER and LUTZE 1976).

Table 3-2: Occurrence of meiofauna taxa with respect to water depth and abundance in
Kiel Bight. Sampling time: June 1980.

taxa	Surendorf medium sand midlitoral (n/100 cm²)	Falkenstein coarse sand midlitoral (n/100 cm²)	"Hausgarten" coarse sand 12 m water depth (n/100 m²)
Nematoda	1,245	2,729	29,400
Harpacticoida	450	1,922	7,844
Turbellaria	125	3,939	398
Oligochaeta	100	1,556	52
Polychaeta	-	-	110
Acari	15	96	373
Ostracoda	-	12	226
Gastrotricha	95	1,119	228
others	5	46	427

3.2.2   Annual Cycles

The macrobenthos of Abra alba communities proves to have an annual cycle of abundance
with a minimum standing stock in the period May to July and a maximum during autumn
and winter (Fig. 3-4). This holds true for regions with favorable oxygen conditions,
while unfavorable situations modify the cycle.

Generally, the long-term dynamics of dominant macrobenthos species are represented in
three types (ARNTZ 1981, THEEDE 1981):

1. species with long life-span and relative long-term stability
   (Arctica, Astarte, Nephtys);
2. species with regular seasonal fluctuations (Diastylis);
3. species with irregular seasonal fluctuations (Abra, Terebellides).

Investigations of KÜHLMORGEN-HILLE (1963, 1965), ARNTZ (1971) and ARNTZ and BRUNSWIG
(1976), showed that the annual cycle of macrobenthos in Kiel Bight is subject to distinct seasonal variations in relation to abundance, diversity and biomass. The typical reasons for the occurrence of fluctuations are defined by ARNTZ and BRUNSWIG
(1976), KÜHLEMORGEN-HILLE (1965), BANSE (1956), ARNTZ, BRUNSWIG and SARNTHEIN (1976)
and ARNTZ (1981):

1. mass mortality in extremely cold winters;
2. variations in the transport of larvae from Kattegat and Great Belt;

BIOMASS, ABUNDANCE [rel. units]

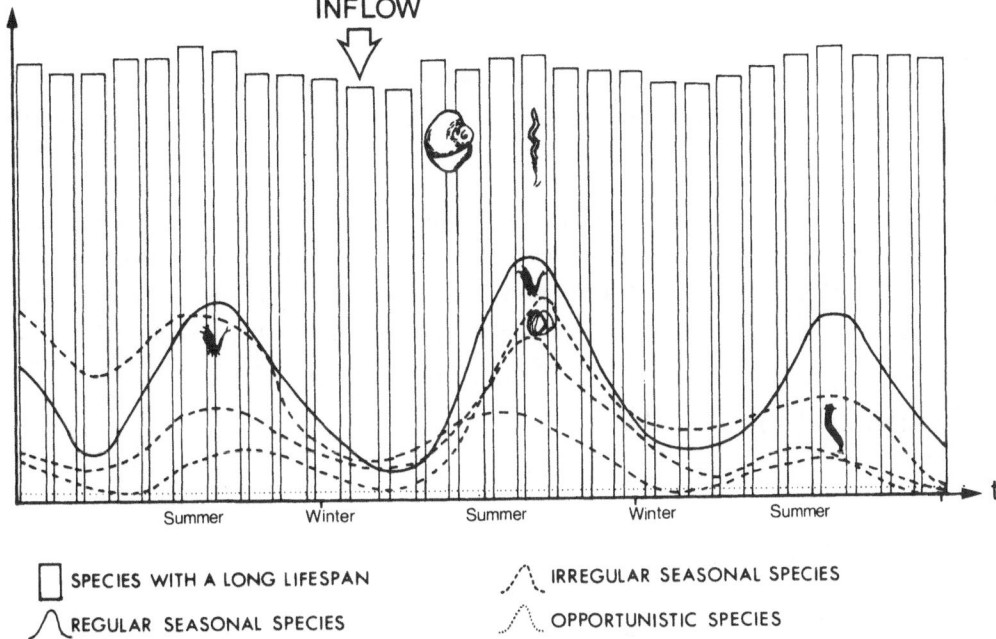

Fig. 3-4: Diagram illustrating macrobenthic community dynamics in the influx zone (north-eastern Kiel Bight). Against a fairly stable background of long-living species (columns) there are seasonal and longterm oscillations of benthic species distribution. The figure refers to ≯ 20 m depth: the time-span considered in this graph is 3 years (modified after ARNTZ 1981).

BIOMASS, ABUNDANCE [rel. units]

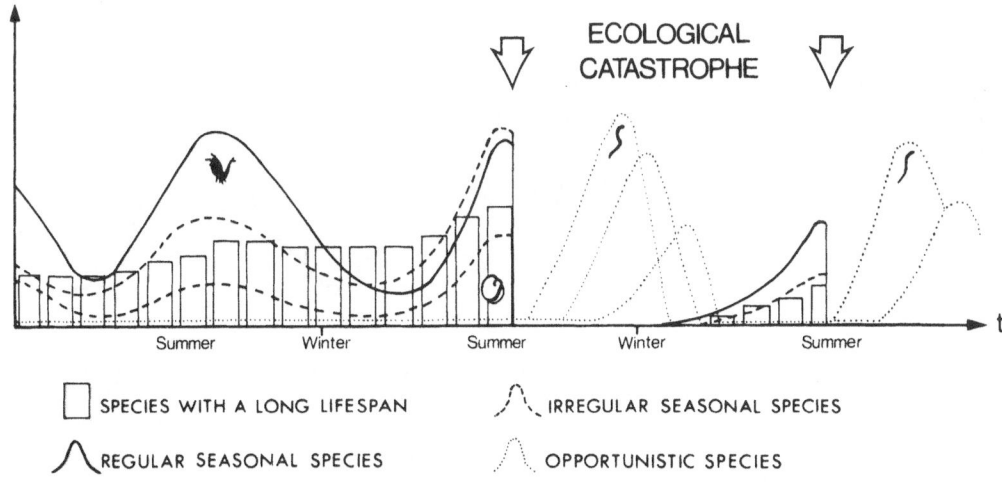

Fig. 3-5: Diagram illustrating macrobenthic community dynamics in the "Hausgarten" area (Kiel Bight). The figure refers to < 20 m depth and to a time-span of 3 years (modified after ARNTZ 1981).

3. stagnation conditions in late summer in the southern and south-western parts of
   Kiel Bight (Fig. 3-5).

Thus fluctuations play an important role, for example demonstrated by the enormous
extent in range of the polychaete <u>Pectinaria koreni</u> (ARNTZ and BRUNSWIG 1976, NICHOLS
1977).

From shallow waters a typical dynamic development of macrofauna abundance was obtai-
ned as displayed in Fig. 3-6 (WORTHMANN 1975) differing from that of deeper waters.
The maximal increase of abundance in April is based on settlement of young larvae.
Typically, in times of high reproduction, the divergence becomes obvious between the

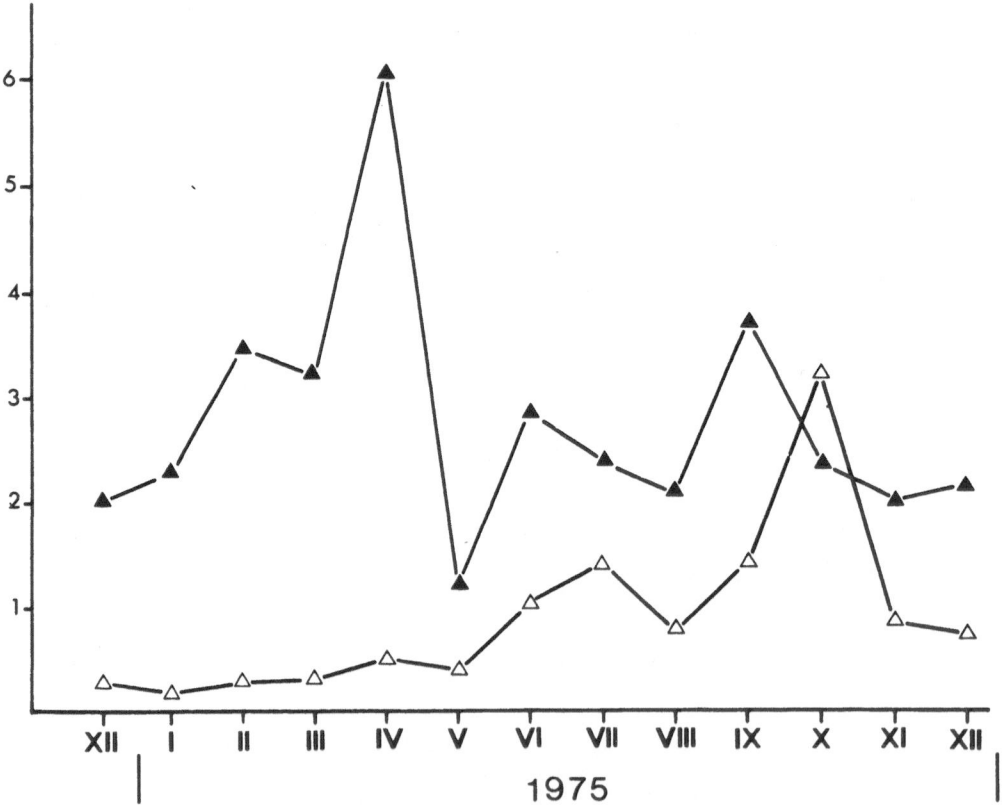

Figure 3-6: Abundance and biomass of macrobenthos in shallow areas of the Kiel
Bight (according to WORTHMANN 1975).

abundance and biomass curves. In the late winter months and in spring (Fig. 3-6) the curves diverge, and in summer and autumn they are more or less parallel. This is in accordance (Fig. 3-3) with the annual abundance and biomass development of meiobenthic organisms based on results from the "Hausgarten" area (10 m water depth) in 1979 and 1980. Periods of highest reproductive activity occur from April to July as demonstrated in the maximal divergence of the two curves presented.

The community development in deep channel systems (Eckernförde Bay) with periodically prevailing oxygen deficiency was investigated by KÜLMEL (1977, 1979) and REIMERS (1976). As a result, it may be stated that the annual cycle of abundance is controlled by oxygen availability in the near-surface sediment. This environmental factor is responsible for immigration and larval settlement and thus for species dynamics. The impoverishment of macrobenthos and meiobenthos, i.e. their mortality, during summer and autumn is also controlled by oxygen deficiency (GRAF et al. 1983). Only some nematodes, foraminifers, ciliates, flagellates and bacteria have adapted to reduced oxygen conditions. In 1972 to 1973 proportional to the decrease in the macrofauna standing stock, a prominent increase in the Cyanobacteria and Beggiatoa species was observed (REIMERS 1976). A process of succession was detected in the Schwentine estuary (Kiel Fjord), a biotope more severely subjected to oxygen deficiency than the channel of Eckernförder Bucht which demonstrates succession steps of communities proportional to distinct O_2 saturations. Three steps of species reduction can be recognized (REIMERS 1976):

"1. a macrofauna association tolerating O_2 concentrations down to 80 % (Polydora ciliata, Capitella capitata);
 2. a cyanophycean-dominated association tolerating O_2 concentrations of about 80 to 40 % (Beggiatoa associated with Monhystera disjuncta (Nematoda), Prorodon discolor, Pleuronema coronatum (Ciliata));
 3. a Protozoa-Sphaerotilus association tolerating O_2 concentrations of about 40 to 1 % (Ciliata: Cyclidium sp.)."

3.2.3   Biomass and production

To estimate the annual production of animal communities, it is essential to obtain information on species biomass development, population dynamics and mortality. But there is a great lack of information regarding autecological and population-dynamic data for most species which play a dominant role in the communities of Kiel Bight. Avoiding this dilemma, ARNTZ (1971) introduced the term "minimum production" which can be reliably ascertained. The annual minimum production is defined as the difference between the lowest and the maximum biomass development per annum. But some insufficiencies of this approach have to be taken into consideration (ARNTZ 1971, WORTHMANN 1976):

"1. only a few species show distinct fluctuations in density;
 2. most of the macrobenthic species have a life span of several years (nearly all molluscs);
 3. the actual production must be somewhat higher because of larval mortality and through predation pressure."

In Table 3-1, the composition of macrobenthic biomass (1968-1971, and 1975-1978) is compiled for the Abra alba community of Kiel Bight. About 85 % (total wet weigth incl. shells) of the standing stock is provided by Arctica islandica, followed by other molluscs (5.4 - 8.0 %), polychaetes (3.5 - 5.6 %), and by crustaceans and other faunal groups (1.5 - 2.6 %) (ARNTZ 1978a, 1980). The latter group includes the starfish, Asterias rubens, the mean wet weight of which runs to 14.3 ± 1.9 g m^{-2} in muddy sand/sandy mud areas (NAUEN, 1978).

Table 3-3: Composition of biomass (BM) and minimum production (MP = $BM_{max.}$ - $BM_{min.}$) in 1968 (ARNTZ and BRUNSWIG 1975a)

	wet weight (g·m^{-2})	ash free dry weight (g·m^{-2})
(a) Abra alba community		
BM of macrobenthos without Arctica	93.3	9.6
MP of macrobenthos without Arctica	98.0	10.05
minimum turnover rate	1.05 y^{-1}	1.05 y^{-1}
BM of macrobenthos incl. Arctica	729.3	30.5
MP of macrobenthos	628.0	27.43
minimum turnover rate	0.86 y^{-1}	0.9 y^{-1}
(b) shallow water communities		
BM of macrobenthos	77.4	
MP of macrobenthos	298.4	
minimum turnover rate	3.85 y^{-1}	

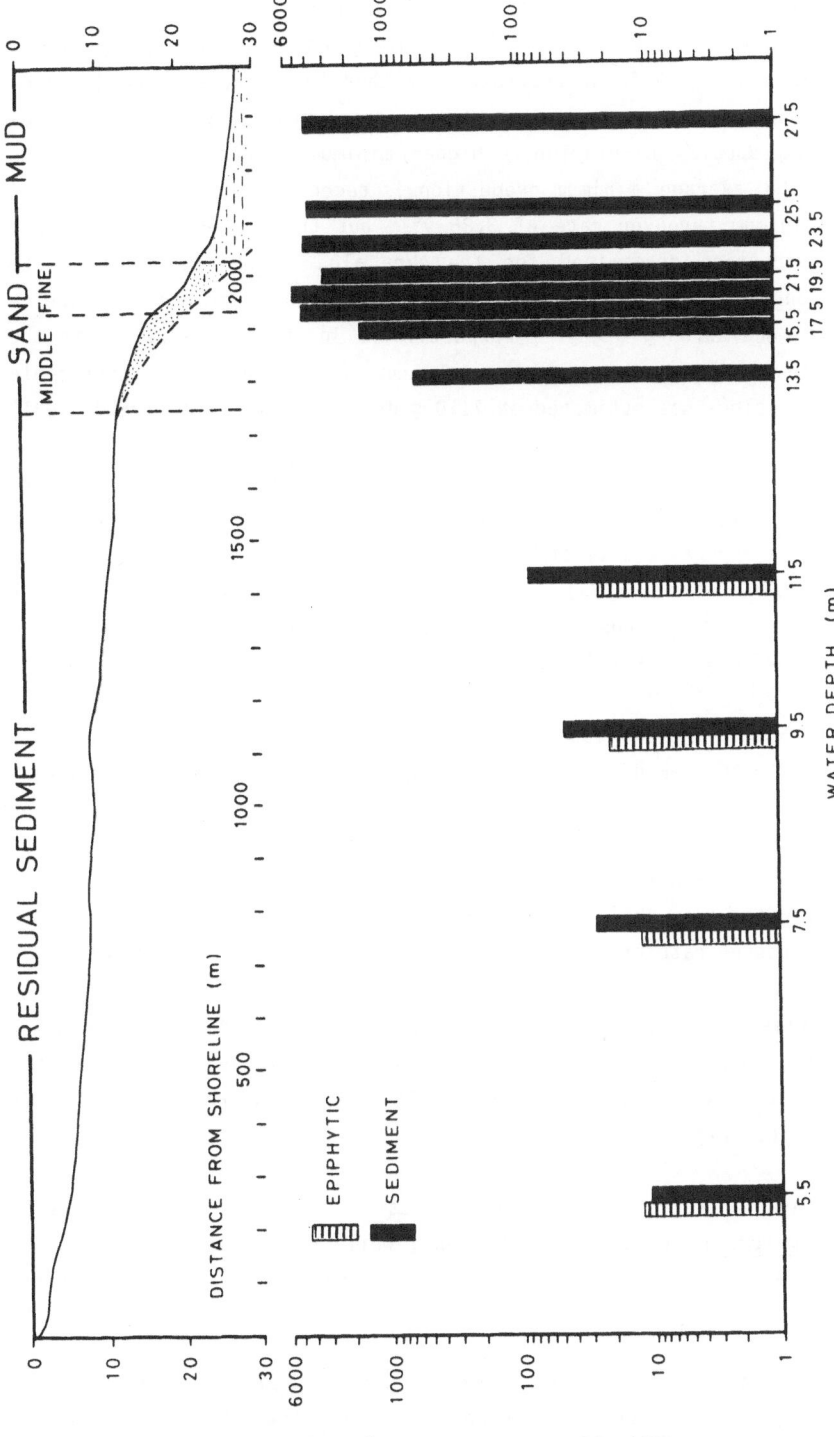

Figure 3-7: Foraminiferal biomass production at the Hausgarten area in the Eckern-förder Bucht. At the top profile showing the erosional pattern with low production and the edge of a small basin with extremely high production has been given (according to WEFER and LUTZE 1976).

The minimum production in the Abra alba community recorded by ARNTZ and BRUNSWIG (1976) amounts to 628 g wet weight $m^{-2} y^{-1}$, and to 98.0 g wet weight $m^{-2} y^{-1}$ omitting Arctica islandica (Tab. 3-3a). This demonstrates a rather low turnover rate for Arctica islandica as well as for the other part of the macrobenthos, half of which consists of fast-growing species displaying a higher minimum production. In shallow waters (Tab. 3-3b) the average minimum production is recorded as 298.4 g $m^{-2} y^{-1}$ thus exhibiting a minimum turnover rate of 3.85 y^{-1}, but this is essentially lower than the minimum production determined for the Abra alba community. Species like Cerastoderma edule and Macoma baltica yielded values of highest minimum production with 286.7 g $m^{-2} y^{-1}$ and 25.03 g $m^{-2} y^{-1}$, respectively. In shallow phytal areas the biomass values are lower than in deeper waters. The minimum production of utilizable food in these phytal regions was estimated at 71.0 g dry weight $m^{-2} y^{-1}$ (THEEDE 1981).

When investigating the biomass of meiobenthos in the "Hausgarten" area (20 - 22 m water depth), SCHEIBEL (1976) recorded a mean of 560 mg dry weight m^{-2} (approx. 250 mg C m^{-2}). This represents approx. 2 % of the macrofauna biomass. The production of meiobenthos was assessed by ARNTZ (1978a) to be in the order of magnitude of 1.5 g C $m^{-2} y^{-1}$. Meiobenthos production thus amounts to 10 to 30 % of the production of the macrofauna which is 5 - 14 g C $m^{-2} y^{-1}$.

The mean meiofauna biomass of the coarse sand community ("Hausgarten", 10 m water depth, Fig. 3-3) amounts to 1.18 g dry weight m^{-2} (approx. 0.71 g C m^{-2}). This determination of biomass is performed by counting the specimens in distinct size classes and by multiplying the numbers with an empiric factor (FAUBEL 1982). The proportion of carbon in meiobenthic dry weight was ascertained to be 57 to 63 % (FAUBEL et al. 1983). In the investigated area the production of meiobenthos amounts to 4.26 g C $m^{-2} y^{-1}$, if the annual turnover rate of meiofauna is assumed to be 6 y^{-1} (GERLACH 1971).

Because of their shells, benthic Foraminifera appear in the size fractions of meiofauna and were counted together with these in some recent investigations. Special attention was paid to this group by WEFER and LUTZE (1976) who tried to quantify biomass production (Fig. 3-7).

In the "Hausgarten" area three sediment-living species dominate the standing stock of Foraminifera: Ammotium cassis is the most abundant, while Elphidium incertum and E. excavatum clavatum are less frequent. Total production for 6 species is calculated (Table 3-4).

Substantial production with 3 - 5.5 g wet weight $m^{-2} y^{-1}$ is found at depths of more than 17 m while the values decrease to insignificant contributions in the shallow sandy sediments. A comparison with meiofauna (SCHEIBEL 1976) on wet weight basis shows the Foraminifera to amount to 6 - 63 % of the total size fraction.

Table 3-4: Total Foraminifera production in the "Hausgarten" area of Kiel Bight

Depth in m	Total production wet weight (mg m^{-2} y^{-1}) (WEFER 1976)	Foraminifera wet weight in % of total meiofauna (SCHEIBEL 1976)
5 - 6	11	6
7 - 8	26	
9 - 10	49	
11 - 12	89	
13 - 14	581	41
15 - 16	1610	
17 - 18	4687	
19 - 20	5411	45
21 - 22	3162	
23 - 24	4478	
25 - 26	4123	63
27 - 28	4361	

3.3 Benthic Response to the Annual Sedimentation Patterns
(G. GRAF)

3.3.1 Introduction

Organic particles originating from pelagic production form the main basis for heterotrophic life in marine sediments. Preliminary models of this pelagic benthic coupling
assumed a continuous and steady vertical flux of particles (STEELE 1974). Recent investigations, however, indicate that in boreal seas this hypothesis requires modification. Concomitant with the seasonal succession of the pelagic organisms, based on
the annual fluctuation in light energy and stratification of surface waters, an annual sedimentation pattern was established (SMETACEK 1980b, HARGRAVE 1980). This pattern is characterized by high sedimentation rates of freshly produced organic matter
during spring and autumn, high resuspension during winter, and low sedimentation rates during summer (see also chapter 2, this volume).

In Kiel Bight, fluctuation in this type of food supply for the benthos is especially
pronounced for two reasons: i) the water column is rather short (< 28 m) and ii) only
a small amount of zooplankton lasts through the winter. The result is that both the
spring bloom and the last phytoplankton autumn bloom are only affected by minor grazing
pressure and thus a large amount of freshly produced cells settles to the sea floor.

The spring bloom especially was studied in detail during recent years (SMETACEK et
al. 1978; PEINERT et al. 1982). These studies demonstrate that up to 1/3 of the yearly pelagic input can reach the sea floor within 1 or 2 weeks. Corresponding results
from a Norwegian fjord and from the north-east Atlantic Shelf demonstrated that this
process is also important in boreal regions with deeper waters and a larger zooplankton population in the winter (SKJOLDAL and LÄNNERGREN 1978, WALSH 1981).

The question arises as to the extent to which benthic organisms respond to this sudden
food supply with high nutritious value. Is a possible benthic reaction buffered or
hidden by the annual temperature cycle or by the amount of organic matter already
stored in the sediments?

3.3.2 Sedimentation of Plankton Blooms and Incorporation in the Sediment

The two main sedimentation events in spring and autumn in Kiel Bight were reported by
KOELMEL (1977) from two annual cycles of the seston content. However, in order to distinguish between the freshly produced input and older material, more specific analyses are needed, for instance proteins and carbohydrates, or C/N ratios (LENZ 1977).

Fig. 3-8 depicts sedimentation of the spring bloom in 1980. While the chlorophyll a equivalents were measured in particles collected in a sediment trap (2 m above the sea floor), ATP and EC ratios were analysed from water samples taken immediately above the sea floor. The ATP content of these particles indicates that a high amount of living cells had just settled, which at least at the beginning of this process were still in a viable state. A greenish-brown layer apparently consisting of such cells was visible on top of the corresponding sediment samples (GRAF et al. 1982).

However, from the data presented in Fig. 3-8, it is not possible to quantify the real input to the benthal of the "Hausgarten" area, for it is uncertain how much of this material will be carried to deeper areas by horizontal advection.

Several factors determine how much of the particulate organic matter arriving at the sea floor is finally incorporated into the sediment. Currents and wave-induced bottom turbulence, the topographical features of the area such as slopes and depressions and the type of macrofauna inhabiting the investigation area determine the amount of organic matter incorporated into the bottom that becomes available for microheterotrophs and meiofauna living in deeper sediment layers. Differences in the composition of macrofauna and the corresponding differences in intensity of bioturbation cause a difference in the carbon consumption of sediments in the "Hausgarten" area (21 m water depth) by a factor of 4 (GRAF et al. 1983). This finding fits very well with the bioturbation model given by YINGST and RHOADS (1980), i.e. a higher bioturbation rate will increase the carbon consumption of microheterotrophs.

For these reasons, the actual input of sedimenting plankton blooms to the benthal requires investigation not only in sediment traps, but also in the sediment itself. Fig. 3-9 demonstrates the increase of protein and carbohydrate content in the top sediment layer (0 - 1 cm) following sedimentation of the autumn bloom 1981 and the spring bloom 1982. Additionally the figure depicts the importance of resuspension and macrophyte debris input during winter. If the enrichment of organic compounds in the sediment is compared to the matter collected in a sediment trap, in the corresponding time interval, large discrepancies may be observed, caused by the processes mentioned above. GRAF et al. (1983) introduced the term advection factor for a quantitative description of this phenomenon. It is defined as the ratio between a) the rate of organic matter accumulating in the top sediment layer plus the rate of organic matter being consumed by organisms and b) the rate of organic matter sedimenting into a near-bottom sediment trap. In a slope area such as "Hausgarten" an advection factor of 7 was found for the settling spring bloom 1982.

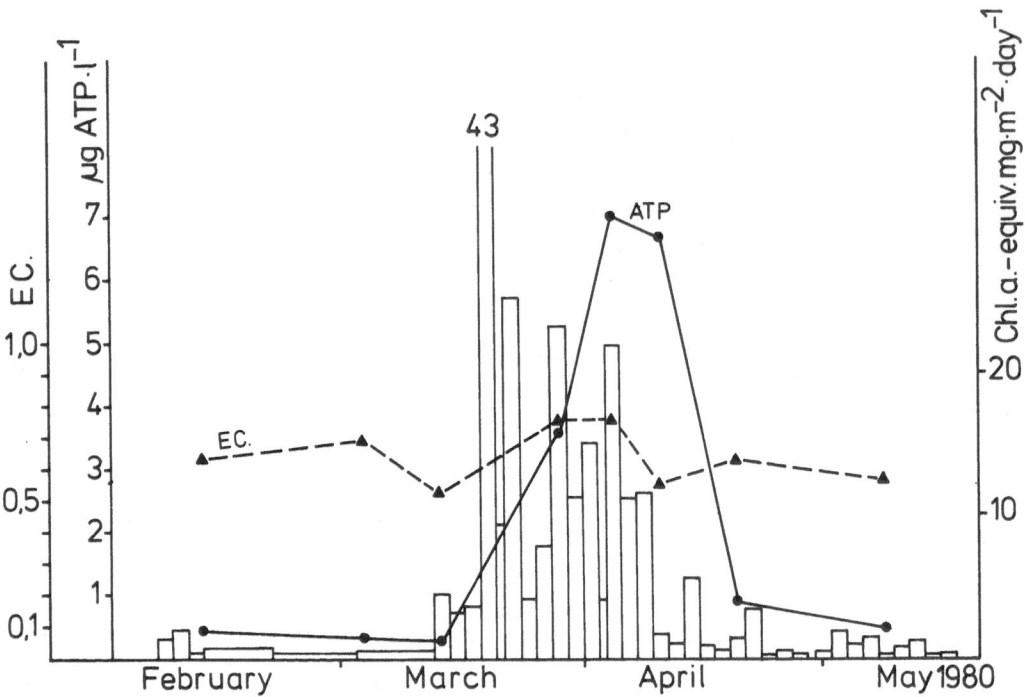

Figure 3-8: Sedimentation of the spring phytoplankton bloom as described by chlorophyll a equivalents (collected in a sediment trap), ATP biomass and adenylate energy charge (EC) of particulate organic matter in the bottom water (water depth 21 m). According to PEINERT et al. 1982, GRAF et al. 1982).

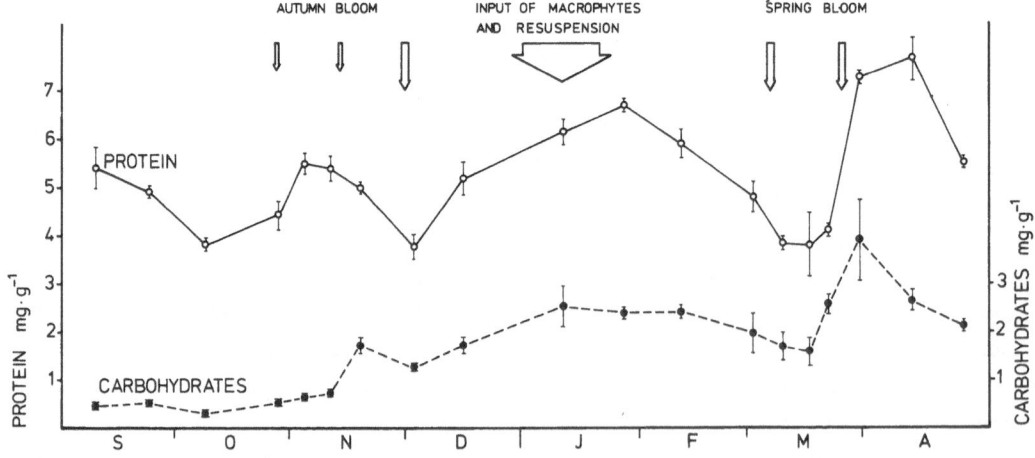

Figure 3-9: Variations in the concentration of carbohydrates and proteins in the top layer of the sediment caused by sedimentation events during the period autumn 1981 to spring 1982 (18 m water depth). According to GRAF et al. 1983b, MEYER-REIL 1983.

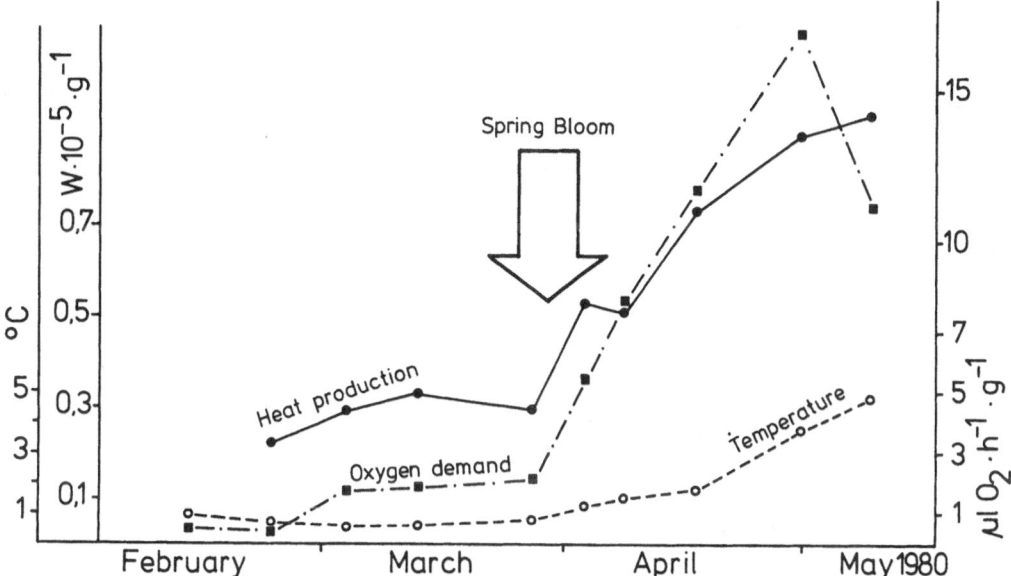

Figure 3-10: Heat production, oxygen demand (based on the activity of the electron transport system), and temperature in the top layer of the sediment in the "Hausgarten" area (21 m water depth) during spring 1980. According to GRAF et al. 1982.

3.3.3   Response to a Settling Spring Bloom

For three consecutive spring blooms it has now been established that sedimentation caused an outburst of benthic activity (GRAF et al. 1982, 1983b, SCHULZ 1983). Fig. 3-10 depicts the doubling in heat production during spring 1980 within 12 days and during 1982 the response, in terms of heat production is even more impressive (Fig. 3-11a). All three food inputs during the period autumn 1981 to spring 1982 caused a temporary shift from aerobic towards anaerobic metabolism, especially towards fermentation. This is indicated by the increase in the ratio heat production/ETS-activity (Fig. 3-11c). Such a shift of benthic metabolism was already postulated by PAMATMAT (1977) and HARGRAVE (1980). In spring 1980 the opposite was found, most likely induced by the strong bioturbation effect caused by the very abundant polychaete Pectinaria koreni, which aerated the sediment. Another effect of the food input during spring is a strong increase in benthic ATP-biomass (Fig. 3-11b). Especially bacteria show a rapid response by an increase in their dividing activity and biomass (MEYER-REIL 1983), whereas meiofauna response is a little delayed.

The vertical flux of particles during a spring bloom ranges between 10 - 15 g C m^{-2} and the material is consumed within 2 - 3 weeks. But even if an advection factor of 7 is to be postulated as found during spring 1982 and more than 70 g C m^{-2} within 2 we-

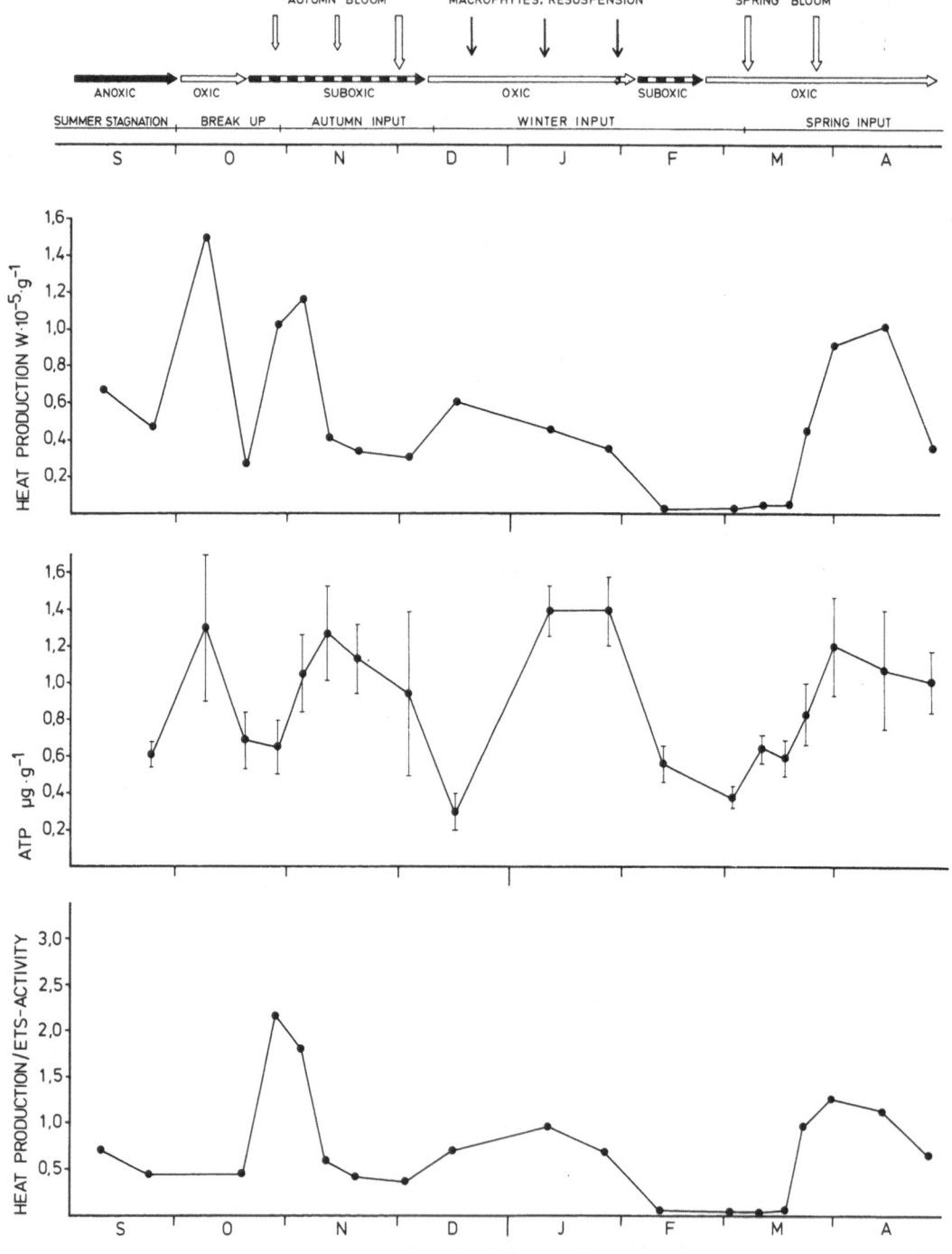

Figure 3-11: Development of a) heat production, b) ATP biomass, and c) change in the type of metabolism following three periods of food input during the period autumn 1981 to spring 1982 in the top sediment layer of the "Hausgarten" area (18 m water depth). According to GRAF et al. 1983

eks are introduced into the sediment, this is not sufficient to supply benthic life throughout the summer. A budget for the spring period 1982 demonstrates that even such a high supply of food is "burned" within 5 - 6 weeks. The strong decrease in heat production during late April 1982 was apparently caused by food limitation (Fig. 3-11a).

3.3.4   Response to the Break-Up of Summer Stagnation and to the Settling of the Autumn Bloom

During autumn an external food supply such as the input from a settling bloom also leads to an immediate response in both activity and biomass production (Fig. 3-11a, b). However, during 1981 a stable stratification of the water caused a long anoxic period in parts of Kiel Bight deeper than 18 m. The break-up of anoxia was followed by a drastic increase in benthic metabolism, which led to a suboxic period lasting several weeks. This sudden explosion of benthic metabolism was fed by protein, which had accumulated during the period of anoxic conditions, which interfered with its decomposition (MEYER-REIL 1983). When oxygen was introduced into the sediment the stored protein became available and was rapidly consumed by micro-organisms. Such a process could be termed internal food supply as opposed to the external food supply by sedimentation events.

A consequence of the anoxic period was a striking reduction of macrofauna and bioturbation and a less effective incorporation of particles into the sediment. A budget for this period indicates that the autumn bloom was not quantitatively consumed by the benthic community and that organic matter of nutritious value reentered the water column by resuspension. This finding may explain the surprising results of SMETACEK and HENDRIKSON (1979), who failed to demonstrate a significant difference in the composition of particulate organic matter during summer, when it mainly consists of phytoplankton, and during winter when it is dominated by resuspended matter.

3.3.1.5 Response to Winter Sedimentation

According to SMETACEK (1980b) the winter period from December to February is characterized by high sedimentation rates which, however, mainly consist of resuspended matter and of inorganic particles. Surprisingly the quality of this matter was nutritious enough to stimulate benthic heat production and an increase in biomass despite decreasing temperatures (Fig. 3-11a). MEYER-REIL (1983) demonstrated that this reaction was first of all a response of benthic bacteria which accounted for more than 50 % of ATP biomass. The high quality of the food supply in winter, which has partly to be explained by the incomplete decomposition of the settling autumn bloom, is also visible in the increase of protein and carbohydrate in the sediment (Fig. 3-9). The accumulation of carbohydrates during winter suggests that refractory material originating from deeper parts of the channel system is retransported into shallower areas (GRAF et al. 1984, PEINERT et al. 1982).

Particles deriving from benthic primary production are also involved in the cycling of particulate organic matter described above, especially particles from macrophytes. Macrophyte production in the shallow parts (< 13 m) of the "Hausgarten" areas was found to be 37 - 387 g dry weight m^{-2} y^{-1} depending on water depth (M. MEYER pers. comm.). These macrophytes are torn off during the winter and fall to the ground in deeper areas by bed load motion. A chlorophyll $\underline{a}$ maximum in the sediment at 18 m water depth was to be seen during late January 1982 (SCHULZ 1983). Quantitatively, however, the role of macrophytes in sediment metabolism is not yet fully understood. In any case they contribute mostly during winter, especially during January, in the deepest part of the channel system (SCHULZ 1983). KÜLMEL (1977) reported an expressed maximum in meiofauna abundance during winter.

3.3.6   Coupling of Pelagic and Benthic Processes

Overall benthic activity including all types of metabolism is depicted in Fig. 3-12 as a preliminary, simplified model for the slope of the "Hausgarten" area (18-22 m) water depth). An attempt is made to categorize the rate of metabolism according to the sources of the organic matter fueling the observed activity.

The solid line (——————) indicates the part of metabolism based on stored organic matter in the sediment, i.e. mainly refractory material. Only after an anoxic period can this material be of higher nutritious value. The decomposition rate parallels the annual temperature cycle and is influenced by anoxic conditions. The area "A" and "B" reflect the fact that more refractory material is decomposed when fresh organic matter is supplied during periods of plankton bloom sedimentation.

The dashed line (— — — —) indicates the part of metabolism based on actual sedimentation. It is basically independent of temperature and shows maximum values during spring and autumn and has a less pronounced maximum during winter. During the summer period such matter is decomposed as fast as it is provided from the pelagial. Only during periods of anoxia is the settling matter partly stored and activity based on this source is reduced.

Summarising, it can be stated that benthic overall metabolism (——————) is very much similar to the annual sedimentation pattern (cf. SMETACEK, chap. 2, Fig. xy, this volume). Only the part of metabolism shown as area "C", based on material that has been actively incorporated into the sediment by the macrofauna (e.g. filter feeders), and special situations of anoxia give some particular characteristics to the benthic activity pattern.

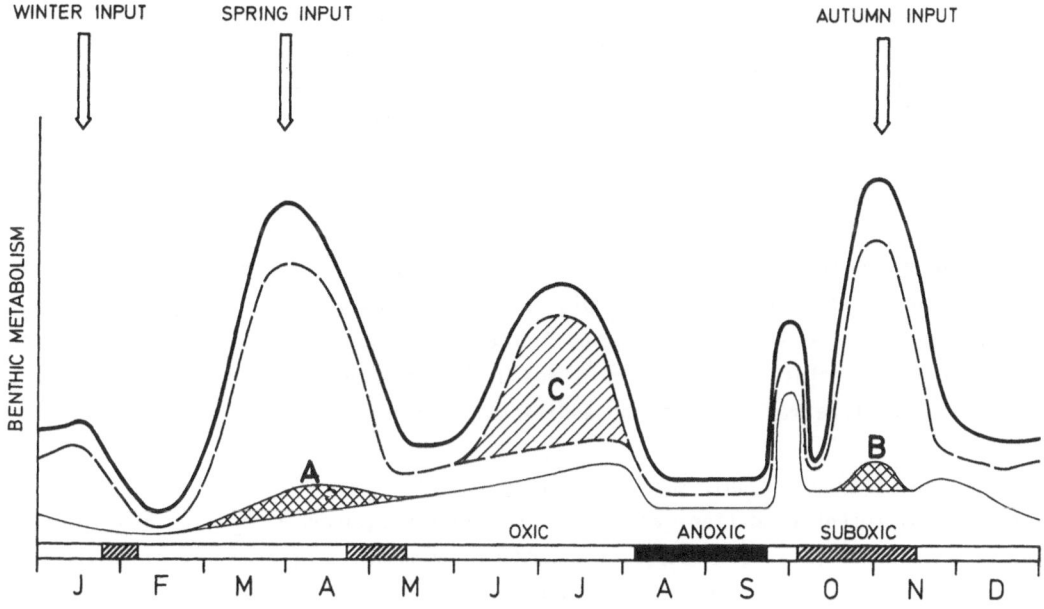

Figure 3-12: Tentative model of benthic metabolism in the sediment at the slope of the "Hausgarten" area, Kiel Bight.

The solid line (———) gives overall metabolism. It consists of the part of metabolism based on stored organic matter (———) and the one based on actual sedimentation (-----). Areas A and B represent the increased consumption of old and refractory material in the presence of an excess of freshly produced matter. Area C is based on the material actively incorporated into the sediment by benthic filter feeders.

Some differences should be considered prior to direct coupling of the pelagic to the benthic system during spring and autumn. The spring diatom bloom starting at a zero point of the pelagic system is one of the few predictable events in the ecosystem of Kiel Bight. It's sedimentation and impact on the benthal functions as a trigger for benthic activity. However, the starting level was found to be very different in different years. The final development of the pelagic system in autumn is much more variable. The last autumn bloom is either a _Ceratium_ bloom in September or a mixed diatom bloom in November. In some years both occur. Their input to the benthal is thus variable. The only existing example, analysed in autumn 1981, was probably a very extreme one. The long anoxic period led to a strong decrease in macrofauna, and the remaining benthic community could not fully utilize the food supply. An early drop in temperature may have been an additional reason.

However, the fate of organic matter deriving from the last autumn blooms may be different if different conditions are given. For instance, when a _Ceratium_ bloom during September settles to a well-established benthic community at prevailing high temperature, there is only little chance that much nutritious food will be left for the next annual cycle.

Thus, both the timing of certain periods in our model as well as the absolute values of minima and maxima must be expected to vary extremely in different years. Especially the development of the benthic system in spring is highly dependent on the development of the whole ecosystem during the preceeding year. It will be very difficult to determine a zero point for the benthic system even if this is possible for the pelagic system. The different levels of food supply in spring also provide one factor that influences the long-term fluctuations of macrofauna, i.e. the cycles over several years (cf. 3.2, this chapter).

3.4   Biomass and Activity of Benthic Bacteria*
        (L. MEYER-REIL)

3.4.1   Introduction

Coastal sediments play an important role in nutrient regeneration for marine ecosystems (cf. chapter 4, this volume). These processes are governed by the activity of
benthic bacteria. Microbiological work in sediments has been concentrated on the investigation of nutrient cycles as well as on the spatial and temporal distribution of
bacteria. In the studies of nutrient cycles, the activity of benthic bacteria was
indirectly concluded from changes in the concentrations or turnover rates of inorganic
and organic chemical parameters. Generally, corresponding information on the development of the bacterial populations themselves could not be included.

This paper summarizes information on the spatial and temporal distribution of benthic
bacterial populations from the brackish water Kiel Bight. Existing data and concepts
illustrating the spatial and temporal distribution of bacteria in shallow water surface sediments are organized under the following topics:
- sediments as habitats for bacteria,
- bacterial number and biomass,
- seasonal development of bacterial communities,
- diurnal fluctuations of bacterial populations,
- estimates of bacterial production.

3.4.2   Sediments as habitats for bacteria

Sediments represent a complex environment consisting of particles which are more or
less densely packed and surrounded by interstitial water. In beach sediments which
are extremely exposed to wave action and tidal activity more than 95 % of the bacteria are attached to particle surfaces (MEYER-REIL et al. 1978a). Deeper sediments
carry a much higher percentage of bacteria "free-floating" in the interstitial water
(up to 50 % of the total number; WEISE and RHEINHEIMER 1978, 1979).

Bacteria colonize only a small proportion of the available particle surface area
(0.01 to 5 %; range of the data obviously dependent upon the method used for the calculation of surface area; cf. HARGRAVE 1972, RUBLEE and DORNSEIF 1978, WEISE and
RHEINHEIMER 1978, DeFLAUN and MAYER 1983). Deep pores of weathered feldspar and clay
grains do not appear to be inhabited by bacteria (DeFLAUN and MAYER 1983). Bacteria

---

* presented during the SCOR-Seminar "Biogeochemical Processes at the Land-Sea Boundary"
  in Roscoff, France, October 22-24, 1984; this section has also been accepted for
  publication in the proceedings of this seminar.

94

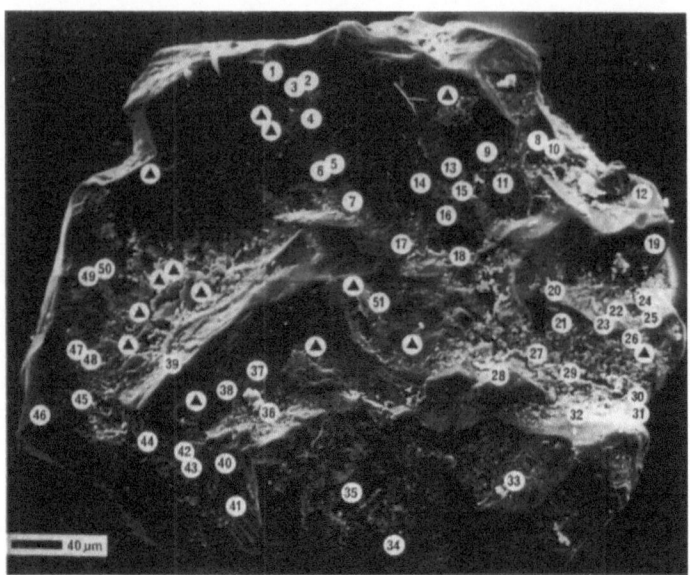

Fig. 3-13: Scanning electron micrograph of a 0.5 mm quartz grain. Note that organic material is concentrated in protected areas. The numbers on the quartz grain mark sites of specific investigation (from WEISE and RHEINHEIMER 1978).

preferentially colonize areas of low relief such as depressions and crevices of particles where the cells are protected against grazing and mechanical demages (NICKELS et al. 1981; cf. Fig. 3-13). The habitation in these areas is an expression of preferential survival rather than preferential colonization as it could be demonstrated by DeFLAUN and MAYER (1983).

The benthic bacterial flora is of a high diversity: rods, cocci and curved cells of different sizes. Single cells or small colonies prevail. Most of the bacteria are found in extracellular slime layers consisting of bacterial fibrous webs and mucus produced by diatoms (WEISE and RHEINHEIMER 1978, MORIARTY and HAYWARD 1982). Through the formation of this organic material of a considerable structural complexity, bacteria may influence the texture of sediments, as demonstrated for fungi mycelium in sand dune soils (CLOUGH and SUTTON 1978).

3.4.3   Bacterial Number and Biomass

As basic members of the food chain, benthic bacteria represent an important nutrient source for meio- and macrofauna. The distribution of bacterial number and biomass was shown to be closely related to sediment properties, from which organic material and grain size are probably most important.

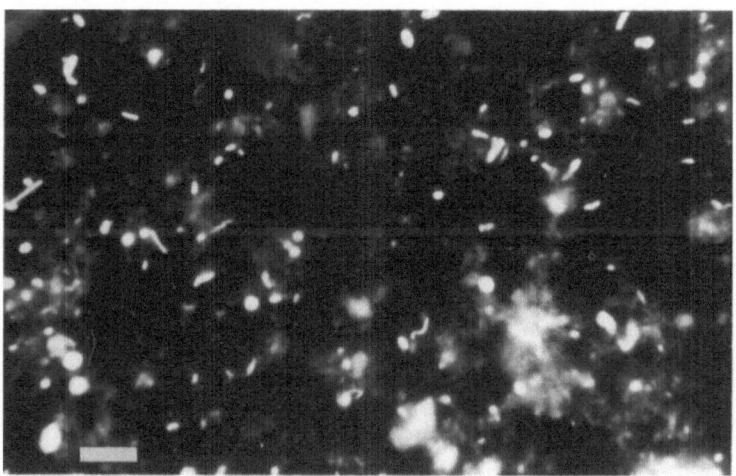

Fig. 3-14: Epifluorescence photograph of bacteria from a sandy sediment of Kiel Bight. The bacteria were liberated from the sediment by sonication. The sample was diluted and stained with acridine orange. Bar represents 3 μm (from MEYER-REIL 1984a).

Beside scanning electron microscopy, epifluorescence microscopy permits a reliable estimate of number and biomass of bacteria in sediments. However, prior to counting bacteria have to be liberated from the particle surfaces. Different extraction techniques have been applied: treatment of sediments with surface active agents, homogenization (MEYER-REIL et al. 1978a) and sonication (WEISE and RHEINHEIMER 1978, ELLERY and SCHLEYER 1984). From these techniques, sonication gave the most reliable results with the highest percentage of bacteria (approximately 95 % of the total number) liberated from the particle surfaces (MEYER-REIL 1983). However, most recently this technique has been discussed with special regard to a possible destruction of bacterial cells (ELLERY and SCHLEYER 1984).

After sonication sediment samples are diluted, filtered onto prestained Nuclepore polycarbonate filters (pore size 0.2 μm) and stained with fluorescence dyes (e.g. acridine orange). Bodies with clear outline, bacterial shape and distinct fluorescence (orange or green) are counted as bacterial cells. For biomass determinations slides of characteristic microscopic fields can be prepared. The slides may be analysed by means of an image analyser which reduces the uncertainty of the visual grouping of the bacteria into arbitrary size classes (KRAMBECK et al. 1981, MEYER-REIL 1983). Using conversion factors, data can be extrapolated to biomass in terms of carbon (cf. FERGUSON and RUBLEE 1976, CAMMEN 1982, BAKKEN and OLSON 1983).

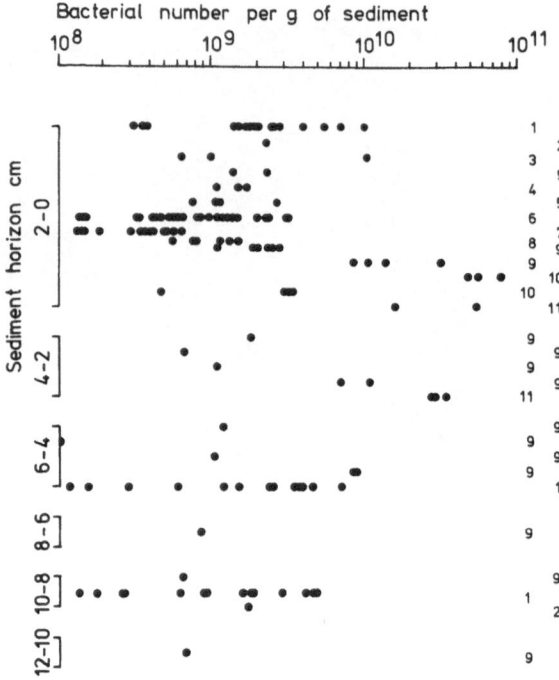

Fig. 3-15: Summary of data concerning bacterial numbers in sediments of different areas. Each of the dots represents one data point taken from the literature listed on the right panel: 1-DALE 1974; 2-KEPKAY et al. 1979; 3-GRIFFITHS et al. 1978; 4-MEYER-REIL et al. 1981; 5-MEYER-REIL et al. 1978; 6-MEYER-REIL et al. 1980; 7-MEYER-REIL 1981; 8-WEISE and RHEINHEIMER 1978, 1979; 9-MEYER-REIL, unpublished data; 10-RHEINHEIMER, unpublished data; 11-JONES 1980; 12-MORIARTY 1980.

Uncertainties arising from inadequacies of the sonication technique and the conversion factors are obviously minor as compared to the subjectivity involved in the counting procedure. However, up to now, beside scanning electron microscopy epifluorescence microscopy is the only direct approach to gain information on bacterial number and biomass (cf. Fig. 3-14).

Data on local variations in bacterial number and biomass summarized from the literature are presented in Fig. 3-15 und 3-16. The data set comprises sediments from arctic, antarctic, boreal and tropical regions. Generally, bacterial numbers are in the range of 10^8 to 10^{11} cells per gram of dry weight sediment with a maximum frequency around 10^9 cells (cf. Fig. 3-15). Numbers turned out to be much more related to the type of sediment than to the region from where the samples were taken. The lowest cell numbers were found in sandy sediments and the highest numbers in muddy sediments (cf. below). With increasing sediment depths, a slight decrease in bacterial numbers becomes obvious. However, even in 11 m sediment depth of an Antarctic sediment core, 10^9

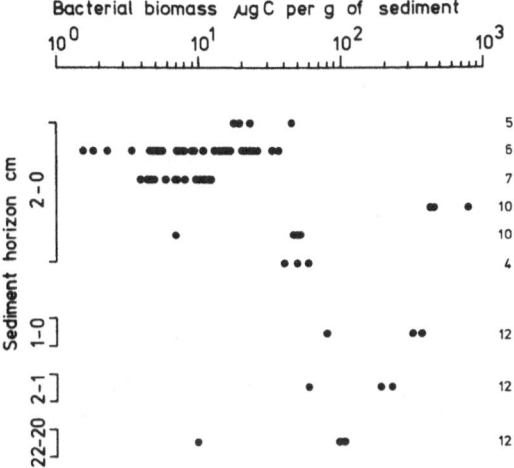

Fig. 3-16: Summary of data concerning bacterial biomass in sediments of different areas. For explanations cf. Fig. 3-15.

per gram of sediment could be detected, which was only insignificantly less than in the surface horizon (MEYER-REIL 1984a).

Because of the time consuming procedure of size fractionation, much less data are available on bacterial biomass (Fig. 3-16). Generally, bacterial carbon is in the range of 1 to 10^3 µg per gram of dry weight sediment thus contributing significantly to the total benthic biomass as determined by ATP measurements (for data cf. GRAF et al. 1982a, GRAF et al. 1983). Again, the lowest bacterial biomass values were measured in sandy sediments and the highest values in muddy sediments.

A horizontal profile from beaches into deeper water (18, 28 m) of Kiel Bight revealed an inverse relationship between benthic <u>bacterial number and grain size</u> of the sediment (Fig. 3-17). Cell numbers increased significantly from sandy to muddy sediments. This general inverse relationship certainly reflects both the greater surface area and the higher organic matter content in fine sediments as compared to coarse ones (DALE 1974, TANOUE and HANDA 1979). However, the question still remains open whether the bacteria simply respond to higher organic matter level originally present in fine sediments or whether higher bacterial numbers in fine sediments result in higher organic matter accumulations (DeFLAUN and MAYER 1983).

Relationships between bacterial biomass and organic matter content of the sediments seem to be very complex. In sediments of Kiel Bight, bacterial carbon increased significantly from sandy beach sediments to sandy mud sediments. However, in muddy sediments bacterial biomass did not further increase despite the more than twofold increase in organic matter as compared to sandy mud sediments (Fig. 3-18).

On the average, bacterial carbon in sediments of Kiel Bight accounted for 0.7 ± 0.2 % (standard deviation of the mean; 76 observations) of the sediment organic carbon. In the literature, higher values have been reported (1.2 %, DALE 1974; less than 2 %, CAMMEN 1982) which may be partly due to methodological differences in calculating bacterial biomass. Bacterial carbon as percentage of the total sediment organic carbon varied dependent upon the types of sediments in Kiel Bight. The lowest percentage was found in sandy beaches (0.5 ± 0.1 %; 36 observations), a considerably higher percentage in sandy mud sediments (0.9 ± 0.2 %; 21 observations) and a lower percentage in muddy sediments (0.7 ± 0.2 %; 19 observations).

Similarly, the average bacterial cell weight varied in the different types of sediment investigated in Kiel Bight. For sandy sediments an average cell weight of 15.8 ± 2.5 x 10^{-9} µg of bacterial carbon was calculated as compared to 20.4 ± 3.9 x 10^{-9} µg for sandy mud sediments and 11.3 ± 1.7 x 10^{-9} µg of bacterial carbon for muddy sediments.

As an explanation for the observations described above, both the concentration and the suitability of the sediment organic material have to be considered. In Kiel Bight, the sandy mud sediments obviously offer optimal conditions for the development of the bacterial populations, sustaining the highest total bacterial biomass, the highest averge cell weight and the highest percentage of bacterial carbon as compared to the total sediment organic carbon. In sandy beach sediments as well as in muddy sediments, conditions for the development of bacterial populations have to be regarded as suboptimal. In sandy sediments the concentration of organic material may be the limiting factor for bacterial growth. In muddy sediments, however, much of the organic material may consist of highly refractory compounds thus restricting bacterial biomass development.

Grouping the bacteria into different size classes revealed a characteristic spectrum of bacterial biomass in sediments of Kiel Bight. Generally, small sized bacteria (volume < 0.3 µm³) accounted for the major part of the total bacterial biomass, followed by medium sized bacteria (volume 0.3-0.6 µm³) and large sized bacteria (volume > 0.6 µm³). This pattern fits into the average "Sheldon" spectrum, a characteristic distribution of benthic biomass with two main peaks in the largest (> 2 mm; corresponding to macrofauna) and in the smallest size classes (< 2 µm; corresponding to bacteria, respectively; cf. SCHWINGHAMER 1981). Deviations from this typical pattern

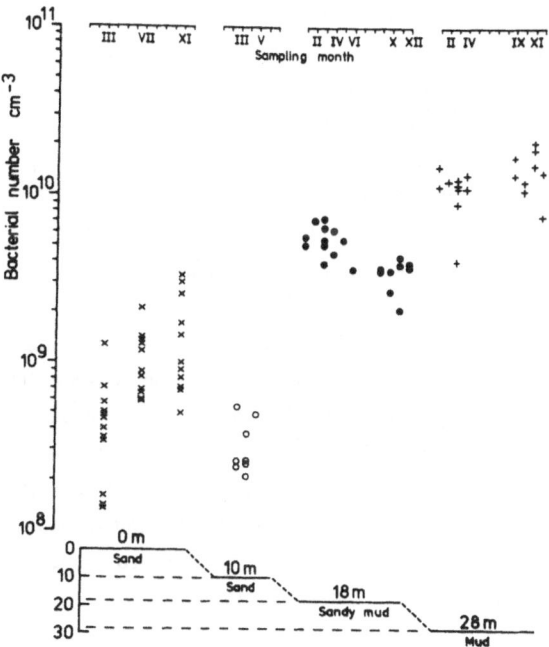

Fig. 3-17: Bacterial numbers per cm³ of sediment on a hypothetical horizontal profile from sandy sediments (water depth 0 and 10 m) to sandy mud sediments (water depth 18 m) and to muddy sediments (water depth 20 m) of Kiel Bight.

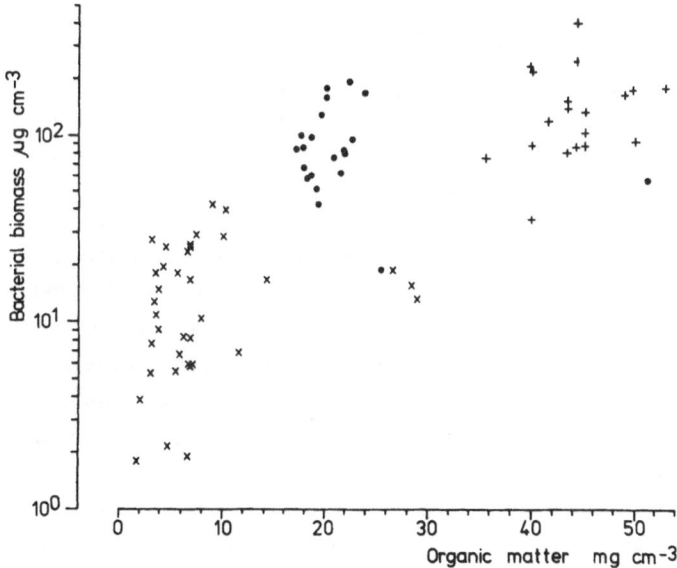

Fig. 3-18: Bacterial biomass versus total organic matter content on a hypothetical horizontal profile from sandy beaches (x) to sandy mud sediments (•) and to muddy sediments (+) of Kiel Bight.

in the distribution of benthic biomass have been interpreted by the author as an expression of the effect of exogenous disturbance factors. This interpretation can be related to deviations from the "normal" bacterial biomass spectrum as well. The input of the phytoplankton blooms into the sediments in autumn and spring as well as mass abundances of predators led to drastic shifts in the size spectrum of bacterial biomass in sediments of Kiel Bight (cf. below).

3.4.4   Seasonal Development of Bacterial Communities

In the literature information on the seasonal development of benthic bacteria is limited. From the investigations of MONTAGNA (1982) and CAMMEN (1982) no general trend for seasonal variations of benthic bacteria could be detected. The studies of RUBLEE (1982) and DeFLAUN and MAYER (1983) revealed a positive correlation between benthic bacteria and temperature. Detailed investigations with high sampling rates in a sandy mud sediment of Kiel Bight reveated strong seasonal variations in the development of bacterial populations closely related to specific ecological situations and sedimentation events (MEYER-REIL 1983, 1984b).

In recent years it could be shown that in boreal marine systems the main sedimentation events and thus the main food supply for the benthos occur in autumn and spring. Studies in Kiel Bight have demonstrated that large amounts of pelagic primary production in autumn and spring do not enter the pelagic food web but rather settle onto the sediment (cf. chapter 2, this volume). This material already represents 2/3 of the total yearly input from the pelagic into the benthic system (SMETACEK 1980b). Investigations of PEINERT et al. (1982) and GRAF et al. (1983) followed the biomass development of the autumn and spring phytoplankton blooms, their sedimentation and incorporation into the benthal.

In sediments of Kiel Bight, three periods of accumulation of organic material could be distinguished: in autumn, winter and spring (Fig. 3-19; MEYER-REIL 1983). The enrichment of organic material during November could be traced back to the breakdown and sedimentation of the autumn phytoplankton bloom composed of dinoflagellates and diatoms. Total organic matter, proteins and carbohydrates accumulated in temporally separated peaks. During winter a slow continuous increase of organic material was observed in the sediment surface. The organic matter consisted of resuspended sediment, material from terrestrial origin as well as macrophyte debris eroded by winter storms. The breakdown of the spring phytoplankton bloom (mainly diatoms) led to an enrichment of organic material in the sediment surface during late March to mid-April. Again temporally separated peaks were recorded for total organic matter, proteins and carbohydrates. These accumulation periods termed "autumn"-, "winter"- and "springinput" turned out to be of high relevance for the development of the benthic communities in sediments of Kiel Bight (GRAF et al. 1983, MEYER-REIL 1983, 1984b).

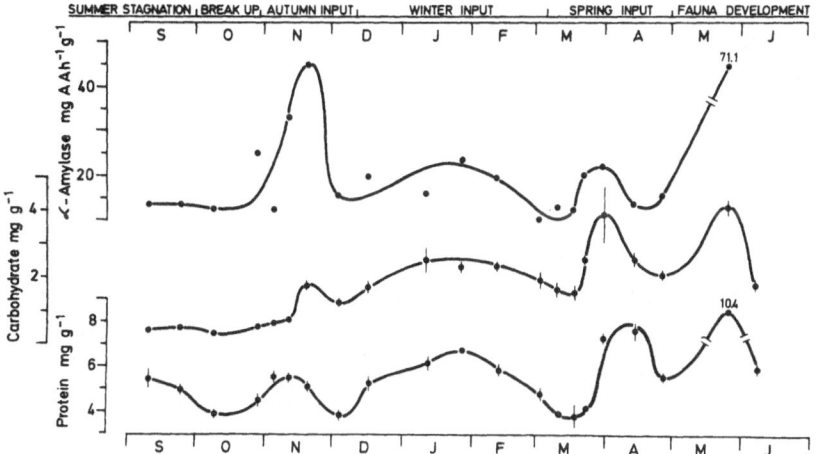

Fig. 3-19: Seasonal variations in concentrations and exoenzymatic decomposition rates of organic material in the 0 to 1 cm horizon of a sandy mud sediment (water depth 18 m) of Kiel Bight. Illustrated are: concentrations of proteins and carbohydrates and activity of α-amylase (mg of amylopectin azure decomposed per g of dry weight sediment per hour).

Direct calorimetry (heat production) represents a reliable method for the estimation of total benthic community metabolism, because it is a direct measurement of the energy flow through the system. Although the heat release from the activity of extra-cellular enzymes and from chemical oxidations are also included, the latter two components are thought to be of minor importance for total heat loss from sediments (PAMATMAT 1982).

As an immediate response to the "autumn"- and "spring-input" in sediments of Kiel Bight, heat production culminated. A less obvious response was found during "winter-input". However, considering the temperature dependence of benthic metabolism, there was still a considerable and long-lasting heat production during this period (GRAF et al. 1983). Whereas heat production comprises all types of benthic metabolism, electron-transport-activity (ETS) relates to the activity of respiratory chains. As pointed out by PAMATMAT et al. (1981), the quotient between heat production and ETS-activity should serve as a qualitative indicator for changes in the type of metabolism. Following the "autumn"- and "spring-input", a strong increase in this quotient was observed, demonstrating a shift in the type of benthic metabolism towards fermentation. This coincided with suboxic conditions in the sediment surface mainly caused by biological oxygen consumption (GRAF et al. 1983, see also section 3.3, this chapter).

The main portion of the input of organic material into the sediment requires <u>enzymatic</u> <u>decomposition</u> prior to the incorporation into cells living in the sediment. In this process, extracellular enzymes are involved which are secreted by living cells (CORPE and WINTERS 1972) or liberated during the lysis of dead and decaying cells. As shown by BURNS (1980) some of these enzymes may retain their activity outside the cells by the formation of humic-enzyme complexes bound to clay particles.

The accumulation of organic material in sediments of Kiel Bight during "autumn"-, "winter"- and "spring-input" led to corresponding stimulations in the enzymatic de-composition rates of carbohydrates and proteins (activity of α-amylase, proteolytic enzymes; Fig. 3-19). Exoenzymatic responses were highest in autumn as compared to winter and spring. This is obviously a reflection of both the higher temperature and the higher benthic biomass in autumn. During "autumn"- and "spring-input" a stimula-tion of enzymatic decomposition rates already occurred when concentrations of parti-culate organic material started to accumulate in the sediment surface indicating an induction of enzymatic activities by increasing concentrations of suitable substrates (MEYER-REIL 1983). There is strong evidence from laboratory and field data that under anoxic conditions the enzymatic decomposition of protein is retarded. During summer stagnation, an anoxic period in which hydrogen sulfide prevails in sediments of Kiel Bight, protein accumulated. The concentration of the stored protein was even compara-ble to that measured during "autumn"- and "spring-input". Following the introduction of oxygen into the sediment (break up of summer stagnation), protein concentrations significantly decreased. Parallel peaks in heat production and ATP imply that the stored protein was rapidly consumed and incorporated into benthic biomass (GRAF et al. 1983, MEYER-REIL 1983; see also section 3.3, this chapter).

Bacterial development in autumn and spring

As shown for sediments of Kiel Bight, bacteria reacted to the "autumn"- and "spring-input" with two separate biomass peaks. The first peak already occurred when concentrations of organic material started to accumulate in the sediment surface. This demonstrates that bacteria almost immediately responded to the availability of decomposable organic material. The second peak in bacterial parameters coincided with the main input of organic material following the final breakdown and sedimentation of the phytoplankton blooms (Fig. 3-20).

The bacterial population faced with the "autumn-input" was derived from an anoxic popu-lation (fermentative bacteria, sulfate reducers) which prevailed during summer stagnation. Within this population the input of freshly produced organic material caused a drastic shift; bacteria primarily reacted with a strong increase in cell vo-lume (biomass production). Deviating from its "normal" distribution (cf. above), the size spectrum was dominated by medium and large sized cells. Following the final break-

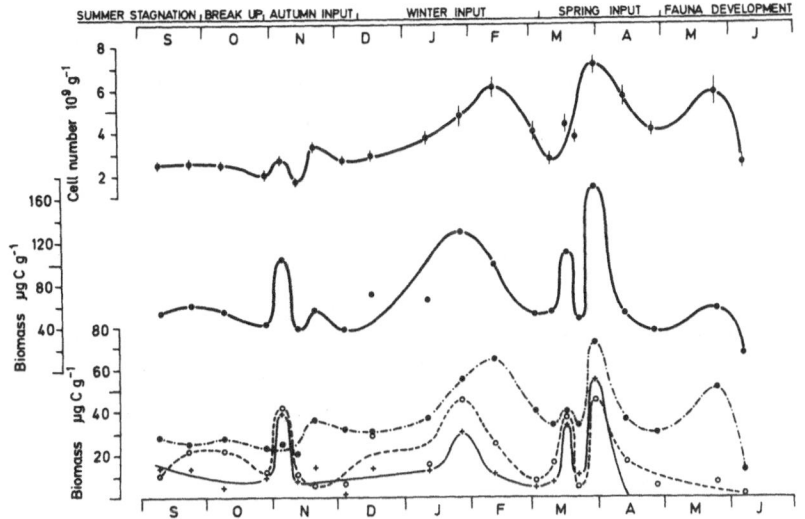

Fig. 3-20: Seasonal variations in microbiological parameters in the 0 to 1 cm horizon of a sandy mud sediment (water depth 18 m) of Kiel Bight. Illustrated are: total number of bacteria, total bacterial biomass and biomass spectrum. The biomass spectrum comprises: small sized bacteria (volume $<$ 0.3 μm^3; closed circles), medium sized bacteria (volume 0.3-0.6 μm^3; open circles) and large sized bacteria (volume $>$ 0.6 μm^3; crosses).

down and sedimentation of the autumn phytoplankton bloom, the bacteria responded with cell division (increase in cell number).

Compared to autumn, the history of the bacterial population faced with the "spring-input" was quite different. During winter oxic conditions prevailed in the sediment. Due to erosion of the sediment caused by winter storms, the bacterial population was declining. The "spring-input" encountered an impoverished bacterial community which immediately reacted with a strong increase in cell volume, but only a small increase in cell number. Again, deviations from the normal size spectrum were observed: small, medium and large sized cells almost equally contributed to the total biomass.

Following the final breakdown and sedimentation of the spring phytoplankton bloom, bacterial number and biomass responded with a second maximum.

Bacterial development in winter

During winter resuspended sediment, terrestrial material and eroded macrophytes represent an additional food supply for benthic bacterial populations of Kiel Bight.

Bacterial number, biomass and cell division activity showed a slow continuous increase up to values that were even higher as compared to those obtained following the "autumn-input" (Fig. 3-20). Taking into account the more refractory nature of the organic material and the low temperature with reduced metabolic activity rates, the accumulation of bacterial biomass during winter is surprising (cf. temperature-dependent development of benthic bacteria reported by DeFLAUN and MAYER 1983). However, the relatively long time available for the "undisturbed" development of the bacterial population and the limited number of grazers have to be considered. With regard to the nature of the organic material and the slow continuous development, the response of the bacterial community to the "winter-input" differed basically from the spontaneous bacterial development following the input of the phytoplankton blooms in autumn and spring.

3.4.5   Bacteria and fauna development

The development of the benthic fauna in spring greatly influenced the composition of the bacterial community in sediments of Kiel Bight. Through the activity of polychaetes the sediment surface was firmly glued together. This ecological situation was reflected by a bacterial population which consisted of a high number of almost exclusively small sized cells with a corresponding low biomass (Fig. 3-20). Since the bacteria actively grew as demonstrated by a high number of dividing cells, nutrient deficiency could not be the reason for the impoverishment of the bacterial population. More likely, preferential grazing on medium and large sized bacteria led to the reduction in the size spectrum of the bacteria. Literature regarding interrelationships between bacteria and fauna components is sparse; however there is some evidence for the stimulation of bacterial activity by grazing (GERLACH 1978; MORRISON and WHITE 1980).

3.4.6   Diurnal fluctuations of bacterial populations

The strong diurnal rhythms of benthic primary production (JORGENSEN et al. 1979, KARG 1979, REVSBECH et al. 1981) imply a coupling between autotrophic and heterotrophic processes in shallow water sediments. Evidence for the existence of diurnal rhythms in benthic bacterial activities obviously closely related to primary production was obtained from investigations of sediments below sea-grass beds in Moreton Bay, Queensland, Australia (MORIARTY and POLLARD 1982) and from studies in sandy sediments of Kiel Bight (MEYER-REIL and GRAF, unpublished data). Since the latter study comprises various parameters related to benthic biomass and activity, the discussion of diurnal fluctuations of bacterial populatins will be based upon these results.

Benthic biomass and activity

Total organic matter, benthic biomass and activities revealed strong diurnal fluctuations in sandy sediments of Kiel Bight sampled during a 36 hour cycle in summer
1980 (Fig. 3-21 and Fig. 3-22). Total organic matter and living benthic biomass derived from ATP measurements accumulated around midnight and decreased during the day.
Bacterial biomass, however, revealed an opposite fluctuation pattern: maxima were
observed in late afternoon and minima between midnight and early morning (cf. Fig.
3-21).

As indicated by the almost constant ratio between total organic matter and protein,
the composition of organic material did not change during the day/night cycle. Assuming a conversion factor of 200 for calculating organic carbon from ATP, and assuming
that carbon represents 50 % of the total organic material, living biomass on an average accounted for 5 % of the total organic carbon. Bacterial biomass made up 5 % of
the total living biomass.

Measured parameters of benthic activity were well correlated among each other despite
the fact that quite different parameters of heterotrophic activity were measured such
as total metabolic activity (heat production), exoenzymatic decomposition of parti-
culte organic material (activity of α-amylase) and bacterial uptake of dissolved organic substrates (^{14}C-labelled glucose). Benthic activities increased during the
morning, culminated around noon and decreased during afternoon and night (Fig. 3-22).
A corresponding diurnal pattern in bacterial activity based upon the incorporation of
thymidine into DNA was reported by MORIARTY and POLLARD (1982) for sediments below
sea-grass beds.

Trophic interrelationships

Shallow water sediments may represent self-supporting systems governed by benthic
primary production, at least in periods with sufficient light supply. It may be speculated that heterotrophic benthic activities are stimulated by the exudation of
substances from the benthic primary production (microphytobenthos) which is initiated
by light in the early morning. Bacteria take up the exudates and respond with a subsequent increase in biomass (note the time lag between bacterial uptake of substances
and increase in biomass; Fig. 3-21, 3-22). Bacteria as basic members of the food
chain are grazed by meio- and macrofauna. As a consequence, total benthic activity
(heat production) and enzymatic decomposition of particulate organic material in the
sediment as well as in selected meiofauna organisms (cf. FAUBEL and MEYER-REIL 1983)
increase. After maximum values are reached around noon, benthic activities decline
towards the afternoon and night obviously due to a decreasing supply with primary-
produced material.

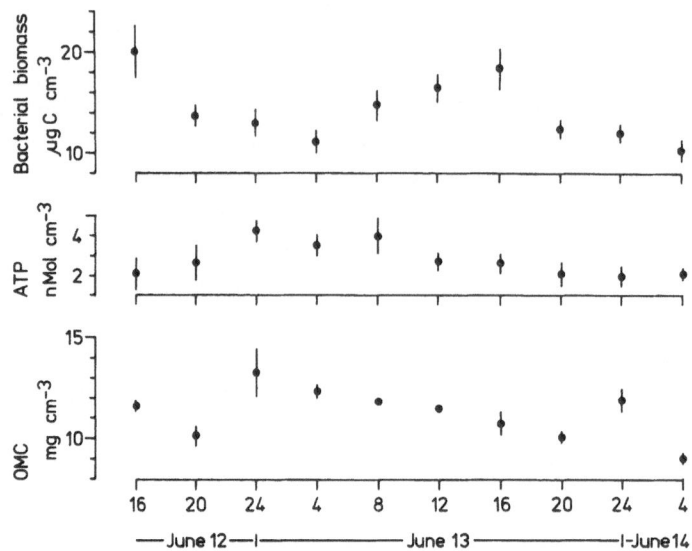

Fig. 3-21: Diurnal fluctuations of total organic matter, ATP and bacterial biomass in the 0 to 1 cm horizon of a sandy sediment (water depth 10 m) of Kiel Bight sampled every 4 hours during a 36 hour cycle (June 12 to June 14, 1980).

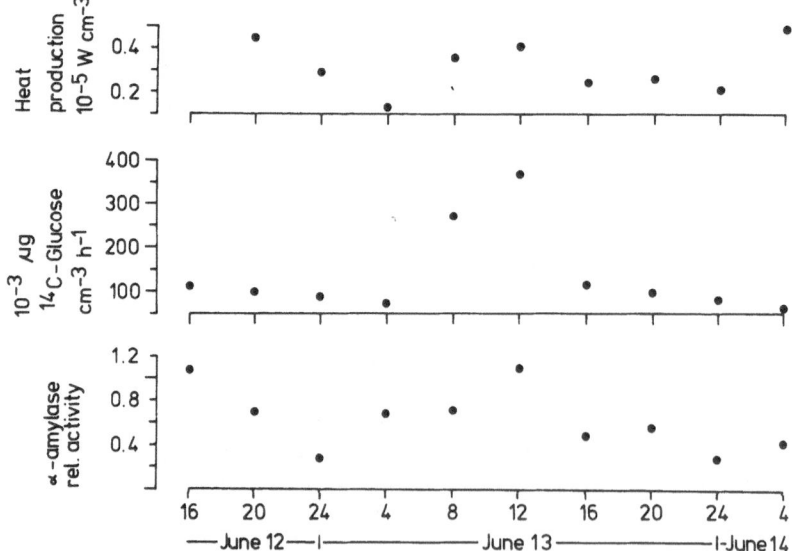

Fig. 3-22: Diurnal fluctuations of exoenzymatic decomposition rates of carbohydrate (activity of α-amylase), bacterial uptake of ^{14}C-glucose and heat production in the 0 to 1 cm horizon of a sandy sediment (water depth 10 m) of Kiel Bight sampled every 4 hours during a 36 hour cycle (June 12 to June 14, 1980).

3.4.7   Estimates of bacterial production

Since bacteria play an important role as basic members of the food chain in sediments (YINGST and RHOADS 1980, GERLACH 1978), information on bacterial production is urgently needed. In the literature, however, reliable data on bacterial production are sparse. Certainly this is a reflection of the methodological problems involved. Different approaches have been applied to gain information on bacterial production in sediments (Table 3-1).

Table 3-1: Summary of data concerning bacterial production in sediments of different areas

Sediment	Method	Production $(mg\ C\ m^{-2}\ d^{-1})$	Literature
Sandy beaches Kiel Bight	Glucose uptake	28	MEYER-REIL et al. 1980
Sediment associated with sea-grass beds Moreton Bay, Queensland, Australia	Thymidine incorporation	12	MORIARTY and POLLARD 1982
Sand Nearshore western Atlantic Ocean, Sapelo Island, Georgia, USA	Thymidine incorporation	100-800	FALLON et al. 1983
Sandy mud/mud Kiel Bight Autumn Winter Spring	Seasonal changes in biomass	140/370 20/ 10 300/120	MEYER-REIL 1983 MEYER-REIL, unpubl.
Sand Kiel Bight	Diurnal changes in biomass	80	MEYER-REIL, unpubl.

Glucose uptake

In sandy beaches of Kiel Bight, actual uptake (flux) of glucose by bacteria amounted to 0.1 µg C (glucose) per gram of dry weight sediment per hour ($g^{-1} h^{-1}$) (summer conditions, average of 12 observations; MEYER-REIL 1978, cf. MEYER-REIL et al. 1980). In these sediments approximately 2 % of the total dissolved organic carbon (DOC) existed in the form of labile carbon (1 % amino acids, 1 % monosaccharides) from which glucose roughly represents one-quarter (MEYER-REIL et al. 1978). Under the assumption that bacteria take up the remaining three-quarters of the DOC with the same rate, total bacterial carbon uptake would amount to 0.4 µg $g^{-1} h^{-1}$. Taking into account respiration (40 % as an average of the respiration of different organic substrates) and excretion (estimate of 10 %, no data available) bacteria would produce 0.2 µg C $g^{-1} h^{-1}$, which is equivalent to 28 mg C $m^{-2} d^{-1}$ (cf. Table 3-1). This means that between 10 and 30 % of the microphytobenthos primary production in this area (KARG 1979) is fixed by bacterial secondary production - a reasonable range with respect to the bacterial production data calculated.

Thymidine incorporation

The incorporation of thymidine into DNA was used by MORIARTY and POLLARD (1982) and FALLON et al. (1983) to calculate benthic bacterial production. For sediments below sea-grass beds in Moreton Bay, Queensland, Australia, the former authors reported a bacterial production of 12 mg C $m^{-2} d^{-1}$. Nearshore sediments off Sapelo Island, Georgia, USA, revealed a much higher baterial production (100-800 mg C $m^{-2} d^{-1}$; FALLON et al. 1983). The lower limit of these data agrees well with the bacterial production extrapolated from seasonal and diurnal changes in bacterial biomass in sediments of Kiel Bight (cf. below and Table 3-1). However, calculating bacterial production from the incorporation of thymidine requires a number of assumptions which are difficult to verify. Among these, the question of isotope dilution and the validity of conversion factors to calculate production from bacterial uptake of thymidine seem to be most important (MORIARTY and POLLARD 1982, FALLON et al. 1983).

Diurnal and seasonal changes in biomass

Calculating bacterial production from changes in bacterial biomass in samples collected frequently during short periods of time (range of hours; cf. Fig. 3-21) at a fixed location offers the most reliable estimates. By this method the addition of exogenous substrates and artificial conditions during the incubation in the laboratory can be avoided. Except for the conversion of bacterial volume to carbon, no conversion factors or assumptions must be applied. However, other problems arise. A sufficient number of subsamples have to be collected by divers within each time interval from a defined sediment area to avoid artefacts due to patchiness. Since bacterial excretion of substrates and grazing on bacteria between the sample intervals can not be accounted for, the values gained certainly present an underestimation of bacterial production.

Based upon studies of diurnal fluctuations at a shallow water sediment in Kiel Bight, a bacterial net production of 80 mg C m^{-2} d^{-1} was calculated. Further estimates of bacterial production were derived from investigations of seasonal variations in bacterial biomass at two sediment stations in deeper waters of Kiel Bight. As a response to the input of the phytoplankton bloom in autumn, bacterial production amounted to 140 and 370 mg C m^{-2} d^{-1} (sandy mud and muddy sediment, respectively). The corresponding values in spring were 300 and 120 mg C m^{-2} d^{-1}, respectively (cf. Table 3-1). As it was pointed out above, these values agree well with the lower range of bacterial production data derived from the incorporation of thymidine into DNA (FALLON et al. 1983).

3.4.8   Conclusion

From the information available it can be accepted that bacteria colonize sediments in high number and biomass. Although they represent less than 1 % of the total sediment organic carbon, bacteria contribute significantly to the benthic biomass, emphasizing their role as an important food source for the benthic fauna. Close correlations exist between the distribution of bacteria and sediment properties such as grain size and organic matter. However, as shown for the relationship between organic material and bacterial biomass, interpretations are difficult.

Processes within the benthic bacterial community may occur in very short time scales. The strong diurnal rhythms observed indicate a close coupling between autotrophic and heterotrophic processes in shallow water surface sediments. In sediments of Kiel Bight the seasonal development of the benthic bacterial populations proved to be strongly influenced by certain ecological situations and events, of which the input of the phytoplankton blooms in autumn and spring, the accumulation of organic material during winter and the development of the benthic fauna in spring were the most important. The enrichment of organic material in the sediment surface led to corresponding increases in the enzymatic decomposition rates of particulate organic material. Bacteria immediately reacted to the availability of organic material with a drastic shift in the size distribution of biomass.

Reliable data on benthic bacterial production are urgently needed. From very few data based upon quite different approaches, production ranges between 10 and some 100 mg of bacterial C m^{-2} d^{-1}. These estimates already illustrate the importance of bacteria in the turnover of organic material in sediments.

Although basic information on the spatial and temporal distribution of benthic bacterial biomass and activity is available, a number of questions still remain open. Among these, the mesurement of bacterial production, seasonal variations in the metabolic activity of bacterial populations and interactions between bacteria and the benthic fauna need further attention.

Chapter 4:   DIAGENESIS AND EXCHANGE PROCESSES AT THE BENTHIC BOUNDARY

W. BALZER, H. ERLENKEUSER, M. HARTMANN, P.J. MÜLLER
and F. POLLEHNE

4.1   INTRODUCTION

Benthic decomposition determines the portion of detrital organic matter that is utilized by organisms for their biomass production and energy requirements, the remainder being buried in the sediment column. Benthic decomposition on the other hand also determines to a large extent the nutrient release to the overlying water column which - in the case of shallow sea areas - may be large enough to satisfy up to 100 % of the nutrient requirements for pelagic primary producers (ZEITZSCHEL and DAVIES 1978, ZEITZSCHEL 1980). For an evaluation of the role of the bottom sediment in the cycling of organic matter constituents, an area of investigation such as Kiel Bight appears particularly suited since (i) a decade of interdisciplinary research has yielded results to tackle most of the problems currently under discussion in this field, (ii) data from long-term studies stretching back more than 20 years may be recalled to rate short-term findings and (iii) Kiel Bight comprises widely different sedimentary regimes ranging from newly eroded sands to rapidly accumulating, organic-rich mud sediments.

In this study of early diagenetic processes, degradation of organic matter is considered as the entirety of single steps of decomposition by which the primary product assembled from (mainly) solar energy, carbon and nutrients is transformed back to its equilibrium components at earth's surface conditions, i.e. CO_2 (HCO_3^-), NO_3^- and HPO_4^{2-}. Thus degradation is not confined to bacterial activity, but includes - aside from intramolecular reactions, spontaneous autolysis, synthesis of intermediate products - also the degradational activity of higher members of the food chain. The actual communities of benthic organisms are simply treated here as a "black box" which, in addition, is assumed to be at "steady state" within the seasonal and spatial limits set by temperature, sediment supply and sediment composition, respectively (a detailed analysis of the benthic community is provided in chapter 3, this volume).

In the following a few results on the hydrography and rate and composition of inputs to the sediments are summarized (which are analyzed in detail in chapter 2, this volume) and are compared with the composition and the rate of burial of sedimentary organic matter. The second section (4.3) analyses various processes of decomposition and release, and compares the relative contributions from differing near-surface sites of decomposition, with emphasis on seasonal and spatial variations; it concludes with an attempt to describe quantitatively the role of the bottom in organic matter cycling (4.4). The subsequent two sections (4.5 and 4.6) report some few results on the distribution of trace elements and on parameter assemblages indicative for pollution. Seeing its main task in a critical and comprehensive presentation of the research efforts of the SFB 95, it was beyond the scope of this chapter to review the presently available literature on all fields which are addressed; thus, relatively few, more general citations are given in the text.

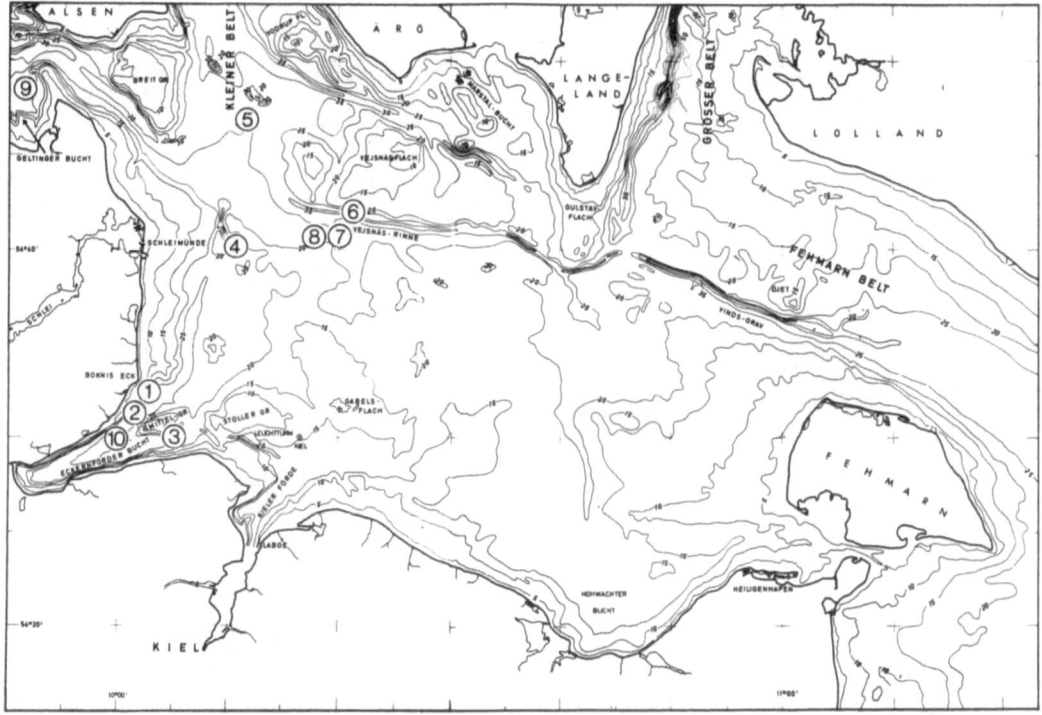

Fig. 4-1: Kiel Bight, Western Baltic. Principal investigation area (No. 1: "Hausgarten") and the location of cores pertinent to accumulation rate studies
(for numbers: see Tab. 4-1). (Map taken from SEIBOLD et al. 1971).

4.2   SEDIMENTARY ORGANIC REGIME

4.2.1   Hydrographical Regime

In Kiel Bight (Fig. 4-1) the episodic inflow of high saline water at the bottom
through the deep channels leads to large short-term salinity fluctuations (surface
water: 9 to 22 °/oo, bottom water: 13 to 3o °/oo) and the build-up of a halocline. The
density stratification is intensified by the regular formation of a seasonal thermocline from April to October at about 15 m water depth (v. BODUNGEN 1975, ITURRIAGA
1979), which separates surface water with an annual temperature range of 0 - 22°C
from bottom water with a temperature range of 2 - 11°C at 20 m. A more detailed hydrographical description is provided in chapters 2 and 5 of this volume. Most investigations for this study were carried out in the "Hausgarten", a 0.6 by 1 km wide area
in the southwestern part of Kiel Bight that is set aside for research purposes. An
important hydrographical feature for the cycling of organic matter is the almost regular migration of the redox discontinuity layer during summer from the sediment upwards into the bottom water of the deep channel adjacent to the "Hausgarten". The resulting H_2S-milieu causes the macrofauna to disappear almost completely from this
depth zone (KÜLMEL 1976).

Tab. 4-1: Accumulation of organic carbon in Kiel Bight sediments

No. in Fig. 4-1	Sed. type x)	Locality	Water depth (m)	Porosity	Sedimentation rate (mm y^{-1})	Dry matter flux (g m^{-2} y^{-1})	Organic carbon Accumulation (range in brackets) (% dW)	(g C m^{-2} y^{-1})	Time interval (AD)	Dating method	GIK-core-no. and references
1	fS	"Haus- garten"	20	0.37	1	1580	0.5 (0.38-1.0)	8 (6-16)	1900 - 1980	^{210}Pb	14 261-1, 3, 4, 5, 6
1	sM	"	25	0.60	1.5$^+$-0.5	1500	1.1 (0.5-1.5)	17 (8-23)	1900 - 1980	"	14 262-2, 3, 4, 5, 6
1	M	"	28	0.86	1 / 3.2	350 / 1100	4.5	16 / 50	1850 - 1920 / 1930 - 1978	"	14 263-5
1	M	"	28	0.85	1.2 / 3.1	520 / 1220	4.5 (4-5) / 4.5 (4-5)	23 / 55	1870 - 1930 / 1940 - 1980	"	14 263-1, 2
2	M	near "	29	0.86	1.4 / 1.4	484 / 484	6.0(5.5-6.6) / 4.8 (4-5.5)	29 / 23	400 - 1400 / 1400 - 1900	^{14}C on org. C	11 883; ERLENKEUSER et al. 1974
3	M	South of Mittelgrund	21	0.81	0.2	100	4.5(4.1-5.2)	4.5	-2000 - present	"	13 949; WHITICAR 1978
4	M	off Schleimünde	28	0.76	0.3	195	3.7(2.7-4.7)	7	-2500 -	"	13 704-1
5	M	SW of Ärö	26	0.83	1.7	710	4 (3-5)	28	1850 -	" indirect	11 777-3; ERLENKEU- SER et al.1974
6	M	Vejnäs Rinne	27	0.82	0.4	190	3.2 (3-5)	6	1000 -	"	12 888
7	M	"	23	0.79	0.3	160	4(3.5-4.5)	6	1000 -	"	12 887
8	M	"	23	0.79	0.3	150	3.8(3.5-4.1)	5.6	1000 -	"	12 886
9	M	Geltinger Bucht	20		2.3	920	3.5^{xx}(3-4)	32.5		^{210}Pb	MÜLLER et al. 1980
10	M	Eckernförd. Bucht	21		3.2	1030	5 ^{xx}(4-6)	51.5		"	"
	fS	Lübecker B.	15		1.6	540	4.5 ^{xx}	24.3		"	"

x) fS: fine sand; sM: sandy mud; M: mud   xx) calculated from total nitrogen assuming C/N = 10 by weight

4.2.2   Composition and Supply of Organic Matter to the Bottom

In a long-term study it was shown that despite the large scale water exchange processes the "Hausgarten" area exhibits remarkably constant seasonal patterns of nutrient
and biomass distribution. An advection of water masses, although a predominant feature
in this region, has essentially little effect when compared to the vertical stability of the water column and the interaction between bottom and overlying water column (v. BODUNGEN 1975): starting from a well mixed water column at the end of winter
and an annually recurring nutrient threshold of 20.2 - 22.4 mMol PO_4 m^{-2} and 234 - 254
mMol tot-N_{inorg} m^{-2} in a 20 m water column, nutrients are entirely incorporated into
the spring phytoplankton bloom which quantitatively settles out to the bottom. In late
spring the nutrient source for the euphotic zone is benthic regeneration, while during summer production is based on recycling within the euphotic zone due to stratification with contributions from lateral advection of shallow sediment derived nutrients
(POLLEHNE 1980). Although primary production is highest during summer, sedimentation
is low. After the breakdown of the thermal stratification due to strong westerly
storms, nutrients accumulated in the bottom water are mixed to the euphotic zone and
initiate an autumn bloom which might be even larger than the spring bloom. The annual
primary production was estimated to be 158 g C m^{-2} y^{-1} (v. BODUNGEN 1975).

Quantitative annual figures of the organic carbon sedimentation rate for the
"Hausgarten" area are given by ZEITZSCHEL (1965), v.BROECKEL (1975), HENDRIKSON
(1976), ITURRIAGA (1979) and POLLEHNE (1980); the most detailed description of sedimentation is provided by SMETACEK (1980b): in a 3 years study he found the seasonal pattern
of sedimentation to be alike each year although considerable differences were observed
in the quantity of material collected (see also chapter 2, this volume). The main
contributors to the trapped material were primary produced matter originating from
the pelagic system (phytoplankton cells and detritus) and resuspended sediment giving
rise to high sedimentation rates during the non-growth season.

If the organic carbon trapped during the growth season (March 1 to November 15) is
considered representative for the sedimentation from the pelagic system (the potential sedimentation of 6.3 g C m^{-2} primary production during the winter months being
assumed to be counterbalanced by resuspension during the growth season), one would
reach a yearly figure for primary organic carbon sedimentation of 40 g C m^{-2} y^{-1}
(calculated from data of SMETACEK 1980b). A comparably low value for the growth season (30.9 g C m^{-2} y^{-1}) is given by POLLEHNE (1980) along with rates of nitrogen (3.7
g N m^{-2} y^{-1}) and phosphorus sedimentation (0.8 g P m^{-2} y^{-1}). During the growth season
phytoplankton forms the bulk of particulate organic carbon in the water column, much
less being present as detritus. The average composition of the trapped organic carbon
at 20 m water depth was 38 % as proteins, 23 % as carbohydrates, 25 % as lipids and 3.4
% as humic acids (SMETACEK and HENDRIKSON 1979).

From a comparison of fixed and non-fixed sediment trap material HENDRIKSON (1976) inferred degradation rates of 15 % to 25 % per year; the degradation intensity decreased
from lipids over water-soluble carbohydrates, proteins, carbohydrates to humic acids,
the value for proteins being low due to the build-up of saprophyte protein. The degradation processes in the water column are partly reflected in an increase of the
C/N ratio (by atoms as used throughout the text) in trapped material from 8.1 at 10 m
to 8.4 at 18 m water depth averaged over the year (ITURRIAGA 1979). From the data of
SMETACEK (1980b) a mean value at 18 m can be calculated of C/N = 9.8 for the growth
season and C/N = 10.2 for the winter season; this author reported C/N ratios well below 9 only for the spring and autumn bloom. For the material trapped at 18 m during
the growth season POLLEHNE (1980) reports a C:N:P-composition (by atoms) of
152:14.6:1 (C/N = 10.4).

4.2.3   Sedimentary Regime and Sedimentation Rates

According to PRATJE (1939, 1948) and BARNER (1964), recent depositional environments
of the Baltic may be divided into four zones: nearshore sand accumulation, erosion
areas with lag sediment, offshore sand accumulation and mud areas. A detailed description of the sediment cover of Kiel Bight is given by SEIBOLD et al. (1971) (see
also chapter 5, this volume). A topographic map of the "Hausgarten" together with information on sediment distribution was constructed by WEFER und TAUCHGRUPPE (1974). Fig.
4-2 shows areas of lag sediments (exposed areas of glacial till) down to 13.5 m, of medium and fine sand, and of mud mainly accumulating in the deep channel below 23 m water
depth. Additional information is provided for a station (11 m) with coarse sand by BOJE
(1974) and a station with fine and muddy sand at 20 m water depth (BALZER 1978). In the
central basin of the Eckernförder Bucht () 20 m water depth) the fine grained, organic-
rich Holocene muds reach a maximum thickness of 10 m (WHITICAR 1978). Surface sediments
at that depth are composed of muddy silt to clay (median diameter 13 - 25 µm). For
the cycling of organic matter in the entire system from the shore to the maximum
depth it is important to recognize that the basin area receives most organic matter
but is least colonized by macrofauna.

Sediment accumulation rates pertinent to Kiel Bight are compiled in Tab. 4-1. Radiocarbon studies of these sediments have shown that the organic matter in nearshore
clay sediments is derived from a mixture of various fractions of different ages
(ERLENKEUSER 1979); the admixture of older eroded and redistributed organic carbon to
the recently produced carbon results in an overall radiocarbon age of about 850 years
for surface deposits of the area, if extrapolated from pre-industrial sediment strata
(cf. Fig. 4-24).

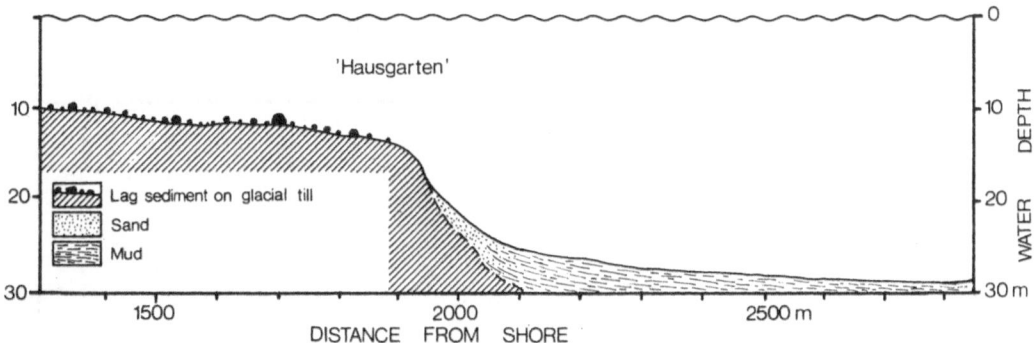

Fig. 4-2: Sediment distribution in the "Hausgarten" region (after WEFER and TAUCH-
         GRUPPE 1974).

The introduction of large amounts of fossil carbon by man's use of fossil fuel has resulted in an "ash-effect": the dilution of the natural ^{14}C-distribution of the sediments through fossil carbon causes an apparent age reversal in surface sedimentary
layers (SUESS and ERLENKEUSER 1975a, ERLENKEUSER 1979). The input of up to 0.8 % dry
weight fossil carbon into bottom sediments started about 1860 in Kiel Bight and later
in more easterly Baltic basins (SUESS and ERLENKEUSER 1975b). During recent decades
nuclear bomb-produced radiocarbon has entered the sea, providing a tracer for identi-
fying modern deposits. Modelling the "ash"- and "bomb"-effect on radiocarbon distribution, ERLENKEUSER (1979) deduced the sources of the organic carbon in Bornholm
Basin surface deposits: during the last 70 years recent carbon varied from 0.5 to
1.3 % sediment dry weight, the "eroded" fraction amounted to 3.6 % dry weight with an
apparent age of about 1500 years and fossil carbon ranged between 0.2 and 0.3 % dry
weight. In view of the construction of diagenetic models it appears noteworthy that
the old organic carbon is not digested by living bivalves typical of the Kiel Bight
sedimentary habitat (ERLENKEUSER 1976). Neither does Nephtys ciliata incorporate the
old organic fraction. If the same holds true for bacterial degradation, the actually
available reactive part of the total organic carbon would in fact be considerably lower.

Inhomogeneities in the recent sedimentation rate become apparent from a combination
of ^{14}C and ^{210}Pb dating of a core taken adjacent to the "Hausgarten" and analyzed
systematically to deduce the origin of organic matter (see 4.2.6). The ^{137}Cs and
^{210}Pb distribution in the upper 28 cm of this core is shown in Fig. 4-3. The ^{210}Pb
depth distribution suggests an average sedimentation rate of 3.1 mm y^{-1} in the upper
section of the core, an exceptionally high rate of 8.3 mm y^{-1} between 9 and 18 cm,

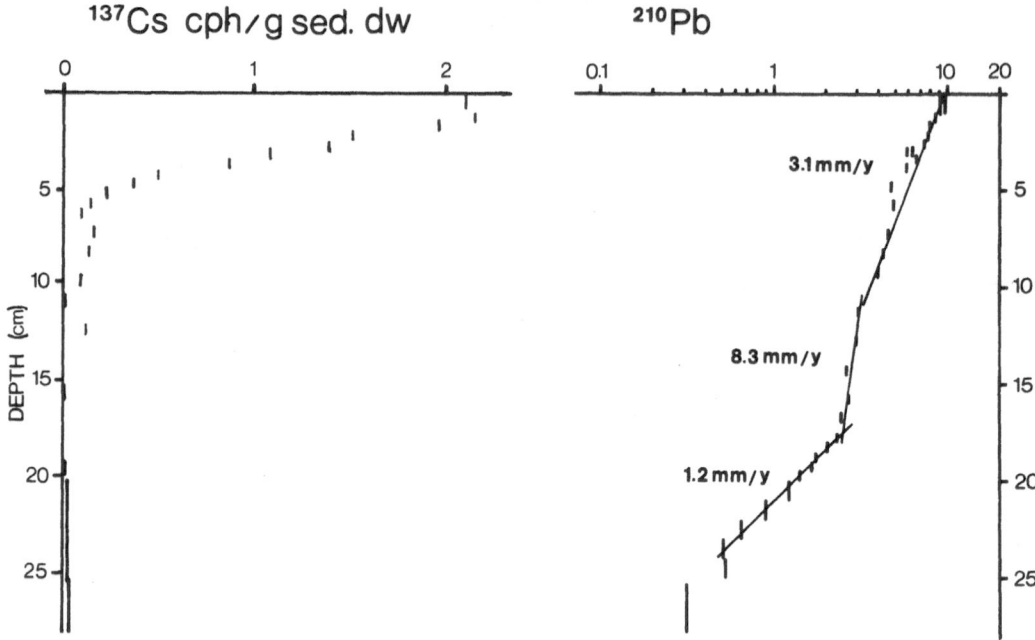

Fig. 4-3: Station at 28 m, near "Hausgarten" (core GPI 14 263-1): distribution of ^{137}Cs and ^{210}Pb activity (ERLENKEUSER unpubl.).

and a lower rate of 1.2 mm y^{-1} below. The sedimentation rate of sediments below 18 cm is in accordance with the rate of 1.4 mm y^{-1} calculated by ERLENKEUSER et al. (1974) from radiocarbon data for sediments between 20 and 160 cm below the sediment/water interface at the same location. Due to the decrease in natural radiocarbon content by fossil carbon fallout from industrial activities in sediments younger than about 120 years, these earlier radiocarbon studies did not allow an estimate of the sedimentation rate for the upper 20 cm. A combination of both ^{210}Pb and ^{14}C studies suggests that the sedimentation rate remained relatively constant during the past 1100 years at this location but increased significantly during the last 50 years. ^{137}Cs derived from nuclear weapon tests in the atmosphere since 1953, with a test activity peak at 1962, appears first in this sediment core at a depth of about 10 cm correlating closely with the depth of 9 cm, as expected from the ^{210}Pb profile. Similar features were obtained for cores from the Skagerrak (ERLENKEUSER and PEDERSTAD 1984), from the Gdansk basin (ERLENKEUSER unpubl.) and the Western Baltic (G. MÜLLER et al. 1980). The latter authors suggested that only a minor fraction was immediately scavenged from the water column, whereas a major ^{137}Cs fraction was first deposited on land and then transported into the basin. However, our studies for different areas of the ocean of various parameters, such as sedimentary ^{14}C, ^{210}Pb, ^{137}Cs, anthropogenic trace metals, seasonal variations of the dry matter yield of sediment traps and of course direct sedimentological evidence emphasize the role of redistribution processes for the deposition and lateral distribution of fine-grained material. Thus, the retarded development of the ^{137}Cs profile in cores from areas of deposition of clay-sized particles seems more likely to result from a temporary storage of this nuclide in higher energetic environments at shallower water depths and subsequent gradual export to deeper sedimentary basins. In addition to the depositional regime controlled by currents and wave action, the availability of the tracer for export from shallower reservoirs should also depend on the rate and the depth range of bioturbation as a retarding factor.

4.2.4   Organic Matter Distribution and Accumulation Rates

To investigate degradation at and within the bottom sediment, the organic matter distribution was recorded both vertically and horizontally. Accumulation rates are important figures for the quantitative understanding of organic matter cycling and of its horizontal transport (see 4.4). In the transect orthogonal from the shoreline to the maximum depth of Eckernförder Bucht as described by WEFER and TAUCHGRUPPE (1974) the non-accumulating lag sediments extend 1750 m offshore down to 13.5 m water depth (Fig. 4-2). According to data obtained at 11 m water depth by BOJE (1974) these sediments are characterized by a C_{org} content of 0.1 % dry weight in the top 5 cm. A higher value ($C_{org} \approx 0.4$ %) was found by MEYER-REIL (1981) in the top millimeter during the settling of the spring bloom (see also chapter 3, this vol.).

The transition zone of medium sand and muddy sand between the lag sediment and the mud region below 23 m may be characterized by stations between 18 and 20 m which have been extensively investigated (BALZER 1978 and 1984, POLLEHNE 1980). The depth distribution of organic carbon, organic nitrogen and organic phosphorus (Fig. 4-4) shows a rapid decrease of C_{org} and N_{org} in the top two centimeters and fairly constant values in the layers below. The observed profile can be generated by an exponentially decreasing rate of decomposition eventually flattened by biogenic mixing, as well as by an increased organic matter load due to pollution (G. MÜLLER et al. 1980). From the intensive remineralization in the youngest layers (RHOADS 1974, see also section 4.3.3) and its reflection in calcium carbonate dissolution (WEFER 1976, BALZER 1978), decomposition was estimated to be the dominating process (BALZER 1984). Accumulation rates (C_A) compiled in Tab. 4-1 for different water depths of the "Hausgarten" were calculated from the percentage organic carbon of dry weight sediment (C %), the sedimentation rate (ω in mm y^{-1}), the porosity ϕ and the dry solid density (taken as $\rho_s = 2.54$ g cm^{-3}) according to

$$C_A \ (g\ C \cdot m^{-2} \cdot y^{-1}) = 10 \cdot (C\ \%) \cdot \omega \cdot \rho_s \cdot (1-\phi) \quad (\text{MÜLLER and SUESS 1979}).$$

120

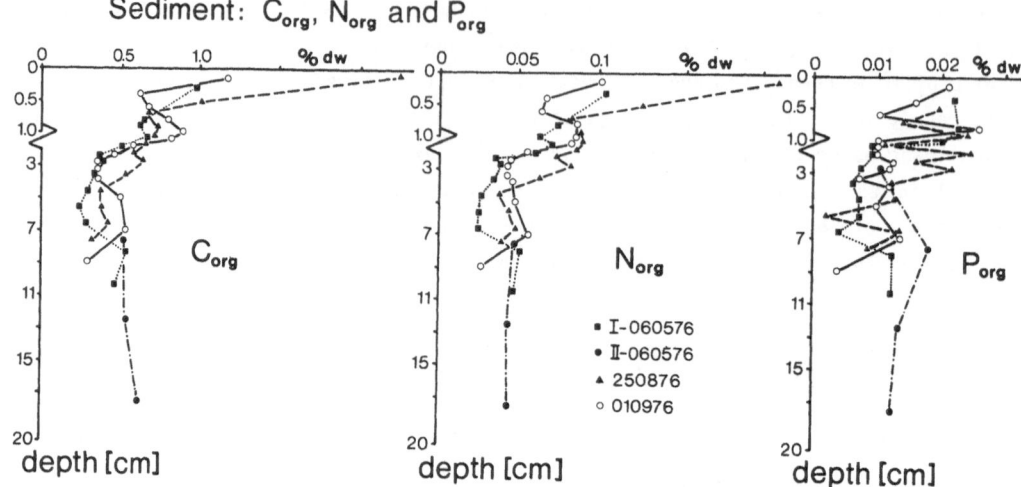

Fig. 4-4: Station at 20 m, "Hausgarten": depth distribution of organic carbon, organic nitrogen, and organic phosphorus (BALZER 1984).

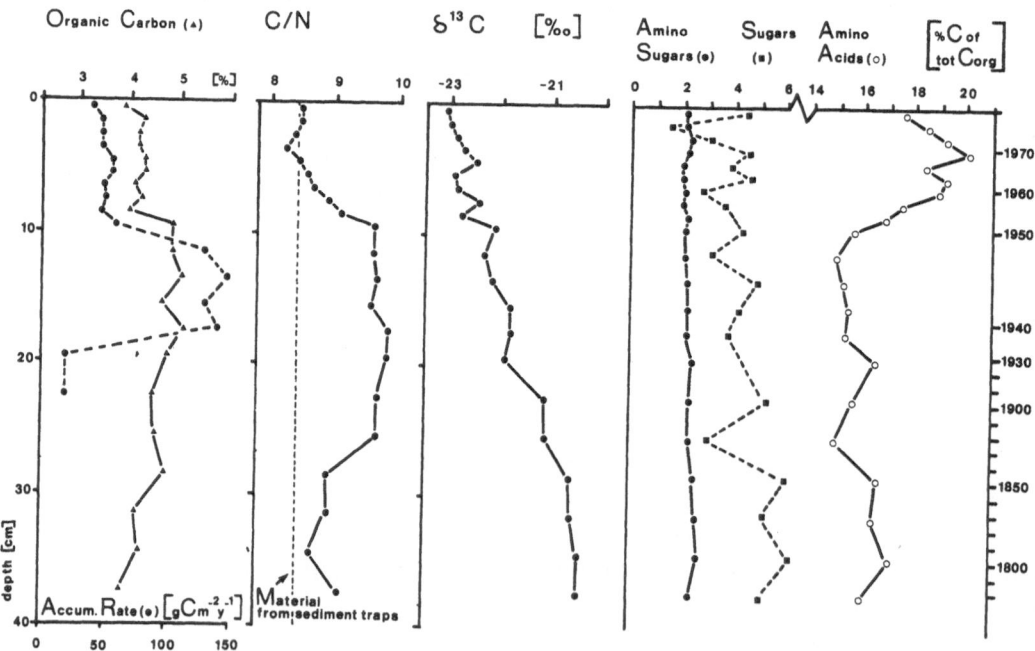

Fig. 4-5: Station at 28 m, near "Hausgarten" ((core GPI 14 263-2): depth distribution of organic carbon (triangels: percent content; solid dots: accumulation rate), C/N ratio, $\delta^{13}C$, and percentages of amino sugars (glucosamine
and galactosamine), sugars and amino acids as carbon equivalents (ERLEN-
KEUSER, LIEBEZEIT, MÖLLER unpubl.).

If the top 10 centimeters are taken as a black box and assumed to be at steady state - an unsatisfactory assumption in view of the seasonality of supply, bioturbation and decomposition - a flux to the surface ($\sim$ 17 g $C \cdot m^{-2} \cdot y^{-1}$) and a "burial" rate below 10 cm can be calculated yielding a mean decomposition rate of 10 g $C \cdot m^{-2} \cdot y^{-1}$ within the bioturbated zone; the respective burial rates for organic nitrogen and organic phosphorus through the 10 cm horizon are 0.60 g $N \cdot m^{-2} \cdot y^{-1}$ and 0.15 g $P \cdot m^{-2} \cdot y^{-1}$, respectively (BALZER 1984).

For the station at 28 m (mud), evidence is obtained from the bulk accumulation rate (Fig. 4-3) and the organic carbon distribution, indicating that over longer periods of time the assumption of steady state supply and accumulation is not valid. The organic carbon contents and accumulation rates are given in Fig. 4-5. Organic carbon contents range from 4 to 5 % in these sediments, the lowest values occurring in the upper 9 cm of sediment and at the base of the core. The abrupt organic carbon increase at a depth of 9 cm matches the change in sedimentation rate just below 9 cm (Fig. 4-3). The organic carbon accumulation rates were about 50 and 20 g $C \cdot m^{-2} \cdot y^{-1}$ in the upper and lower core sections, respectively, and reached an exceptionally high level of 135 g $C \cdot m^{-2} \cdot y^{-1}$ in sediments in between. The organic carbon flux in the upper core section is comparable to the pelagic sedimentation rate obtained from sediment trap deployments (SMETACEK 1980b). However, this accordance is misleading since a major fraction of the organic material is remineralized at the sea bottom and this deep water section also receives laterally transported material.

Evidence for appreciable fluctuations in the sedimentary regime at 28 m over the last 4200 years was also obtained from a 6 m long core (GPI 13 939) for which C_{org} along with other sedimentological parameters is depicted in Fig. 4-6 (WHITICAR 1978). This author has proved that degradation within this organic-rich sediment regularly leads to complete exhaustion of sulfate at depth and the formation of methane partly reaching supersaturation in the sediments (see also section 4.3.5).

For a rough quantitative assessment of the amount of carbon that is withdrawn by burial from the "carbon" system a transect was evaluated extending from the shore to the central part (the end of the transect was taken to be half the distance of the 25 m depth-lines between the "Hausgarten" and the "Mittelgrund") of Eckernförder Bucht. The accumulation rates at the different water depth intervals were multiplied by the respective distance they cover in the transect. The transect (Fig. 4-2) was divided into five zones according to the sediment type yielding a total length of 2500 m. By reasonable interpolation of relevant data (Tab. 4-1) carbon accumulation rates per section were calculated (BALZER et al. 1986); while the lag sediments accumulate no carbon, the 1 m wide and 237 m long transect over the transition zone of medium to muddy sand buries only 876 g C y^{-1} and the mud sediments accumulate roughly 19,800 g $C \cdot y^{-1}$ over a distance of 507 m. These two figures correspond to a mean burial rate of

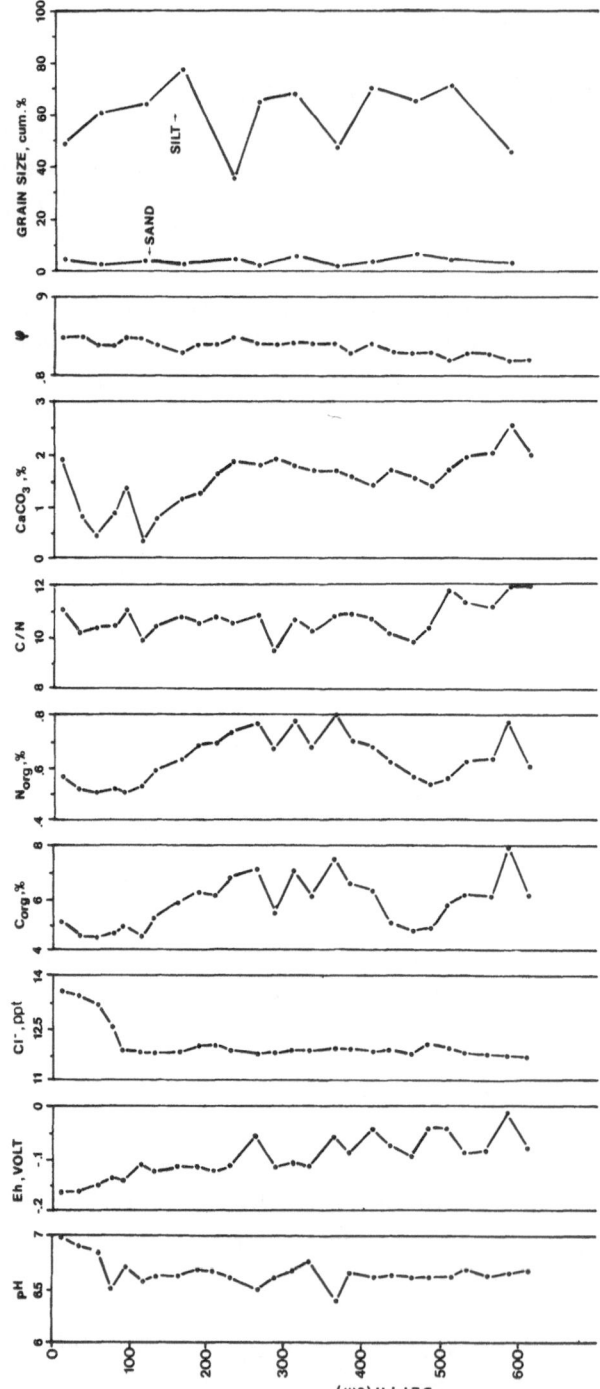

Fig. 4-6: Station at 28 m, near "Hausgarten" (core GPI 13 939; Sta.no. 2, Fig. 4-1): depth distribution of pH, Eh, Cl, C_{org}, N_{org}, C/N, $CaCO_3$, porosity and grain size (WHITICAR 1978).

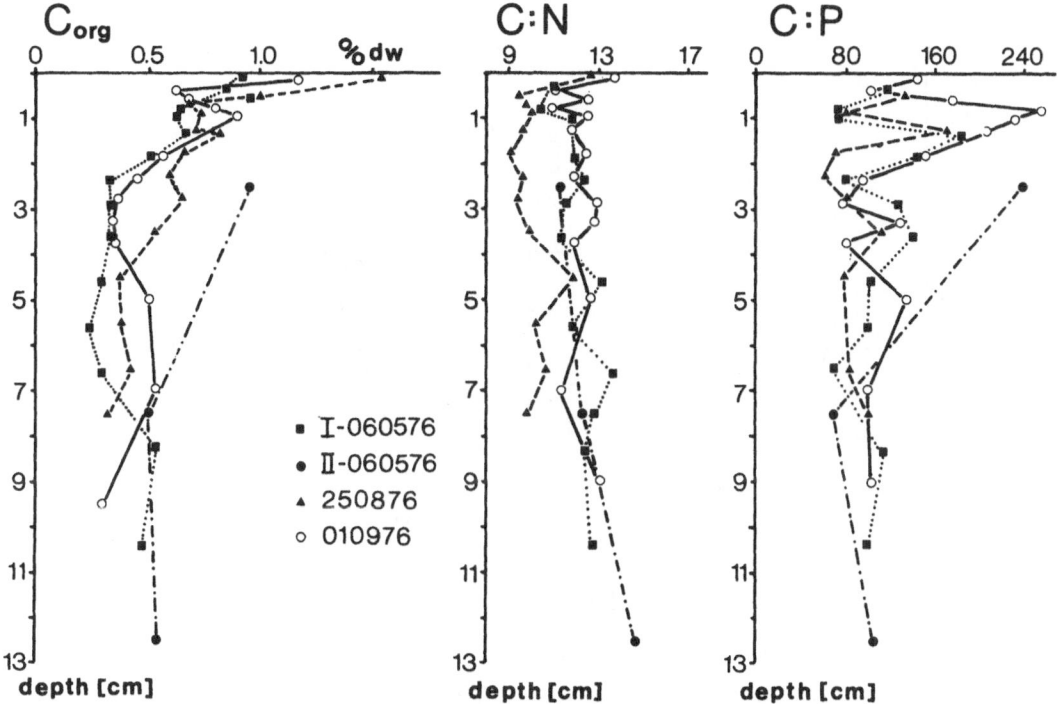

Fig. 4-7: Station at 20 m, "Hausgarten": depth variation of C_{org}, C/N and C/P
ratios of the organic matter (BALZER 1984).

27.8 g $C \cdot m^{-2} \cdot y^{-1}$ over the whole accumulating part of the transect and a mean rate of
8.3 g $C \cdot m^{-2} \cdot y^{-1}$ for the whole distance from the shore respectively (see section 4.4).

4.2.5   Composition and Fractionation of Sedimentary Organic Matter

The organic matter at the sediment surface of the 20 m station in the "Hausgarten"
was reported as being composed of 32.6 % proteinaceous material, 17.5 % carbohydrates,
13.3 % lipids and 23.6 % humic acids in terms of annual mean carbon equivalents, the
rest being unaccounted for (HENDRIKSON 1976). Comparison with settling material collected in sediment traps revealed a slight decrease in the percentages of carbohydrates and lipids and a considerable increase in the proportion of humic acids; nevertheless, HENDRIKSON (1975) stated that autochthonous humification in the water column
and the sediment of this area is slow and that a relatively high percentage of degradable organic matter finds its way into the sediment or is resuspended.

Numerous authors (TRASK 1932, SUESS 1976a, SUESS and MÜLLER 1980) have shown that
analysis of the C:N:P ratio provides an important tool for studying the origin, degradational history and the fractionation of organic matter within the sediment. Bio-

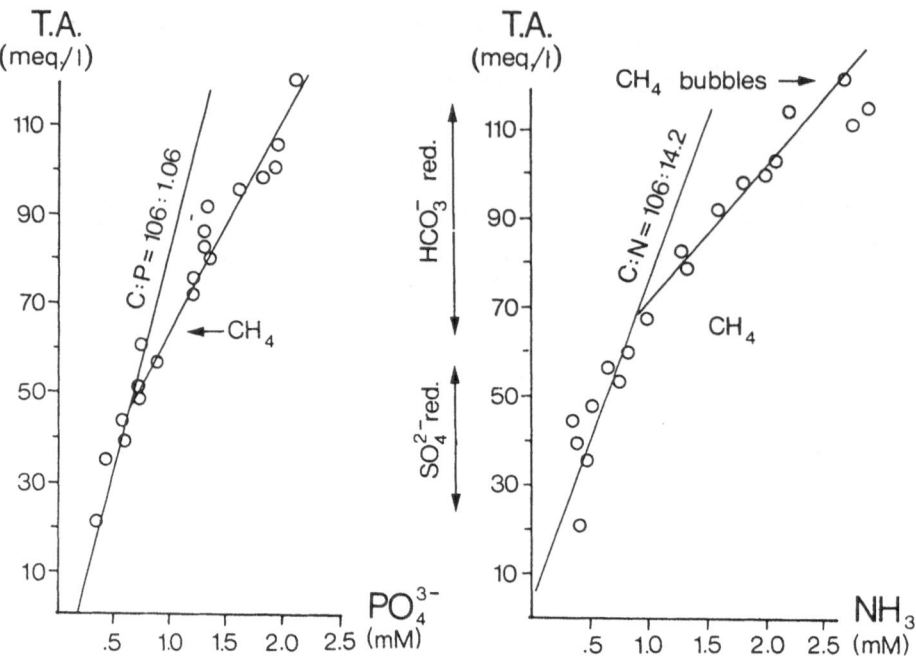

Fig. 4-8: Station at 28 m, near "Hausgarten" (core GPI 13 939; sta.no. 2, Fig. 4-1): plot of interstitial ammonia and phosphate versus titration alkalinity to infer the stoichiometric composition of metabolizable organic matter. Note slope change at depth of the first methane occurrence (WHITICAR 1978).

chemical processes begin selectively to decompose the biogenic material as soon as it leaves the euphotic zone. Fractionation persists throughout the subsequent history of settling, decomposition, reworking, ingestion, resuspension and eventually burial as part of the sedimentary record. C/N and C/P ratios for the station at 20 m are depicted in Fig. 4-7. From visible evaluation a slight increase from C/N = 11 (by atoms) near the surface to values of C/N = 12.5 at 10 cm is obtained, probably reflecting the preferential degradation of proteinaceous material (BORDOWSKI 1965). The decrease of the C/N ratio within the top 5 mm may be due to the build-up of bacterial protein. Comparing average values for trapped particulate material ITURRIAGA (1979) found the C/N ratio to increase from C/N = 8.1 at 10 m to C/N = 8.4 at 18 m water depth. Filtrated particulate matter at 10 m with C/N = 8.1 was considerably lower in the C/N ratio than organic matter at the sediment surface with C/N = 9.1 (HENDRIKSON 1975). After relatively rapid losses of nitrogen during the short time of settling through the water column, there is only a minor difference in the degradation rate of C and N components.

Although it was reported for open ocean sediments (SUESS and MÜLLER 1980) that C/P ratios of organic matter should be even more sensitive fractionation indicators than the C/N ratio, the plot of C/P ratios in Fig. 4-7 does not reveal a preferential loss of phosphorus with depth in sediment. HENDRIKSON (1975) found that the C/P ratio of the particulate matter decreased from C/P = 262 at 10 m over C/P = 195 at 20 m water depth to C/P = 85 at the sediment surface. POLLEHNE (1980) reported a mean C/P ratio of 152 during the growth season for organic matter trapped at 18 m. This behavior - contrary to the general finding of preferential P remineralization (e.g. SEKI et al. 1968, KRAMER et al. 1972, KNAUER et al. 1979) - can be explained in the following way (BALZER 1984): upon death of the organisms phosphate is released autolytically very rapidly (HOFFMANN 1956, GOLTERMAN 1973) without corresponding carbon loss, thus producing high C/P ratios in the suspended matter in the water column at 10 m water depth. For subsequent remineralization only more slowly degradable P compounds are available and organic carbon is preferentially decomposed down to the top sedimentary layers. It is only below these horizons that the proportion of refractory carbon material increases to such an extent that the rates of C and P degradation become more or less equal as observed.

From a careful inspection of all surface sediment data available for the 20 m station BALZER (1984) reached at an average composition for the (model) organic matter of $(CH_2O)_{125}(NH_3)_{10.9}H_3PO_4$ which may serve as a starting point for the decompositional processes at the sediment surface and within its top layers. Formulation of a model detailing the decomposition of organic matter or, put otherwise, the regeneration of nutrients, orginated from REDFIELD et al. (1963), who advanced that nutrients are regenerated in proportions equal to those from which they were assembled. Conversely the C:N:P composition of the metabolizable organic matter may be inferred from the molar ratio of the terminal degradation products contained in the pore water. By this approach (having some serious limitations as discussed by WHITICAR 1978) this author arrived at an average composition of C:N:P = 100:13.4:1 for core GPI 13 939 (28 m) from the pore water constituents (Fig. 4-8) in the layers above methane occurrence.

4.2.6   On the Origin of Organic Matter

Knowledge of the relative amounts of marine and terrestrial organic matter in marine sedimentary basins is of immediate environmental and sedimentological interest but serves also as a basis for the reconstruction of depositional paleo-environments both for scientific and economic purposes. Terrestrially derived organic matter is usually more refractory with respect to microbial decay since it is primarily derived from higher plant vegetation and has been recycled several times before entering the marine environment. SWEENEY and KAPLAN (1980), for example, have shown that microorganisms primarily mediate the oxidation of organic compounds derived from marine organisms, thereby leading to a relative enrichment of the terrestrial component during diagenesis.

The distinct compositional differences between both types of organic matter are partly preserved within sediments and can be used to reconstruct depositional environments, as indicators of diagenesis and catagenesis, and for correlations between oil
and source rocks (e.g. EGLINGTON and MURPHY 1969, SIMONEIT 1978, TISSOT and WELTE
1978, DURAND 1980). Any evaluation of biological markers in ancient deposits, however, requires a thorough knowledge of recent depositional environments and of the
fractionation effects during early diagenesis. In the following we summarize yet unpublished organic geochemical and isotopic results (sugars: LIEBEZEIT unpubl., amino
acids and amino sugars: MÜLLER unpubl., isotopes: ERLENKEUSER unpubl.) obtained from
sediments of the 28 m station.

The atomic ratio of organic carbon to organic nitrogen varies from 9.6 to 11.4 at the
28 m station (Fig. 4-5). This is within the range of C/N ratios of sediment trap material (6.5 - 11.3) as reported by SMETACEK (1980b) for material collected during different seasons. The average C/N ratio of sediment trap material (9.8) is indicated by
the dotted line in Fig. 4-5. The relatively low C/N ratios in the upper core section
and the striking accordance with the average C/N ratio of sediment trap material
seem to rule out a significant contribution of terrestrial higher plant residues
which have atomic C/N ratios well above 17 (GIESEKING 1975). However, material derived
from marine macroalgae may be present in addition to planktonic organic matter. VINO-
GRADOV (1953) and SCHMIDT (1978) reported C/N ratios of 8 - 12 for red and brown algae, suggesting that both sources are indistinguishable by means of the C/N ratio.
The nitrogen-poor organic matter in the central core section could contain about 25 %
of terrestrial higher plant material, assuming C/N ratios of 8.4 and 20 for marine
algae and terrestrial higher plants, respectively. It is more probable, however, that
this signal was produced by the admixture of up to 14 % fossil carbon from fossil fuel
burning as shown by ERLENKEUSER et al. (1974). Neither does the sugar composition
(see below) suggest a significant terrestrial input.

The organic carbon percentages occurring in the form of amino acids, hexosamines and
sugars are depicted in Fig. 4-5. Amino acids account for the highest proportion (15 -
20 %), followed by sugars (1.5 - 6.1 %) and hexosamines (2.0 - 2.8 %). The total recovery of organic carbon in the form of these compounds is lower than in plankton and
sediment trap material (e.g. WEFER et al. 1982) but higher than in oceanic deep-sea
sediments (MÜLLER unpubl.), pointing to a relatively low degradation state of the embedded organic material. This conclusion is further corroborated by the distribution
of non-protein amino acids (β-alanine and γ-aminobutyric acid) which were present in
absolute amounts similar to those found in continental slope and deep-sea sediments.
However, they account only for about 1 Mol % of the total amino acids present as opposed to their relative enrichment in highly degraded sedimentary organic matter,
where they may constitute up to 90 Mol % of the total amino acids (SCHROEDER 1975,
WHELAN 1977, MÜLLER unpubl.).

127

The data plotted in Fig. 4-5 and 4-9 clearly reveal that the upper core section contains a significantly higher proportion of nitrogen compounds relative to total organic carbon, to amino sugars and to sugars. This supports the assumption that the decrease of C/N ratios discussed above is actually due to an enrichment of nitrogen compounds rather than to a decrease in fossil carbon contents.

Fig. 4-9 also shows a few selected ratios of organic compounds which all reflect the different composition of organic matter deposited above and below about 9 cm depth. The molar glucose/ribose ratio (GLC/RIB) is of particular interest since this ratio has been proposed as a criterion to distinguish marine (< 4) and terrestrial (> 20) organic matter sources (MICHAELIS et al. 1976). The highest value recorded in "Hausgarten" sediments was 2.8 (Fig. 4-9) and there is no need to infer a terrestrial source for the carbohydrate fraction. The high relative amino acid content points in the same direction.

The significant compositional differences in sedimentary organic matter above and below 9 cm in the core from 28 m water depth may be explained by different proportions

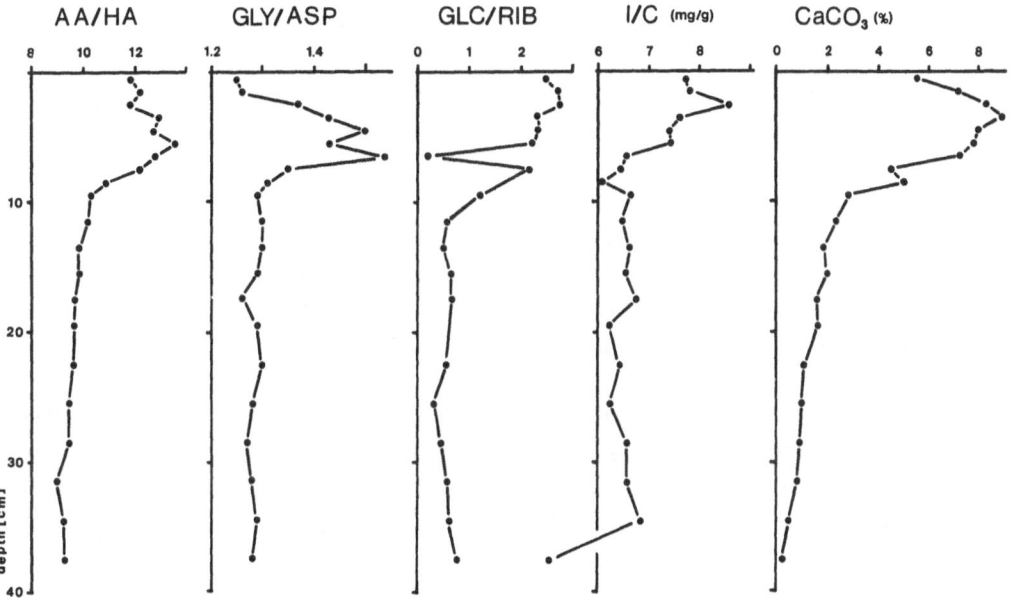

Fig. 4-9: Station at 28 m, near "Hausgarten": molar ratios of amino acids/hexosamines (AA/HA), glycine/aspartic acid (GLY/ASP), and glucose/ribose (GLC/RIB), iodine per g of organic carbon, and calcium carbonate (HARTMANN, LIEBEZEIT, MÜLLER unpubl.).

of organic matter from planktonic and macroalgal sources. The iodine and calcium carbonate distribution, also included in Fig. 4-9, point to an increased input of ma-croalgal detritus in the youngest core section. Marine macroalgae may contain up to 1 % iodine per dry weight primarily bound to peptides and proteins (SHAW 1962). The higher iodine/organic carbon ratios are in accordance with the view that the youngest core section has received a higher proportion of macroalgal material. The concomitant increase of calcium carbonate supports this interpretation since macro-algae dry mat-ter from the "Hausgarten" area may contain up to 25 % $CaCO_3$ produced by animals (e.g. Bryozoa, Foraminifera, Serpulida) growing on them (see chapter 6, this volume); a qu-antitative treatment, however, would suggest an additional carbonate source, e.g. glacial till and marl.

Stable carbon isotopes ratios of the organic material in "Hausgarten" sediments range from $\delta^{13}C$ = - 20 to - 23 °/oo and show a more or less continuous change from hea-vy values in deeper sediments to lighter values in surface sediments (ERLENKEUSER unpubl., SCHMIDT 1981). Fig. 4-5 shows the distribution at 28 m water depth. Organic matter of marine planktonic origin should have a $\delta^{13}C$ value of about - 20 %o as esti-mated from a ^{13}C content of zero °/oo for the bicarbonate fraction of recent Baltic Sea surface water at about pH 8 (ERLENKEUSER et al. 1975) taking into account an isotope fractionation of about - 19 °/oo during plankton formation (DEGENS et al. 1968); $\delta^{13}C$ of terrestrial organic matter should be about 25 °/oo. The $\delta^{13}C$ values of sediments de-posited prior to 1870 center around - 20.5 °/oo (Fig. 4-5), in accordance with a plank-tonic origin of the organic material, as also suggested by the parameters presented above. The organic material deposited since 1870, on the other hand, shows a rather continuous depletion in ^{13}C with the lightest values (- 22.4 to - 23.1 °/oo) occurring in the upper 9 cm of sediment, which might indicate a significant proportion of ter-restrial organic matter. The low glucose/ribose ratios, however, and the fact that sediments with the lightest $\delta^{13}C$ values have the lowest C/N ratios and the highest proportion of amino acids are not compatible with a terrestrial organic matter source.

As explained above there is good evidence that the upper core section has received a somewhat higher portion of macroalgal detritus in relation to planktonic organic mat-ter. This can partly explain the coincidence of relatively light $\delta^{13}C$ values and low C/N ratios since red algae from the "Hausgarten" area were found to have compara-tively light $\delta^{13}C$ values ranging from - 26 to - 36 °/oo (STOFFERS 1976) and a low C/N ratio of about 8 (SCHMIDT 1978). Hence, a red algae contribution to organic mat-ter of planktonic origin would shift the $\delta^{13}C$ values towards lighter values but would not significantly alter the C/N ratios. The fossil carbon ($\delta^{13}C$ ≃ -25 °/oo) input would also lower the $\delta^{13}C$ ratios leading to a maximum shift of 0.7 °/oo.

Summing up, it is suggested that the bulk of organic matter originates from marine planktonic and macroalgal sources. A minor fraction of the total organic carbon appears to be derived from fossil fuel burning and subsequent atmospheric fallout. There is no evidence that the sediment contains significant amounts of organic matter from terrestrial higher plants. Distinct differences in the amino acid and sugar composition of the organic material seem to reflect different proportions of both types of algae but an improved knowledge of the chemical composition of source organisms is certainly desirable. A fairly continuous change in sedimentary organic carbon $\delta^{13}C$ values (not exceeding 2.5 °/oo) from heavier values in older sediments to lighter values in surface sediments is not attributed to an increased terrestrial organic matter contribution. The organic parameters rather suggest that this trend is due to an input of fossil fuel carbon since about 1870 and to an increased supply from macroalgae during the past 50 years, both of which superimpose the planktonic $\delta^{13}C$ signal.

4.3   BENTHIC DECOMPOSITION AND REGENERATION

Degradation of organic matter commences immediately upon death of the organism. For the open ocean, it has been shown that a considerable portion of the nutrients is regenerated within the water column (e.g. GRILL and RICHARDS 1964). In shallow water areas, however, regeneration involves the sediment surface and the sediment, since settling times for organic detritus are relatively short and the supply is large (SUESS 1980). Organic-rich debris associated with high productivity in nearshore areas is deposited at the benthic boundary layer in a relatively labile state (SUESS 1976 b). The relative importance of the three decompositional environments (water column, sediment surface, sediment) is controlled by the residence times of the organic particles within each environment, an approximation of which may be deduced from settling velocity and sediment accumulation rate (MÜLLER and SUESS 1979, SUESS 1980, SUESS and MÜLLER 1980).

4.3.1   Characterization and Stoichiometry of Decompositional Processes

Excessive demand for electron acceptors and restricted supply in stagnant waters or sediments may result in rapid oxygen depletion and remineralization via a sequence of anaerobic modes according to the greatest available free energy yield (e.g. REDFIELD et al. 1963, SUESS 1976a, MARTENS 1978). The composition of the organic matter subjected to decomposition by various oxidants differs in the sediments encountered in Kiel Bight (e.g. at 20 m C:N:P = 125:10.9:1 as discussed in the previous sections) from the mean oceanic ratio of C:N:P = 106:16:1 (REDFIELD et al. 1963, RILEY 1956). Thus, the stoichiometric relationships in Tab. 4-2 refer to an organic matter of schematic composition: $(CH_2O)_a(NH_3)_b(H_3PO_4)_c$.

Since the different oxidants consumed refer to the same organic matter, they are equivalent to each other in certain ratios with respect to phosphate, and "oxidation equivalents" as a standardized unit may be defined for stratified systems:

Ox eq = ΔO_2 + 2.5 ΔNO_3^- + 4 ΔSO_4^{2-} (BALZER et al. 1983)

where the Δ_i denotes absolute concentration changes in $\mu gat \cdot l^{-1}$, and nitrate is not incorporated before it decreases during denitrification. Concepts of this kind as well as the application of Redfield-ratio-decomposition to conditions at and within

Tab. 4-2: Schematic representation of stoichiometric relationships of oxidation of organic matter during subsequent stages of oxygen consumption (a), denitrification (b), sulphate reduction (c), and carbonate reduction (d).

$(CH_2O)_a (NH_3)_b (H_3PO_4)_c +$

a) $\quad + a\ O_2 \quad = a\ CO_2 + b\ NH_3 + c\ H_3PO_4 + a\ H_2O$
$\quad + (a+2b)O_2 = a\ CO_2 + b\ HNO_3 + c\ H_3PO_4 + (a+b)\ H_2O$

b) $\quad + \dfrac{4}{5} a\ HNO_3 = a\ CO_2 + b\ NH_3 + c\ H_3PO_4 + \dfrac{7}{5} a\ H_2O + \dfrac{2}{5} a\ N_2$

c) $\quad + \dfrac{1}{2} a\ SO_4^{2-} = a\ CO_2 + b\ NH_3 + c\ H_3PO_4 + a\ H_2O + \dfrac{1}{2} a\ S^{2-}$

d) $\quad + \dfrac{1}{2} a\ HCO_3^- = a\ CO_2 + b\ NH_3 + c\ H_3PO_4 + \dfrac{1}{2} a\ OH^- + \dfrac{1}{2} a\ CH_4$

the sediments rely on the assumptions that (i) all components of organic matter decompose at the same rate, (ii) oxidants are not used for the oxidation of sedimentary components other than organic matter and its intermediate products of degradation and (iii) inorganic end products do not interact with sedimentary solids in any form.

The environmental conditions under which different heterotrophic organisms destroy the non-equilibrium state of organic matter for their own metabolic requirements may be represented by the redox potential (BALZER 1980b). It has been shown, however, that thermodynamic equilibrium calculations applied to natural systems are somewhat problematic for reasons of reversibility, buffer capacity, kinetics and homogeneity of the system considered (e.g. STUMM and MORGAN 1970, WHITFIELD 1974). For comparison with directly determined redox potentials within the pH and concentration limits of the bottom sea water enclosed in a bell jar, equilibrium calculations were conducted for the redox couples: O_2/H_2O, O_2/H_2O_2, NO_3^-/NO_2^-, NO_3^-/N_2, MnO_2/Mn^{2+}, $Fe(OH)_3/Fe^{2+}$, and SO_4^{2-}/HS^- (BALZER 1978).

4.3.2   Sites of Decomposition and Vertical Transport Processes

The release of remineralized constituents and their net flux to the overlying water column are the result of a complex set of processes which occur both at the depositional interface prior to burial and within the sediment column following burial (KLUMP and MARTENS 1981, REIMERS and SUESS 1983): at the interface rapid remineralization of

labile organic matter may interact with removal of products via autotrophic or
heterotrophic uptake, precipitation, complexation or adsorption; below the interface
diagenetic regeneration may be coupled with transport processes such as diffusion,
pore water advection, bioturbation, flushing and bubble ebullition.

4.3.2.1 Molecular Diffusion

The non-biogenically mediated flux across the sediment/water interface of a constituent
dissolved in sediment pore water arises from diffusion and pore water advection, the
latter being less than 1 % of the diffusion term in the sediments considered here.
For flux calculations Fick's first law is mostly applied in the form (BERNER 1971):

$$F_S = - \varphi_0 \cdot D_S \cdot \left(\frac{\delta c}{\delta z}\right)$$

where F_S is the flux to the overlying water, φ_0 is the porosity at the sediment/water
interface, where D_S denotes the bulk sediment diffusion coefficient at the interface
and $(\delta c/\delta z)$ is the pore water gradient. Uncertainties in those calculations derive
from the assumption of a plane surface and insufficient resolution of the actual
gradient at the sediment/water interface. Appropriate numbers for the diffusion
coefficients of the different components have been critically revised in a number of
recent publications (KROM and BERNER 1980, SAYLES 1979, KATZ and BEN-YAAKOV 1980,
ALLER 1980a, KLUMP and MARTENS 1981). In order to avoid uncertainties in assessing
the sediments' inhibition of molecular diffusion, it was measured (DICKE in prep.)
by comparing diffusion in the sediment to self-diffusivities. For the different
sediment types of the "Hausgarten" correction factors were established to transform
the diffusion coefficients at infinite dilution into bulk sediment diffusion
coefficients. The factors obtained vary from 0.39 for medium sand to 0.60 for mud
and are within the range of values reported from LI and GREGORY (1974) for red clay
or KROM and BERNER (1980) for anoxic mud.

4.3.2.2 Physical Exchange Processes

Besides current and wave action stirring up sediment and interstitial water, an
additional physical mechanism of pore water exchange with overlying bottom water was
detected by SMETACEK et al. (1976): when enclosing a water column of 30 m³ over 3 m²
coarse sand sediment they observed considerable water exchange with outside waters
through the sediment, driven by salinity fluctuation in the surrounding water. Due to
changes in the density of the bottom water, the interstitial water of coarse grained
sediment is flushed out by gravity displacement. The effect should be most pronounced
in the shallow part of the "Hausgarten" where bottom waters that exhibit large
salinity differences (cf. section 4.2.1) flow over coarse grained sediments deposited
on a slope so that lateral intrusion is possible (POLLEHNE 1980, REIMERS 1976). The

processes of physical stress (current and wave action, density displacement) applied to the sediment surface appear to be at a minimum during summer (POLLEHNE 1980). The potential effect of salinity fluctuation on pore water exchange was also investigated in the 28 m deep channel adjacent to the "Hausgarten" from July 1972 to October 1973 (REIMERS and KÖLMEL 1976). These authors made monthly records of bottom water salinity and that of the pore water at the 2 - 5 cm and the 8 - 10 cm sediment layer as depicted in Fig. 4-10. Up to August the bottom water exhibited a salinity difference of up to 3 °/oo against the pore waters. The adjustment to the higher and lower salinity values of the bottom water occurs with a certain time lag but still appears too rapid to be explained solely by molecular diffusion. The exchange may be supported by the burrow

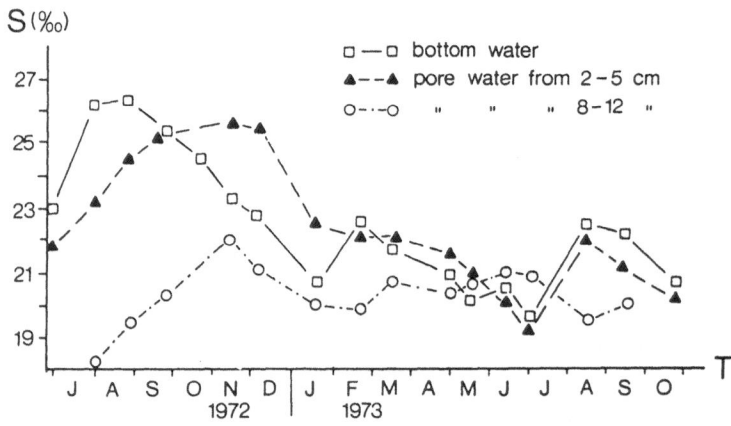

Fig. 4-10: Seasonal changes of salinity in the bottom water and two pore water horizons at 28 m water depth, near "Hausgarten" (after REIMERS and KÖLMEL 1976).

structures created by macrofauna organisms which, however, died off in late summer due to anoxia (KÖLMEL 1976). REIMERS and KÖLMEL (1976) conclude that for this 28 m station displacement by bottom water of higher density is not the main mechanism of pore fluid exchange contrary to findings in two nearshore stations in Kiel Bight (less than 9 m water depth) being influenced also by fresh water intrusion.

4.3.2.3 Biogenic Transport

Macrofauna organisms contribute to the exchange of pore water constituents both by providing a sediment structure for enhanced transport and by actively irrigating within the burrows (ALLER 1978, 1980a). To resolve this complex phenomenon, first the spatial and seasonal distribution of bioturbating organisms along with their feeding and survival habits has to be assessed for the area. Since any kind of net flux is based on the concentration gradients of a dissolved component between bottom and pore water, a model has to be developed that relates organism activity within the sediment to observable fluxes and gradients of the system. Following GOLDHABER et al. (1977), McCAFFREY et al. (1980) and ALLER (1978) the effect of biogenic reworking by motile deposit and suspension feeders on pore water exchange was approximated by an effective diffusion coefficient valid over a fixed depth. It should be kept in mind that simple models of this kind assume random distribution and random activity of organisms (for other models see ALLER 1980b).

POLLEHNE (1980) inferred biogenic exchange rates from a comparison of silica release rates with and without poisoning of the sediment/water system held in vitro. During winter at 0°C he found no difference in release rates which in addition were similar to pure molecular diffusion rates and he concluded that the macrofauna was inactive with respect to irrigation. Temperature elevation to 5°C yielded an increase of SiO_4^{4-} release in the control system by 50 % over the poisoned system when using sediments from 18 m water depth. In a similar experiment during autumn (10°C) he found a 40 % reduction in the release rate after addition of formaldehyde. According to DICKE (in prep.) bioturbation did not seem to be the principal factor in controlling silica release from the sediment. In early summer 1983 she found a silica flux almost as great as in spring after the settling of the plankton bloom (in either case ca. double the winter flux), whereas the contribution of bioturbation in summer (as well as in winter) was reduced by at least 50 % as compared to the spring situation. A detailed study of spatial and seasonal variation in biogenic exchange processes was started in 1980 (DICKE in prep.). The effect of biogenic reworking on pore water exchange was estimated by measuring an effective diffusion coefficient valid over a fixed depth. Taking diffusion in sterilized or poisoned cores as a reference the contribution of bioturbation was evaluated. The model which assumes random distribution and random activity of organisms was confirmed for the "Hausgarten" sediments in general, but other effects were found as well. An almost complete stirring of the sediment was observed in one case; in two others there occurred biogenic pumping of water selectively into deeper sediment horizons. Bioturbation increased transport by factors of 2 to 5 compared to molecular diffusion at the 20 m station in autumn, winter and summer. In early spring bioturbation was enhanced after the settling of the phytoplankton bloom (in one case even twenty fold). In the summer 1980, however, during a period of oxygen deficiency transport rates were reduced to values which could almost completely be attributed to molecular diffusion. In the mud sediments of the 25 m and 28 m stations, the contribution of bioturbation to the transport of solutes was much less than at 20 m, except Halicryptus spinulosus was present which caused considerable irrigation of the sediment.

4.3.3   Total Rates of Oxidant Consumption and Nutrient Release

In order to understand the relationship between the consumption of oxidants and the simultaneous release of decomposition products, to obtain representative rates and to evaluate influencing mechanisms several in vitro experiments were conducted (POLLEHNE 1980, BOJE 1974) as well as in situ (BALZER 1978, 1984, BALZER et al. 1983). Sediments were enclosed together with their overlying bottom waters. The transition zone of the "Hausgarten" between lag and mud sediments (Fig. 4-2) and especially the depth range from 18 to 21 m containing fine to muddy sand sediment was thoroughly investigated. In a recent study on the role of sulfate reduction in sediment decomposition processes (POLLEHNE in prep.) rates of oxygen and sulfate consumption were compared for stations in 20 and 28 m depth during the course of the year.

4.3.3.1 Benthic Decomposition under Oxic and Anoxic Conditions

Results from a typical in situ experiment (20 m water depth) are depicted in Fig. 4-11 and 4-12 (BALZER 1984). This experiment was started in May after precipitation of the spring phytoplankton bloom. With a high initial (233 ml O_2 $m^{-2}d^{-1}$) but slow-

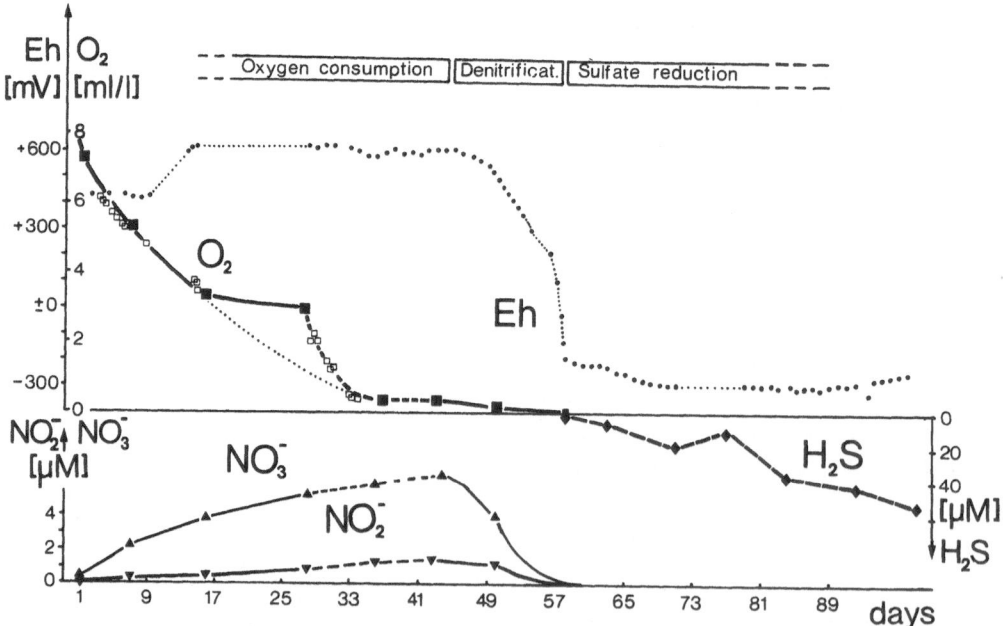

Fig. 4-11: Organic matter decomposition at the 20 m station ("Hausgarten") studied in a bell jar experiment: changes with time of redox potential (Eh), O_2, NO_3^-, NO_2^- and H_2S. (The inhomogeneity in the O_2-curve is due to insufficient stirring and partial water exchange.) (BALZER 1984).

ly decreasing rate, oxygen concentration fell to 0.3 ml l^{-1} on the 36th day. The redox potential during the first two weeks was Eh = + 420 mV as reported for short term
measurements in oxic seawater (COOPER 1937, BAAS-BECKING et al. 1957). The subsequent
increase and stabilization at Eh = + 620 mV after the initial time needed for
"equilibration" of the electrode surface with surrounding oxic water of low redox
buffer capacity were attributed (BALZER et al. 1983) to effective redox control by the
couple O_2/H_2O_2 as proposed by BRECK (1974). When denitrification commenced at an oxygen concentration of 0.25 ml l^{-1} (49th day), Eh started to decrease probably representing mixed potentials since no agreement with calculated values for the redox
couples NO_3^-/NO_2^- and NO_3^-/N_2 could be observed (BALZER 1984). After denitrification and the onset of sulphate reduction (as reflected in the H_2S increase), redox potential further fell to values between -200 mV and -300 mV in agreement with calculated values for the redox pairs SO_4^{2-}/HS^- and SO_4^{2-}/S^0. After reaeration of the anoxic
bell jar water in a similar experiment, a quick response of the electrode to high positive Eh values was obtained, thus ruling out substantial poisoning of the electrode
surface (BALZER et al. 1983).

During high initial oxygen saturation (Fig. 4-12) phosphate was fixed to the bottom
down to a concentration of 0.9 µMol l^{-1} reflecting the interaction between bottom

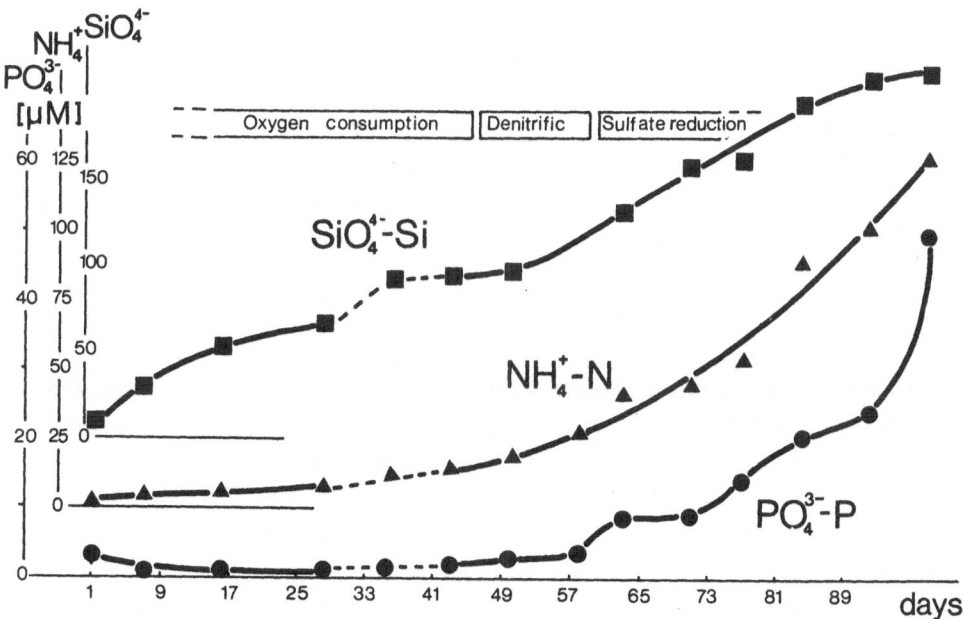

Fig. 4-12: Organic matter decomposition at the 20 m station ("Hausgarten") studied
in a bell jar experiment: concentration changes of nutrients with time
(BALZER 1984).

water phosphate and sedimentary solids as shown by POMEROY et al. (1965). After reaching oxygen concentrations below 3 ml l^{-1} phosphate was released continuously at a slow rate when compared with the twelve times higher (mean) rate of 744 μMol $m^{-2}d^{-1}$ during anoxic conditions (BALZER 1984). A rate of fixation of similar height was observed when reaerating anoxic waters in which large amounts of phosphate had accumulated (BALZER et al. 1983). Among the inorganic nitrogen species in all enclosure experiments conducted, ammonia (Fig. 4-12) always played the main role both in terms of concentration and in terms of release rates, even though the sediment/water interface was found to be the site of most intensive nitrification in the Kiel Bight ecosystem (SZWERINSKI 1981). Like phosphate, ammonia is released at a significantly higher rate after the onset of anoxic conditions contrary to similar experiments in other bottom water/sediment systems (ENGVALL 1973). The increase in the release rate is partly due to fast ammonification from organic matter of meio- and macrofauna organisms dying at the sediment surface when the redoxcline moves up into the bottom water (BALZER et al. 1983); but most of the excess ammonia probably arises from the fact that removal of ionic nitrogen by the combined action of nitrification and denitrification no longer can take place. During oxic conditions after fixation of phosphate, nitrogen and phosphorus are released at a mean ratio of N/P = 6.89 showing a preferential release of phosphate that is - on a long-term view - partly counterbalanced by the fixation during high oxygen tension. With respect to the regeneration of silicate by dissolution from diatoms (as the most abundant siliceous organisms in this area) no direct link to the consumption of oxidants and the release of nutrients may be expected. The close correlation - nevertheless often found between silicate and N and P release - is normally attributed to the same kinetics for organic matter degradation and silica dissolution (e.g. GRILL and RICHARDS 1964). Contrary to the results of v.BODUNGEN (1975), who found a close correlation of silica and phosphate during the transition from oxic to anoxic conditions, no evidence for a redox dependency of silica release was found in the in situ experiments (Fig. 4-12).

4.3.3.2 Seasonal Variation and Total Annual Rates

Oxygen consumption and nutrient release rates during various seasons are compiled for sediments of the transition zone at 18 m (POLLEHNE 1980) and 20 m (BALZER 1984) in Tab. 4-3. Evidence is presented that organic matter decomposition as reflected in the oxygen consumption rate has an absolute minimum during winter and maxima after the spring phytoplankton bloom and in late summer. A similar seasonal cycle with relative minima in January and June was reported by BOJE (1974) for an 11 m station in the lag sediment zone of the "Hausgarten". While she found an annual mean of chemical oxygen demand (taken as O_2 consumption after formalaldehyde addition; discussed and questioned by DALE 1978) of only 4 %, POLLEHNE (1980) supposes that decompositional activity at 18 m during summer in form of anaerobic metabolism must be higher. This author suspects that anaerobic processes create an oxygen debt by accumulation of reduced compounds during summer which has to be compensated for by high chemical oxygen demand during winter.

An extended maximum in oxygen uptake by the sediment could be observed in November and December 1983 following the breakdown of the summer stratification (POLLEHNE in prep). The utilization of sulfate as an alternate oxidant seems to play a minor role for the "Hausgarten"-sediments as long as the sediment surface remains oxidized. Rates of sulfate reduction increase when oxygen transport towards the sediment surface is impeded by water stratification. These rates, however, are more enhanced by a sudden supply of organic material to the sediments as it normally occurs with the sedimentation of the phytoplankton spring and autumn blooms. The fast decomposition of this fresh material leads to temporary anoxic conditions in a microlayer at the surface of otherwise oxic sediments. Longer time-spans of stable stratification and the resulting oxygen depletion may cause a large part of the macrofauna to die off. The decaying macrofauna biomass (about 5 -10 gC m^{-2} at the 18-20 m depth station (as reported by BREY 1983) is subsequently oxidized rather quickly by sulfate reducers resulting in the production of large amounts of poisonous hydrogen sulfide that kills off more aerobic organisms. A feedback mechanism like this may be the reason for dras-

Tab. 4-3: Mean rates of oxygen consumption and nutrient release during different seasons at 18 m station (incubation of cores: POLLEHNE 1980) and the 20 m station in the "Hausgarten" (in situ bell jar experiments: BALZER 1984). Rates after establishment of anoxic conditions in brackets.

Season	Water depth (m)	Mean Temp. (°C)	Oxygen consumption (ml m^{-2}d^{-1})	Phosphate	Release of Nitrogen (µMol m^{-2}d^{-1})	Silicate
winter oxic	18	0	94	fixation	50	112
winter oxic	20	4	93	16	134	690
spring oxic	18	5	289	25	377	1991
late spring oxic	20	8.5	233	63*	476	2078
anoxic	20	11	-	(744)	(1623)	-
summer oxic	20	10	161	73	667	1576
anoxic	20	(12.5)		(677)	(2636)	-
late summer oxic	18	10	212	17	264	1124
autumn oxic	18	10	212	fixation	fixation	1047
annual mean	18		192	17	237	1184
	20		180	54	440	1600

* after period of fixation

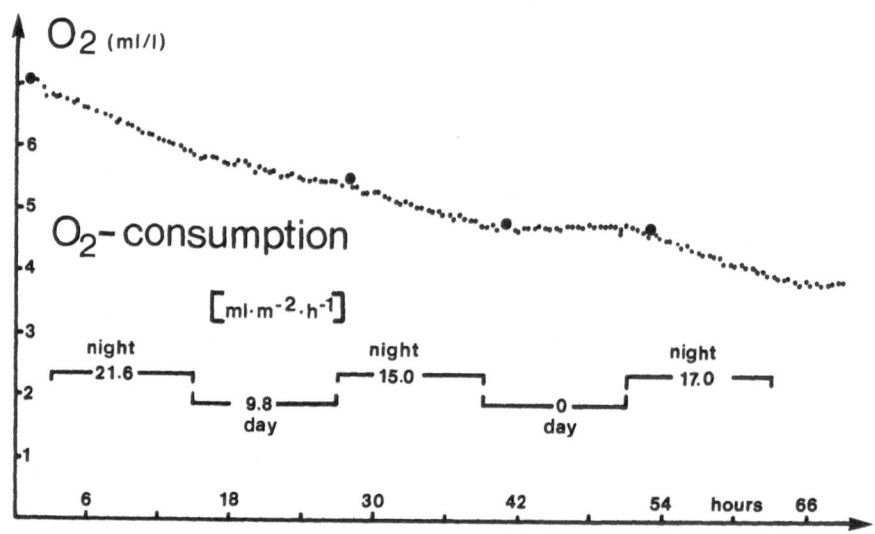

Fig. 4-13: Oxygen consumption (bell jar experiment during spring 1980) at the 10 m station ("Hausgarten") being affected by benthic oxygen production during the day. Large dots denote Winkler determinations, small dots: electrode readings (BALZER et al. 1986).

tic situations like in summer 1981 (EHRHARDT and WENCK 1982), when an unusually large part of Kiel Bight became anoxic and free sulfide could be measured in the whole water column. During the period from September 1983 to September 1984 when such long-term stagnation periods were prevented by wind-induced mixing, the carbon decomposition due to sulfate reduction amounted to a daily mean rate of about 6 mg C m^{-2} d^{-1} at 20 m depth and 30 mg C m^{-2} d^{-1} at 28 m depth. This corresponds to about 5 % and 25 %, respectively, of the carbon decomposition as calculated from oxygen uptake during the same period. To a considerable extent this difference between the two types of sediment may reflect the higher aerating and conditioning activity of bioturbating macrofauna at the 20 m station. For the transition zone sediments at 18 and 20 m water depth POLLEHNE (1980) and BALZER (1984) arrive at annual mean rates which agree reasonably well for oxygen consumption and silicate release (Tab. 4-3). Since POLLEHNE (1980) observed long periods of phosphate fixation in the bottom sediment at his experimental site, which were assumed to be counterbalanced by high release rates during short anoxic periods as shown by BALZER et al. (1983), a large discrepancy persists in the annual mean rate of phosphate release. Besides this, it should be kept in mind that high release rates immediately following inputs of fresh material (POLLEHNE 1980) and the physical processes of bottom current action and density displacement are both unaccounted for; thus the rates presented in Tab. 4-3 may be considered as total release rates from the bottom by different modes of decomposition and transport excluding physical processes.

Tab. 4-4: Rates of oxygen consumption at different water depths in the "Hausgarten".
Note that for the 18 m and the 20 m station annual mean rates are given
calculated from several experiments during different seasons.

Date of experiment	Water depth	Sediment type	Oxygen consumption ($ml\ m^{-2}\ d^{-1}$)	References
May 1980	10 m	sandy, lag sed.	234 (during day) 518 (during night)	BALZER et al. (1986)
Feb.-Dec. 1973	11 m	sandy, lag sed.	528 (dark respiration)	BOJE (1974)
April/May 1980	15 m	sandy, lag sed.	405 (dark respiration)	BALZER et al. (1986)
Feb.-Oct. 1979	18 m	medium to fine sand	192 (dark; annual mean)	POLLEHNE (1980)
Feb.-Sept. 1975,76	20 m	fine sand, sandy mud	180 (annual mean)	BALZER (1984)
May 1980	20 m	fine sand, sandy mud	227 (dark respiration)	BALZER (unpubl.)
May 1980	25 m	silty mud	318 (dark respiration)	BALZER et al. (1986)
May 1980	28 m	clayey mud	426 (dark respiration)	BALZER et al. (1986)

4.3.3.3 Spatial Variations

In order to evaluate the benthic utilization of sedimentary organic matter along the
slope of the "Hausgarten" in relation to the distribution of organic carbon (section
4.2.4), oxygen consumption rates at different water depths are compiled in Tab. 4-4.
There is clear evidence for two maxima of benthic activity in the shallow and the deepest region rather than a continuous increase concomitant with the increase in organic matter content of the sediment. The non-detritus-accumulating site at 11 m is
characterized (BOJE 1974) by a high percentage of living organisms in total organic
matter which have to utilize the organic input immediately before it is transported

to deeper parts of the basin by wave action. In addition to the supply by settling of pelagic primary and secondary products at the shallow sites (10 m, 11 m and 15 m) there is an input of organic matter by benthic primary production. The course of oxygen consumption obtained at the 10 m station by use of a transparent bell jar shows a clear day/night rhythm: benthic metabolism is partly (and on the second day entirely: Fig. 4-13) counterbalanced during day-time by a corresponding oxygen input from benthic primary production. In the deeper region where mud sediments accumulate relatively high amounts of organic matter from various sources (section 4.2.6), the oxygen consumption rate increases again. In these sediments, however, a considerable additional contribution from anaerobic metabolism must be suspected which is not compensated for during the annual cycle. This may be presumed from the high organic carbon burial rate and the intensive decomposition in deep sediments (WHITICAR 1978) and is proven by increasing contents of sulfur in the sediment with water depth, which can be regarded as a permanent storage of reduced compounds (HARTMANN unpubl.).

4.3.4   Molecular Diffusive Flux from Near-Surface Pore Waters

Particularly in coastal areas of rapidly accumulating sediments heterotrophic breakdown processes do not exhaust the reactive fraction of the organic matter while it rests at the sediment/water interface or in the top sedimentary layers. Consequently products of decomposition are released into the interstitial waters and transported to the overlying water by diffusion, if positive gradients are maintained in the sediment as shown in Fig. 4-14. For components which approach a "saturation" value with depth in sediment (e.g. PO_4^{3-} and SiO_4^{4-} in Fig. 4-14), BALZER (1984) and POLLEHNE (1980) use a mathematical expression from TESSENOW (1972) for diffusion-controlled transport making allowance for (depth independent) adsorption. From the interstitial gradients depicted in Fig. 4-14 (obtained during May at 20 m water depth) and bulk sediment diffusion coefficients from the literature (KROM and BERNER 1980, ROSENFELD 1981, WOLLAST and GARRELS 1971), BALZER (1984) calculated a molecular diffusive flux through the interface of 13 µMol PO_4^{3-}-m^{-2}d^{-1}, 51 µMol NH_4^+ m^{-2} d^{-1} and 240 µMol SiO_4^{4-} m^{-2}d^{-1}.

4.3.4.1 Seasonal Influence

A thorough investigation of the seasonal cycle of pore water gradients and diffusive fluxes was provided by POLLEHNE (1980) who inserted teflon frit samplers permanently at 2 cm, 7 cm and 12 cm in a coarse sand (13 m), fine sand (18 m) and muddy sand sediment (21 m). Diffusive fluxes of silicate, ammonia and phosphate at the 18 m station are depicted in Fig. 4-15. It was found that the fluxes (more or less closely) followed the variations in concentration at 2 cm sediment depth while at 12 cm there were only minor changes over the year.

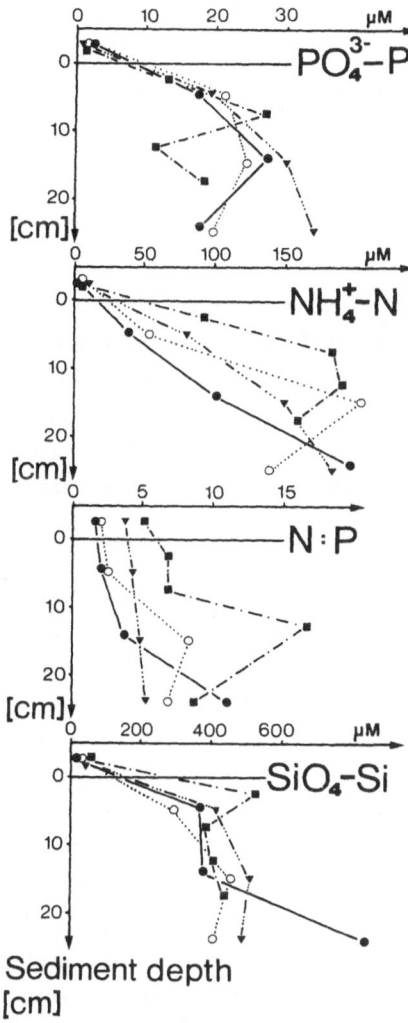

Fig. 4-14: Near-surface pore water concentration profiles of nutrients at the 20 m station in the "Hausgarten". To reduce short-range spatial heterogeneity four cores were taken and processed (BALZER 1984).

Silicate pore water concentration at 2 cm showed a continuous increase from April to December (1978), the decrease of flux after the summer (Fig. 4-15) being caused by a reduction of the gradient due to higher contact water concentration (0 - 1 cm above the interface). Phosphate gradient and flux were neither related to biological events (heterotrophic uptake) nor to sedimentation (POLLEHNE 1980) but showed an inverse relationship to the course of oxygen concentration in the bottom water (KÖLMEL 1976, RUMOHR 1979). Besides a peak during summer stagnation, ammonia flux revealed a clear increase over the year; the restoration of low fluxes at the beginning of spring is assumed to be due to physical exchange processes during winter (POLLEHNE 1980). This study also arrives at the conclusion that diffusive flux contributes only a minor portion to total release of nutrients.

4.3.4.2 Spatial Variation

In order to estimate the relative importance of decomposition within the sediment and of interstitial fluxes through the sediment/water interface in comparison to direct degradation and release at the sediment surface, near-bottom pore water gradients along with exchange coefficients were determined for all sediment types of the "Hausgarten" by means of an in situ pore water sampler. When directly comparing interstitial gradients from different environments it should be kept in mind that these gradients reflect a complex balance between production rates, transport rates and reaction rates with interstitial solids. Since transport mechanisms (e.g. physical processes of flushing and density displacement) are highly variable in the sediments investigated, it is impossible to infer decompositional intensity offhand from pore water gradients. Thus the ammonia and silicate gradients measured during May at different water depths depicted in Fig. 4-16 (BALZER unpubl.) should be interpreted solely in terms of potential molecular fluxes. There is clear evidence from the figure that over a short horizontal distance in the "Hausgarten" (Fig. 4-2) a very broad range of gradients is maintained. Their correspondence in molecular fluxes is reinforced by the simultaneous increase of porosity with water depth. Molecular fluxes based on the observed gradients range from near zero at 15 m where, however, the most intensive sediment flushing occurs, to 700 µMol NH_4^+ $m^{-2}d^{-1}$ and 275 µMol SiO_4^{4-} $m^{-2}d^{-1}$ at 28 m when bulk sediment diffusivities from the literature are applied (KLUMP and MARTENS 1981, ALLER and BENNINGER 1981). During the recording of annual cycles at 13 m, 18 m and 21 m POLLEHNE (1980) also observed increasing steepness of interstitial gradients with water depth for all components studied. These patterns varied with season. Contrary to the uniform behaviour of concentrations at the 12 cm sampling port in 18 m and 21 m, profiles at the 13 m station showed high variance at 2 cm, 7 cm and 12 cm, reflecting the frequent sediment flushing that may reach down to 30 cm (KOHR pers. comm.). Despite intensive short-term fluctuations SiO_4^{4-} is slightly accumulated in the pore water while NH_4^+ and PO_4^{3-} peak during summer, pointing to anoxic microenvironments below a bottom water which never becomes anoxic during the year.

The pore water concentrations of the different components at 21 m follow - at a higher level - more or less the trends of the 18 m station. Although the pore waters at 21 m are more influenced by long-term geochemical processes, POLLEHNE (1980) assumes that even this station loses its high nutrient content by sediment flushing during winterly storms. The tendency to steady changes in gradients along with temperature

143

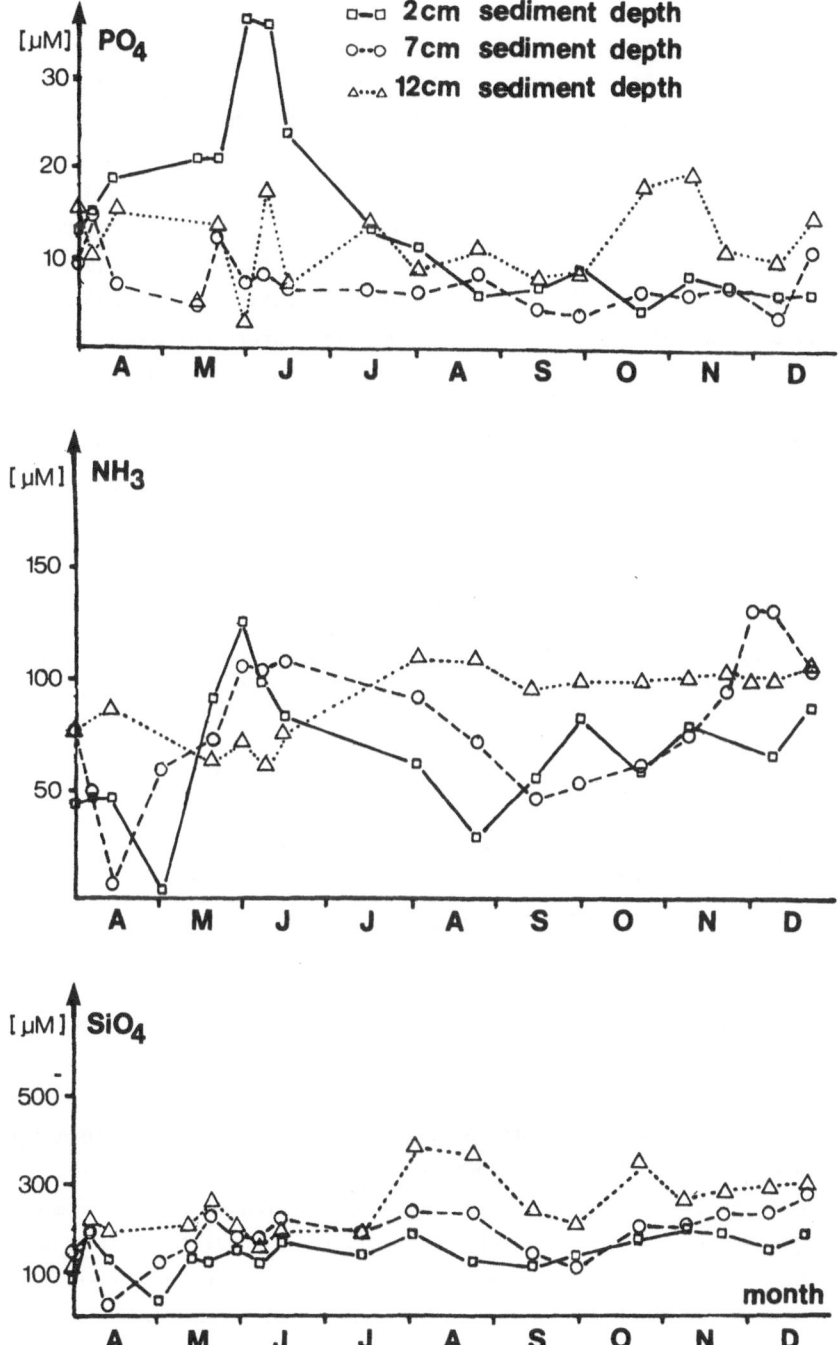

Fig. 4-15: Seasonal pattern (1978) of the pore water concentration of ammonia, phosphate and silicate at the 18 m station in the "Hausgarten" (POLLEHNE 1980).

and oxygen variations of the bottom water is even more pronounced at 28 m (KÖLMEL 1976). For this station, too, slow density displacement was shown to occur (REIMERS and KÖLMEL 1976). In Table 4-5 the mean atomic ratios of the nutrients are compiled for the three stations investigated. It can be seen from the results that the two nutrients involved in exchange equilibria with inorganic solids increase relative to nitrogen with water depth. In contrast to the 13 m and 18 m stations the same holds true for increasing sediment depth at 21 m. This was regarded by POLLEHNE (1980) as an indication for the increasing importance of interaction with sedimentary solids relative to "biological" decomposition. Since most data were obtained during the growth season the low N:P and N:Si ratios in contact waters may be explained by preferential uptake by primary producers and heterotrophic benthic organisms under certain conditions of sedimentation (POLLEHNE 1980). The relatively high N:P ratios at all depths of the 13 m station may be regarded as an effect of sediment flushing which "treats" all interstitial components in the same way leaving no time for P to become equilibrated with solids.

Tab. 4-5: Mean atomic ratios of Si:N:P dissolved in the contact and the pore waters at three stations of the "Hausgarten", Kiel Bight (POLLEHNE 1980).

	Water depth		
Sediment depth	13 m	18 m	21 m
0 cm, contact water	19.2/ 4.3/1	18.8/ 5.1/1	15.7/5.2/1
2 cm, pore water	16.8/ 9.8/1	14.5/ 6.4/1	18.2/6.7/1
7 cm, pore water	23.4/13.5/1	26.3/17.2/1	8.4/3.2/1
12 cm, pore water	23.7/13.2/1	24.6/ 9.5/1	14.8/3.3/1

4.3.5 Early Diagenesis in Deeper Sediment Strata

Geochemical investigations of early diagenetic processes in Eckernförder Bucht deep sediments aim to throw some light on two problem areas: firstly, since interstitial waters are sensitive indicators of post-depositional reactions (SUESS 1976a) an attempt was made to assess extent and relative proportion of nutrient regeneration, the time required for and the factors controlling interstitial distribution and potential authigenic mineral formation (SUESS 1979) in an organic-rich, rapidly accumulating anoxic sediment. Secondly, it was found that Eckernförder Bucht sediment deeper than the sand/mud facies boundary contains large areas which are acoustically turbid (WERNER 1968, HINZ et al. 1969). This was supposed to be due to the presence of gases among which methane can be a product of organic matter decomposition.

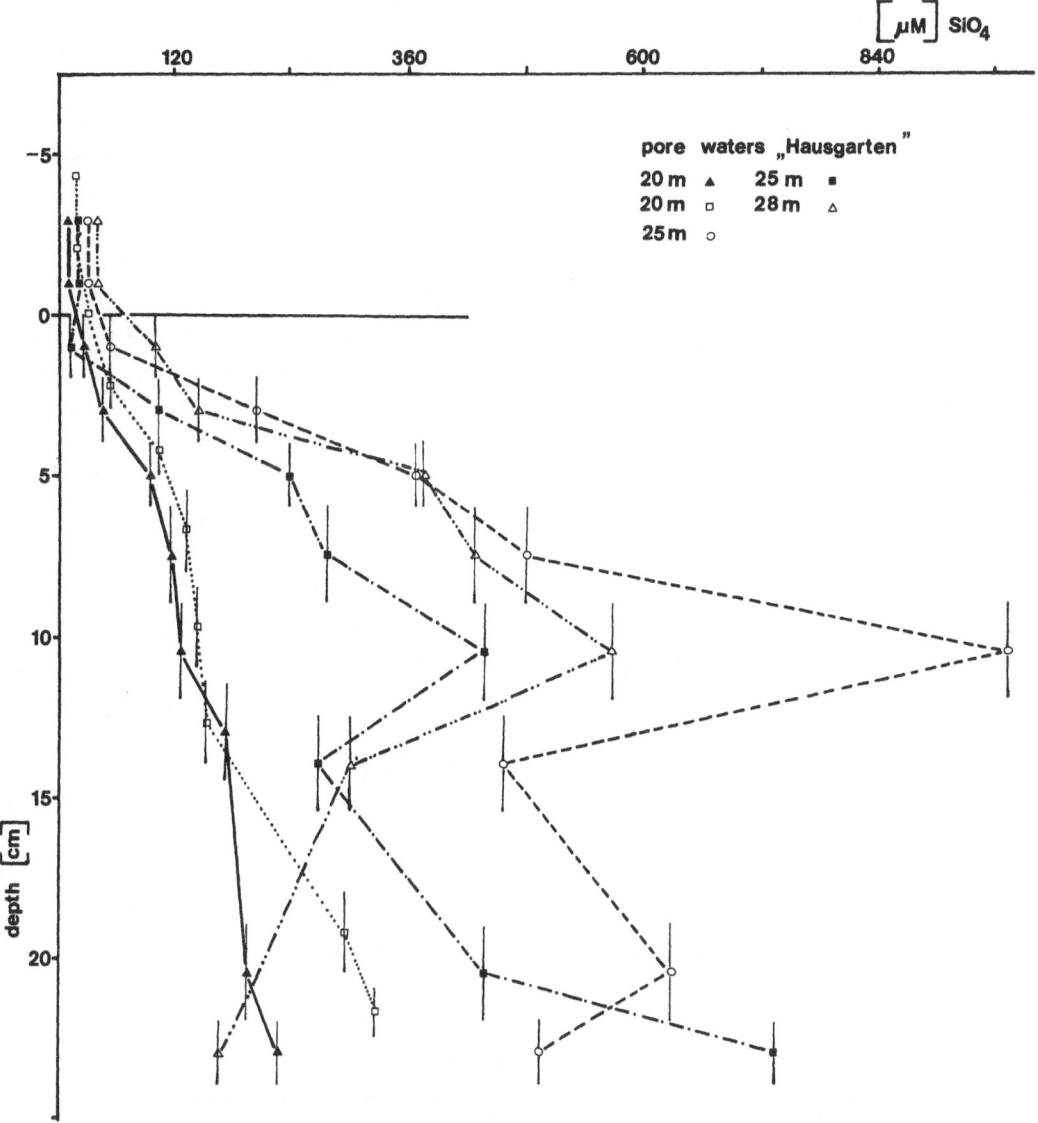

Fig. 4-16: Near-surface interstitial gradients of silicate in "Hausgarten" sediments at 20 m, 25 m and 28 m water depth, taken with an _in situ_ pore water sampler. Duplicate profiles indicate natural variation (BALZER unpubl.).

Up to 7 m long cores from the mud region (for station location see Fig. 4-1) were cut into 20 cm sections and investigated with respect to solid sediment properties and interstitial distribution of nutrients as well as gases (WHITICAR 1978, 1982). Results of solid sediment properties for a typical station (GPI 13 939) were presented in Fig. 4-6. Distribution of interstitial decompositional products is depicted in Fig. 4-17. The intense breakdown of organic matter below the interface leads to sulphate exhaustion already at about 200 cm sediment depth. In spite of the sulphate depletion, titration alkalinity, ammonia and phosphate concentration in pore water increased almost linearly with depth up to the extreme values of 130 mMol l^{-1}, 30 mMol l^{-1} and 2.1 mMol l^{-1}, respectively. Ammonia and phosphate correlate closely at a ratio of N:P = 10.8. Reference to the plot of titration alkalinity versus ammonia (Fig. 4-8) illustrates a distinct break in the C:N ratios from 7.5 - 9.6 in the upper sediment layers to about 3.1 below, probably arising from bacterially mediated carbonate consumption as suggested by SUESS (1976a). Support for this explanation is provided by the sulphate depletion and the methane appearance (Fig. 4-18) at the depth of the C:N ratio break (Fig. 4-8). In order to obtain rate estimates, steady state kinetic modeling of the nutrient release and sulphate reduction was applied to the data, considering effects of deposition and compaction, diffusion, adsorption, ion exchange, mineralization and biological reactions as exemplified in the following simplified equation for sulphate (no adsorption, no dissolution or precipitation):

$$\frac{\delta S}{\delta t} = \left(\frac{\delta S}{\delta t}\right)_{biol} + D_s \left(\frac{\delta^2 S}{\delta z^2}\right) - \left(\frac{\delta S}{\delta t}\right)_{accum}$$

This expression describes the rate of change of sulphate concentration as the combined effect of a bacterial reduction term, a diffusion term and an accumulation (deposition and compaction) term (BERNER 1971). The sulphate depth distribution was estimated using both zero order (near the interface down to SO_4^{2-} = 2 mMol l^{-1}) and first order reaction kinetics for sulphate concentration below 2 mMol l^{-1}. Bacterial reduction rates of up to 0.25 mMol l^{-1} y^{-1} were calculated. Pseudo zero order kinetics applied to ammonia and phosphate revealed average bacterial regeneration rates of $1.9 \cdot 10^{-2}$ and $2.7 \cdot 10^{-3}$ mMol l^{-1} y^{-1}, respectively. The formation rate of alkalinity decreased from $2.2 \cdot 10^{-1}$ to $7.8 \cdot 10^{-2}$ mMol l^{-1} y^{-1} below the sulphate reduction zone (WHITICAR 1978).

Methane accumulation was experienced only at low sulphate concentrations (< 0.4 mMol l^{-1}) or where sulphate was fully depleted (at about 2m below the sediment surface) due to anaerobic bacterial methane consumption (oxidation). In contrast to profiles in other cores investigated where methane remained below saturation, methane concentration at the station described here (Fig. 4-18) increased sharply to and exceeded the saturation limit; under those conditions bubbles may be formed which could be responsible for the acoustic turbidity of the sediment. Upward directed methane flux to

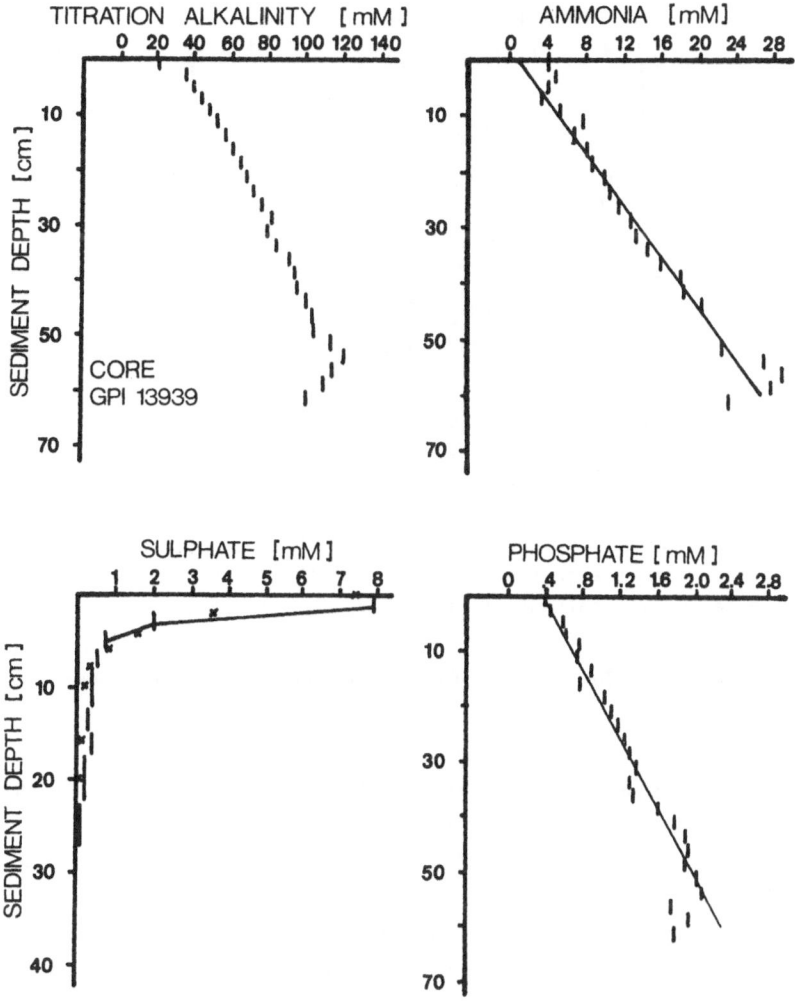

Fig. 4-17: Interstitial concentration profiles of biogenic components in a long core (GPI 13 939) from the 28 m station, near "Hausgarten" (WHITICAR 1978).

the consumption zone of the station described was found to be 4.76 µMol cm^{-2} y^{-1}. For a consumption zone thickness of 50 cm, methane consumption rate was calculated to be 95.2 µMol cm^{-2} y^{-1} (WHITICAR 1981). The production rate of methane was 8.5 µMol l^{-1} y^{-1} which is 25 times lower than the consumption rate, thus precluding escape of methane into the water column. For this reason methane production, transport, redissolution and consumption in deeper sediment layers should have no effect on the nutrient exchange at the sediment/water interface by ebullition as found for other sediments (KLUMP and MARTENS 1981).

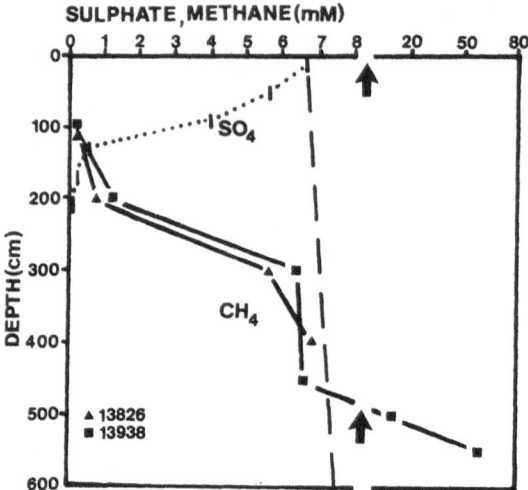

Fig. 4-18: Sulphate depletion and methane formation in pore waters at the 28 m
station. Note scale change for methane concentrtion. Broken line indicates
methane in situ saturation level (WHITICAR 1982a).

4.4 ROLE OF THE BOTTOM IN THE CYCLING OF ORGANIC MATTER: AN ATTEMPT FOR A
BALANCE

To estimate the importance of benthic regeneration for the cycling of organic matter,
decomposition and release rates have to be compared with sedimentation and total annual production. Data available for the 20 m station were used by BALZER (1984) to
calculate a budget for organic carbon, nitrogen and phosphorus at this location
(Fig. 4-19). When combining the carbon and sulfur burial rates with the carbon
equivalent of oxygen consumption a reasonable agreement is obtained with the sedimentation rate during the growth season. The intimate balance between carbon input and
release plus accumulation suggests that there is no net horizontal input or output at
this site, although considerable transport might be expected from the bottom morphology. In contrast to the situation found for carbon the sediment release and accumulation of nitrogen are less than the input (Fig. 4-19) pointing to a removal of
nitrogen by denitrification. The same imbalance is valid for phosphorus when taking
into account the accumulation of P in its organic form only (Fig. 4-19). When, however, the accumulation of P_{inorg} (not shown in the figure) is added, the deficit is
overcompensated for: obviously, part of the phosphate from organic matter decomposition is bound to the bottom in inorganic form and is not immediately available for
new primary production.

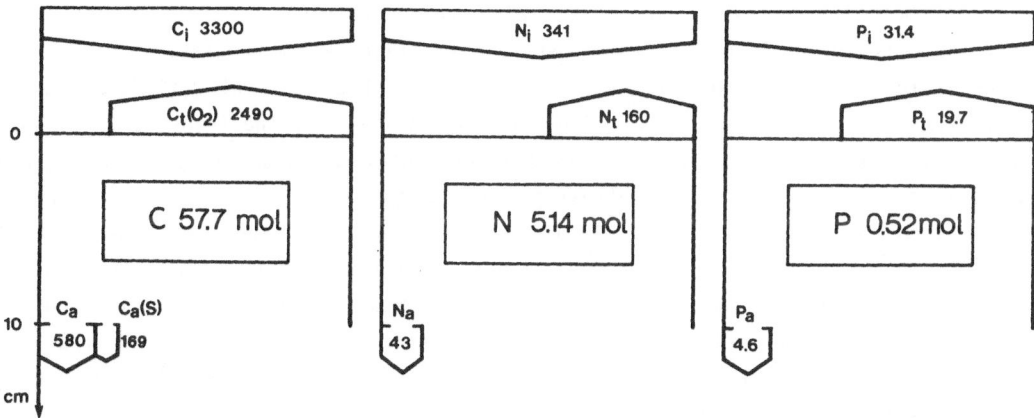

Fig. 4-19: Balance of organic matter cycling at the 20 m station: carbon, nitrogen and phosphorus fluxes (in mMol m^{-2} y^{-1}) in terms of input by sedimentation (X_i), total release (X_t), and accumulation rate (X_a). Total organic carbon degradation was calculated from oxygen consumption ($C_t(O_2)$); c_a (S) denotes the carbon equivalent of reduced sulfur burial (after BALZER 1984).

Tab. 4-6: Budget of organic carbon along the "Hausgarten" transect (Fig. 4-2): organic carbon combustion (as calculated from O_2-consumption rates) and accumulation rates for the different water depth sections. For the calculation of annual means see text (from BALZER et al. 1986).

Depth of measurement (m)	Water depth segment (m)	O_2-consumption rate (annual mean) (ml m^{-2} d^{-1})	O_2-consumption (Mol y^{-1})	C_{org} combustion*** (Mol y^{-1})	C_{org} accumulation (Mol y^{-1})
10	0 - 13.5	454*	12 970	11 030	0
10	0 - 13.5	205**	5 855	4 980	0
15	13.5-17	356	753	64	11
18	17 - 23	177	311	282	62
20	17 - 23	180			
25	23 - 27	279	963	920	300
28	27 - 29	310	1 489	1 452	1 350

* measured during night-time (Fig. 4-13)
** measured during day-time (Fig. 4-13)
*** incl. sulfur-accumulation equivalents

For the whole "Hausgarten" transect a carbon budget in terms of input to the sediment, degradation of organic matter in surface sediments and accumulation in sediment layers below recent bioturbation was constructed by BALZER et al. (1986) including data from recent experiments. Data from the six stations of the transect (Fig. 4-2) were evaluated including both vertical and horizontal transport rates. Results are depicted in Tab. 4-6 for the following five zones i) the lag sediment extending 1750 m from the shoreline down to 13.5 m water depth, ii) a sandy sediment zone in the depth range from 13.5 - 17 m, iii) a section covered by medium and fine sand (17 - 23 m), iv) a zone of sandy mud and mud sediments from 23 m to 27 m and v) a mud accumulating section between 27 m and 29 m water depth. Burial rates for the transect were already presented in section 4.2.4 and are included in Tab. 4-6. Although providing sites for extensive growth of macrophytes and microalgae the shallow zone (0 - 13.5 m) is an erosive area and does not accumulate organic matter. The relatively short deepest section, on the other hand, accumulates the most significant portion of the whole transect. The utilization of organic carbon was calculated from total sediment oxygen consumption by assuming quantitative oxidation of the bulk sediment organic matter having a composition of C:N:P = 125:11:1 (for inherent limitations of this kind of calculation see BALZER 1984). The carbon equivalents of reduced sulfur storage (non-oxidized remains of sulphate reduction) could not be separated from the combustion rate of organic carbon. The annual mean of O_2-consumption of depth intervals, for which direct measurements exist only for one or two seasons, were calculated by assuming the same seasonal cycle as for the 17 - 23 m section. The shallow zone subsystem, however, cannot be balanced in this way, because an oxygen comsumption rate is available only for a sediment not covered by macrophytes (Fig. 4-13) and the extent of sediment coverage by those plants and their rates of production and community respiration are fairly unknown. Since rates for the macrophyte zone cannot be assessed accurately, the transect was restricted to areas below 13.5 m. For the remaining four deeper zones data of carbon input by sedimentation can be balanced against rates of utilization and accumulation and a horizontal transport rate from the shallow to the deeper zones can be derived. The total accumulation of C_{org} in the region between 13.5 m and 29 m water depth amounts to 1 723 Mol C y^{-1}. The total combustion of organic carbon in top sediments below 13.5 m is 3 294 Mol C y^{-1} including reduced sulfur equivalents. From the vertical sedimentation rate, a pelagic input to the sediment of 2 480 Mol C y^{-1} is calculated. To balance the budget between these rates a horizontal input of 2 537 Mol C y^{-1} from the shallow zone to the deeper regions of the transect (below 13.5 m) must be assumed (Fig. 4-20). The data show that the shallow water system is comparatively closed in terms of carbon cycling: carbon loss rates from this system (1.45 Mol C m^{-2} y^{-1}) are small compared to the high metabolism (Tab. 4-6). The deep system is fed by equal amounts of carbon from pelagic sedimentation and horizontal advection, from which two-thirds are recycled and the rest is accumulated in sediment depths below 5 cm (BALZER et al. 1986).

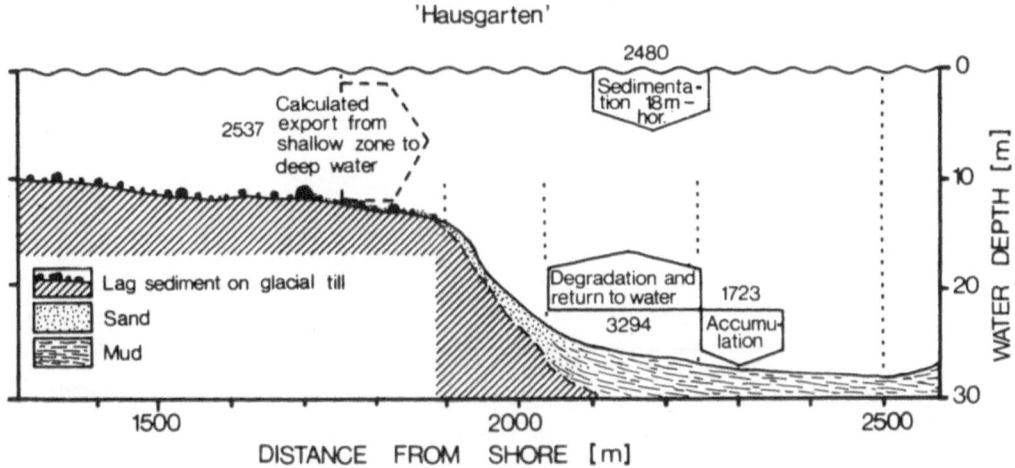

Fig. 4-20: Organic carbon balance for the sediment accumulating part of the "Haus-garten"-Transect. Note the export from the shallows area (after BALZER et al. 1986). Values in Mol C y^{-1} for a 1 m wide transect.

Due to uncertainties in the model assumptions and the seasonal and spatial variations of the observed rates, the estimates should be considered a first approximation.

4.5   TRACE ELEMENT DISTRIBUTION

Heavy metals in the sediment floor are involved in the biogeochemical cycling of or-ganic matter to a varying degree. Among these, iron and manganese hold a prominent position because they are not only dependent on the redox state created by metabolic decomposition of organic matter (TUREKIAN 1977, MURRAY and BREWER 1977), but they may also contribute oxidative capacity during the breakdown of organic matter due to their relatively high sedimentary abundance (FROELICH et al. 1979). In addition, Fe and Mn influence the concentration and distribution of many trace metals in seawater by adsorption or coprecipitation with their oxides (KRAUSKOPF 1956, JENNE 1968, MUR-RAY and BREWER 1977). In the following a few findings are given concerning heavy me-tal distribution in Kiel Bight and the occurrence of concretions along with mecha-nisms of their formation. Iodine is included because of its potential relationships to organic matter and heavy metal cycling.

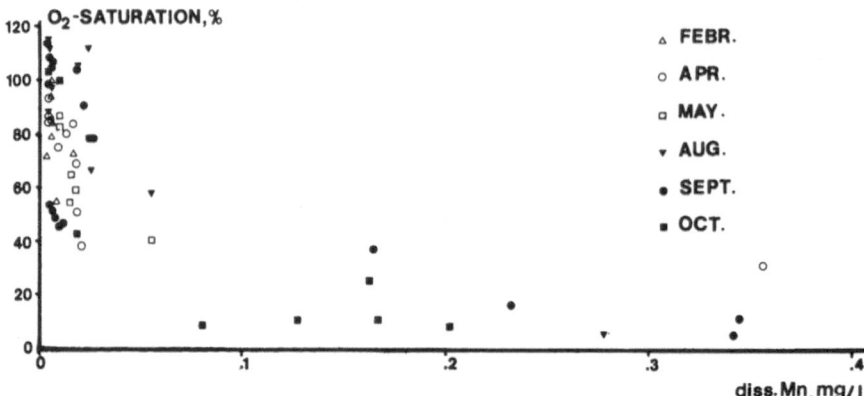

Fig. 4-21: Distribution of manganese in the water column in relation to the oxygen
saturation state. Samples were taken in Kiel Bight between February to
October, 1974 (after DJAFARI 1976).

4.5.1   Trace Elements in Water Columns, Sediments, and Pore Water

Manganese is known to be fairly mobile only under reducing conditions but should be
present in extremely low concentration when in the +IV oxidation state (MURRAY and
BREWER 1977, BENDER et al. 1977). Thermodynamically the transition to higher concentration is controlled by the environmental redox state. During the investigation
of the oxic/anoxic transition zone of northwest African shelf sediments HARTMANN et
al. (1976) found the interstitial manganese concentration to rise when the redox
potential fell below Eh = 300 - 350 mV (at pH ~ 7.5) which is very near to the value
of Eh = 370 mV calculated by HEM (1978) for the redox couple Mn^{2+}/Mn_3O_4. In a study
of environmental conditions favouring the formation of ferromanganese concretions
DJAFARI (1976) found a close relationship between increased manganese concentrations
(up to 7 µMol l^{-1}) in the water column and oxygen saturation values below 40 % (Fig.
4-21). He attributed this finding to a slightly lowered redox state as mentioned above
for pore waters. During a bell jar experiment simulating the transition from oxic to
anoxic conditions BALZER (1982a) found evidence of contradictory behavior: despite
nearly constant redox potentials (Fig. 4-11) during the oxic period in the enclosed
bottom water, manganese concentration rose by two orders of magnitude (Fig. 4-22). A
consistent view of all these observations is obtained if - within long-term thermodynamic control - allowance is made for kinetic effects which may lead to metastable
high Mn^{2+} concentrations in oxic environments. When the redoxcline moves upwards or
lies very close to the sediment/water interface, high concentrations of Mn^{2+} occur in
the pore water of subsurface layers from where diffusive flux into the bottom water
is fast enough to cope with dissolved Mn losses by oxidation to solid Mn(IV)-oxides.
Such a step-by-step shortening of the pathway for diffusive flux would explain the

exponential increase of Mn concentration (Fig. 4-22) during the oxic phase of the experiment (BALZER 1982a). Extremely high concentrations of manganese in the water column (1.82 µMol l^{-1}) coexisting with oxygen saturation in the range 60 - 98 % (KREMLING et al. 1979) and the observation of DJAFARI (1976) can be explained by turbulent mixing of O_2-saturated waters with water masses which had been in contact with reducing conditions. The high manganese concentrations may survive in a metastable state for a certain time due to slow oxidation kinetics.

The process of reductive dissolution, diffusive flux and slow reprecipitation (SUNDBY et al. 1981) in Kiel Bight leads to a relative depletion of Mn in the regions of mud deposition and an enrichment in sediments of intermediate water depths (DJAFARI 1976). The high fluxes observed during the mentioned bell jar experiment (oxic: 362 µMol m^{-2} d^{-1}, anoxic: 1127 µMol m^{-2} d^{-1}) entail a nearly complete loss of

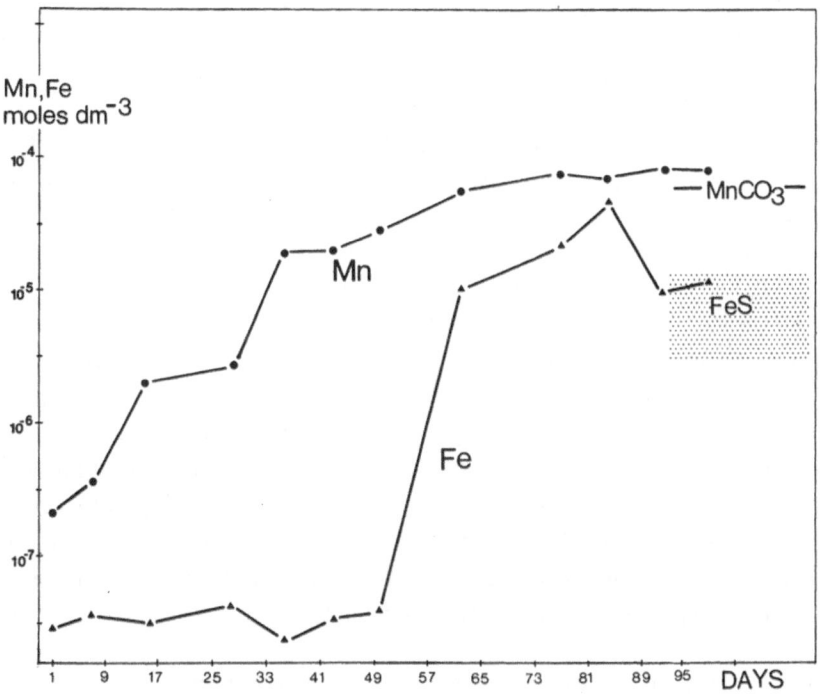

Fig. 4-22: Bell jar experiment: logarithmic plot of manganese and iron in the enclosed bottom water. The shaded area represents the iron concentration in equilibrium with amorphous FeS. The level MnCO$_3$ indicates Mn concentration in equilibrium with rhodochrosite (crystalline Mn CO$_3$) (after BALZER 1982a).

easily dissolvable Mn in the top sedimentary layers (BALZER 1982a). A second consequence of the higher Mn concentration in near bottom waters is the net transport of
Mn into the basins of the Central Baltic according to the general circulation pattern
of the Baltic Sea (HARTMANN 1964). From these basin sediments a high abundance of Mn
oxide nodules and the occurrence of basin sediment sections extremely enriched in Mn-
CO_3 (up to 13 %) have been reported (SUESS 1976a, HARTMANN 1964).

Iron is involved in diagenetic transformations to a lesser extent than manganese.
This is due to the lower redox potential of the Fe^{2+}/Fe^{3+} transition and the low solubility of Fe sulfides on the other hand. Additionally, iron is not subject to slow
kinetics like Mn at the oxic/anoxic boundary: Fe closely follows the abrupt change in
redox conditions during the bell jar experiment (Fig. 4-22 and Fig. 4-11). The observed
maximum in Fe concentration during the transition from low oxic values and the level
determined by FeS precipitation were also reported by ALLER (1980c). This maximum can
be explained by insufficient availability of sulfide ions at the beginning of reducing conditions. In Landsort Deep (Baltic Proper) sediment SUESS (1979) found evidence of the authigenic formation of $FeCO_3$, amorphous FeS and a mixed Fe-Ca phosphate,
the latter two only identified by their respective chemical composition.

Trace metals other than Fe and Mn (Cu, Zn, Ni) in the water column of Kiel Bight were
measured by DJAFARI (1976). KREMLING et al. (1979) provided a seasonal survey data
report on Zn, Cd, Cu, Ni, Fe and Mn of Kiel Bight's surface and bottom water.
Contrary to manganese (and to a lesser extent to Fe and Zn) Cd, Cu and Ni showed no
or only slight increase in concentration with water depth. Strong seasonal variation
basically influenced by biological activity was found for the distribution of Pb, Zn
and Hg at the sediment/water interface (KUIJPERS 1974). Considerable enrichment in
seston and surface sediment was shown for all three elements during late summer. If
these metals were bound to organic matter, a constant ratio of metal/C_{org} should be
the result. Increased ratios during late summer were explained by KUIJPERS (1974) by
delayed remobilization, partly supported by adsorption to or coprecipitation with Fe
and Mn oxides. The increased levels of Zn (DJAFARI 1976) and Cd, Pb, Zn (ERLENKEUSER
et al. 1974) in surface sediments as compared to deeper layers were interpreted as a
consequence of anthropogenic inputs (see section 4.6).

Solid sediments from different water depths contained iodine in excess of 100 ppm,
even reaching 300 ppm in fine-grained, organic-rich muds. Since terrigenic sediments
are characterized by iodine contents well below 10 ppm, marine biological and diagenetic processes must be responsible for the enrichment in recent sediments (HARTMANN
unpubl.). Brown algae known for their ability to concentrate this element (SHAW 1962)
probably contribute to the observed levels (section 4.2.6). A relatively low varia-
tion in the molar ratio I/C_{org} ($4 \cdot 10^{-4}$ - $9 \cdot 10^{-4}$) shows organic matter to be the predominant carrier of iodine in recent sediments. A rapid decrease of this ratio with

depth in the top 20 - 30 cm of cores from the "Hausgarten" (Fig. 4-9) shows preferential release of iodine and suggests a close relationship to easily degradable components of the organic matter. In pore waters from this area iodine increases from its bottom water concentration of about 0.25 μMol l^{-1} up to 8 μMol l^{-1} at 15 cm of a C_{org}-rich sediment core: from the high gradient an iodine flux of 6.8 μMol m^{-2} d^{-1} was calculated (HARTMANN unpubl.), leading to a considerable increase in the I/Cl ratio in the stagnant near-bottom water. Identical release rates obtained during oxic and anoxic periods of a bell jar experiment suggested that the redox state of the bottom water has no influence on iodine release.

4.5.2   Ferromanganese Concretions

A first investigation of the regional distribution, shape and chemical composition of ferromanganese concretions in Kiel Bight was performed by DJAFARI (1976). Three main types of concretions were distinguished morphologically: i) spherical nodules (ca. 1 - 3 cm in diam.), ii) symmetrical and asymmetrical discs (up to about 10 cm in diam. and 1 - 2 cm in thickness) and iii) accretions on shells of living and dead mussels.

These ferromanganese concretions are found in Kiel Bight only in a few restricted areas at certain water depths. In addition to certain chemical conditions the following three main characteristics could be established for the areas of their occurrence: i) low sedimentation rates or slightly erosive conditions, ii) occurrence of pebbles or mussels on the sediment surface and iii) water movement by waves or currents prevailing only for short periods in the near-bottom water. A combination of all these conditions within Kiel Bight is met only at a few places with water depths ranging mainly from 23 m to 26 m (DJAFARI 1977).

The spherical and discoidal concretions normally contain pebbles as substrate in their centers. A counting (HEUSER in prep.) of pebbles in an area rich in ferromanganese concretions showed no preference for a particular kind of material as substrate. There is, however, a clear difference concerning the mussels: while Astarte is normally covered by crusts and coatings of Mn-Fe oxides, Arctica living in the same area is completely free of oxide coatings. This is evidently due to the fact that Astarte lives on the surface of the sediment, buried only with its lower part, while Arctica lives completely buried within the anoxic sediment.

The relations between precipitation of Mn-Fe oxides on naturally occurring substrates and on artificial materials and the variation of hydrographic conditions in the near-bottom water were investigated by DJAFARI (1976), Fig. 4-23 and HEUSER (in prep.). It could be shown that with high O_2 concentration in the water column, precipitation rates are very low (mean value ~ 0.1 μm d^{-1}). This is evidently due to the low Mn content of the sea water during these periods (ca. 2 µg l^{-1}). Such low accretion rates are measured during winter (November - May), when the diagenetic activity within the uppermost sediment section is low due to low temperatures and low supply of fresh organic matter to the sediment. During this time diffusion of Mn^{2+} from deeper sediment sections to the sediment surface and the bottom water is considerably restricted by the oxidizing conditions in the uppermost sediment section. On the other hand additional Mn is suplied to this layer and may be remobilized during conditions of the low redox potential prevailing in summer. This could be shown during the bell jar experiment (BALZER 1982a), where the excess Mn content of the uppermost sediment layer was completely mobilized during several weeks of reducing conditions. Highest accretion rates are found during summer, when the O$_2$ content of the nearbottom water is low due to bacterial remineralization of fresh organic matter and restricted oxygen supply due to thermo-haline stratification (section 4.2.1). During this time the Mn content of the near-bottom water may increase up to several hundred micrograms per liter (Figs. 4-21 and 4-23).

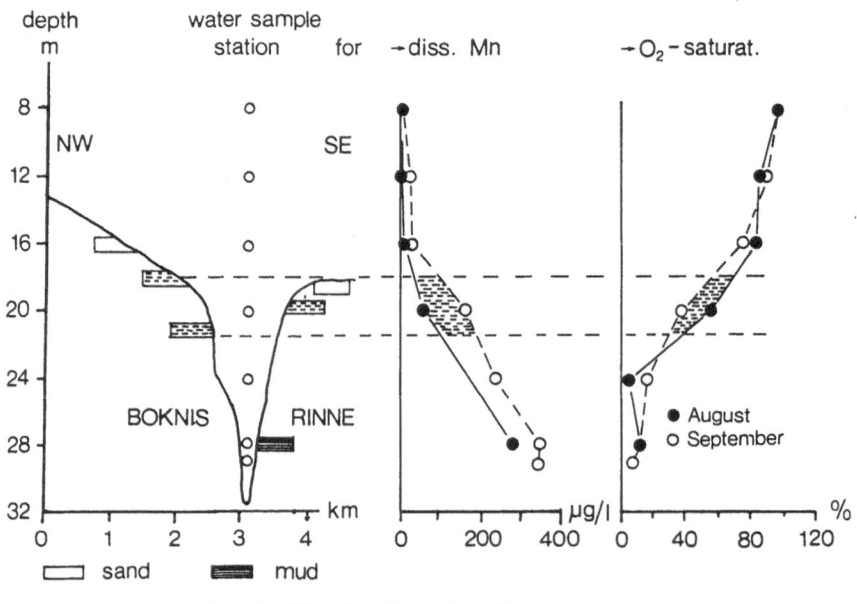

Fig. 4-23: Relation between the zone of increased dissolved Mn levels, the zone of slightly reduced oxygen concentration and the occurrence of concretions in the "Hausgarten" area (after DJAFARI 1977).

In the "Hausgarten" area a maximum accretion rate of 0.4 µm d^{-1} was found between 22.9. - 20.10. 1981. Still higher rates may be expected in areas of naturally occurring crusts and nodules. Short time maxima seem to be reached just before and after an anoxic period in the near-bottom water when highest Mn concentration and still some O_2 are present at the same time. Since the redoxcline is positioned at the immediate interface during these periods and the gradient through the interface is most pronounced, large amounts of Mn cross over into the oxic bottom water and are reprecipitated as concretions at a certain protective distance from reductive dissolution attack (BALZER 1982a).

During anoxic conditions, which normally occur for periods of days up to a few weeks during late summer redissolution of Mn oxides was found at the sites of ferromanganese concretions. While fine-grained hydroxides are dissolved completely within a short time, dissolution attack on crusts and nodules is restricted to their surface as may be seen from thin sections of nodules as well as from artificial substrates sampled during this time. Because Fe oxides are slightly more resistant under reducing conditions, Mn oxide redissolution results in an appreciable decrease in the Mn/Fe ratio of this layer. Exchange reactions between Mn oxides and Fe^{2+} from solution may support this mechanism, giving an additional protection for the deeper layers of Mn oxides within the concretions. Thus the growth of crusts and nodules is subjected to a clear seasonality. Up to now, however, it is still unknown whether the layered structure commonly found in the crusts and nodules is equivalent to an annual structure. Nodules buried within the mud sediment for a certain time show this type of Fe enrichments of the outer crust (HEUSER in prep.).

The chemical composition of the three main types of concretion is shown in Tab. 4-7 (DJAFARI 1976). Only slight differences in Mn and Fe contents are found for the three types. Appreciably higher Zn values and slightly increased values for Cu and Co, however, are found in the concretions on Astarte shells. The same phenomenon of increased Zn contents (as well as of some other heavy metals) was observed in the outer sections (1 - 2 mm in thickness) of discoidal concretions. This was interpreted to be related to the increased anthropogenic input of heavy metals (Fig. 4-26) during the last 100 years (ERLENKEUSER et al. 1974). SUESS and DJAFARI (1977) used this beginning of an increased trace metal content for a first estimation of the accretion rate for Kiel Bight ferromanganese concretions. They found rates between 0.02 and 0.16 mm y^{-1}. Crusts on Astarte shells seem to grow faster than on other substrates (up to several tenths of a mm per year). This is probably due to the activity of the organisms when filtering O_2-containing bottom water and pumping back interstitial water high in Mn at the same time. Active prevention of these organisms from becoming covered by sediments would be a second reason. Microscopic and microbiological examination of the surface layers of ferromanganese concretions from the Breitgrund area (NW Kiel Bight) by GHIORSE (1980) revealed the presence of numerous metal oxide encrusted bacteria suggesting an active involvement of microorganisms in metal oxide deposition. In laboratory cultures of bacteria (Pedomicrobium, Hyphomicrobium, Leptothrix a.o.) isolated from the crusts, active deposition of Mn and Fe oxides could be shown, Leptothrix being the most active species.

Table 4-7: Chemical composition of the three different types of Mn-Fe-concretions
found in Kiel Bight (DJAFARI 1976).

Element	Accretion on shells			Spherical nodules			Disc-shaped nodules		
	average	min	max	average	min	max	average	min	max
Mn, % dw	28.8	20.8	32.0	34.1	31.2	39.2	30.6	23.1	37.2
Fe, % dw	8.1	5.7	13.5	10.0	7.9	12.1	12.0	9.9	16.1
Co, ppm	52	42	62	39	28	56	34	27	43
Ni, ppm	87	69	118	76	62	99	65	51	82
Cu, ppm	27	23	35	11	5	17	17	9	27
Zn, ppm	1128	967	1320	166	120	244	188	128	368
Mn/Fe	3.5	1.5	5.4	3.4	2.8	4.8	2.5	1.8	3.1

4.5.3 Heavy Metals as Indicators of Pollution

Although the study of pollution was beyond the scope of the SFB 95, during our investigations we could not fail to see the interaction between natural and man-made effects in the environment. To tackle these problems SUESS (1978) lists four
approaches: i) in "baseline studies" the concentration of a certain component above a
natural background is determined and the excess is attributed to anthropogenic sources,
ii) the study of entire assemblages is often necessary for a distinction between natural and man-made effects because both anthropogenic input and e.g. diagenetic processes can create the same effect on one component, iii) a more refined approach
includes the evaluation of physical and chemical states of the various materials in
the sediment and iv) further insight is obtained from a synthesis of the various approaches in budget or mass-balance calculations.

The history of the flux of fossil coal residues to the sediment as documented by the
^{14}C age deviation from linear sediment depth-age relation may be compared with the
annual coal mining rate of Northwest Europe and with the accumulation rate of heavy
metals in the sediment (Fig. 4-24). Although the fossil coal deposition rate of 2.4 g
C m^{-2} y^{-1} found for a sediment core at 28 m water depth (No. 2 in Fig. 4-1) (ERLEN-
KEUSER et al. 1974) is higher by a factor of 3 than the average deposition flux calculated from coal production, this is not an argument against the concept of a
dilution of the natural ^{14}C content through ^{14}C free carbon by man's activity. To a
large extent, the bad fit between expected and observed accumulation rates of fossil
fuel residues will be accounted for by physical transport processes which focus
small-sized particulates to the deeper depositories of the fine-grained sediments
and, hence, concentrate particle-associated pollution tracers there (ERLENKEUSER and
PEDERSTAD 1984).

In the same core that was evaluated for the carbon isotope distribution, heavy metals were determined (ERLENKEUSER et al. 1974). The concentrations of Fe, Mn, Co, Al and C_{org} were more or less constant over the entire core length. Cd, Pb, Zn and Cu, however, were constant only at depth in the sediment and gradually increased in concentration from about 22 cm upwards (Fig. 4-25). The enrichment of these metals above their background levels varied considerably: Cd showed the strongest increase of up to 600 % while Pb, Zn and Cu were "only" enriched by an excess percentage of 300, 200 and 100 % respectively. The mobilization of heavy metals through burning of fossil fuels (coal and oil) is between 70 and 200 times higher in the case of burning coal than through combustion of oil (BERTINE and GOLDBERG 1971). From this, the close examination of fine coal pieces and slag fragments, and the fossil carbon input, it was concluded that the heavy metal enrichment represents a certain coal residue assemblage which is added to the surface sediment at rates of 80 mg Zn m^{-2} y^{-1}, 22 mg Pb m^{-2} y^{-1}, 12 mg Cu m^{-2} y^{-1} and 0.58 mg Cd m^{-2} y^{-1} (ERLENKEUSER et al. 1974).

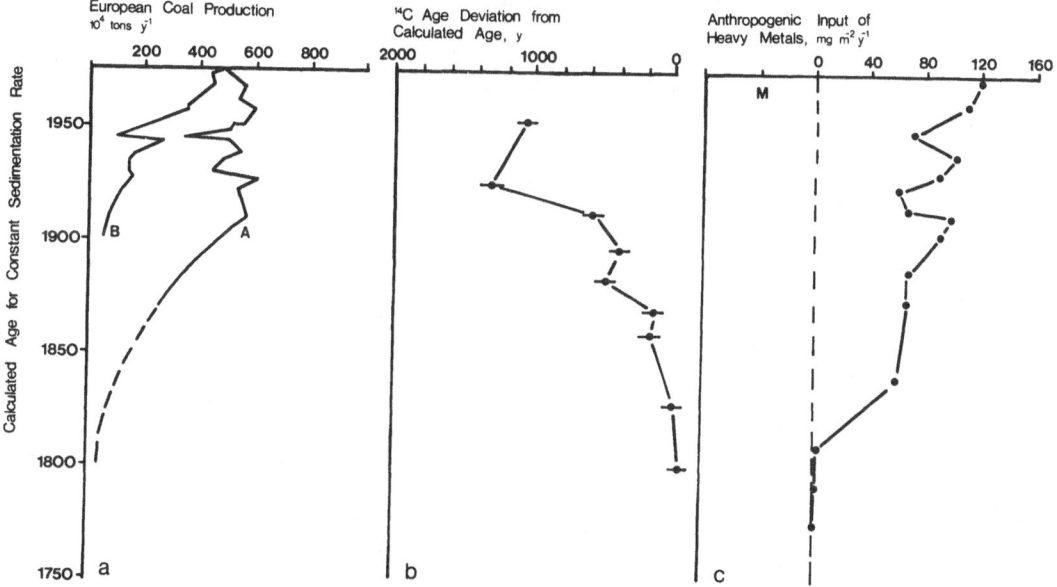

Fig. 4-24: European coal production (a) (A = coal, B = lignite) is compared with the increase of radiocarbon age (b) and with the supply of heavy metals (sum of Pb, Cd, Zn and Cu) to near-surface sediment at 28 m water depth, near "Hausgarten" (c) (M denotes natural flux) (after ERLENKEUSER et al. 1974).

SUESS and DJAFARI (1977) observed the same assemblage concentrated in the outer layers of ferromanganese concretions (which were formed in the same basin) and concluded that this too was caused by the increased metal input during the past century. The analysis of heavy metal distribution, carbon isotopes and sedimentation rates can be used to follow the onset of industrialization effects in different sea areas as demonstrated by SUESS and ERLENKEUSER (1975b) for different basins of the Baltic Sea (Fig. 4-26) and by ERLENKEUSER and PEDERSTAD (1984) for Skagerrak sediments by means of the Zn flux. Rates of sedimentation for the different areas of deposition date the beginning of the increased input at about 1860 for Kiel Bight and the Skagerrak, at about 1900 for the Bornholm Basin and at about 1920 for the Gotland Basin. For a core from the Gdansk Basin, the onset of an increased Zn content (SUESS, unpubl.) was dated by the ^{210}Pb method at 1850 to 1860 (ERLENKEUSER unpubl.). Based on the characteristic assemblage of Cd, Pb, Zn, Cu and carbon isotopes the increased metal and carbon inputs could be identified as being due to the burning of fuel and the smelting of ores. Other metals such as Fe, Ni and Co are far less affected by this input partly because natural mobilization is high and partly because of different diagenetic behavior of the metals.

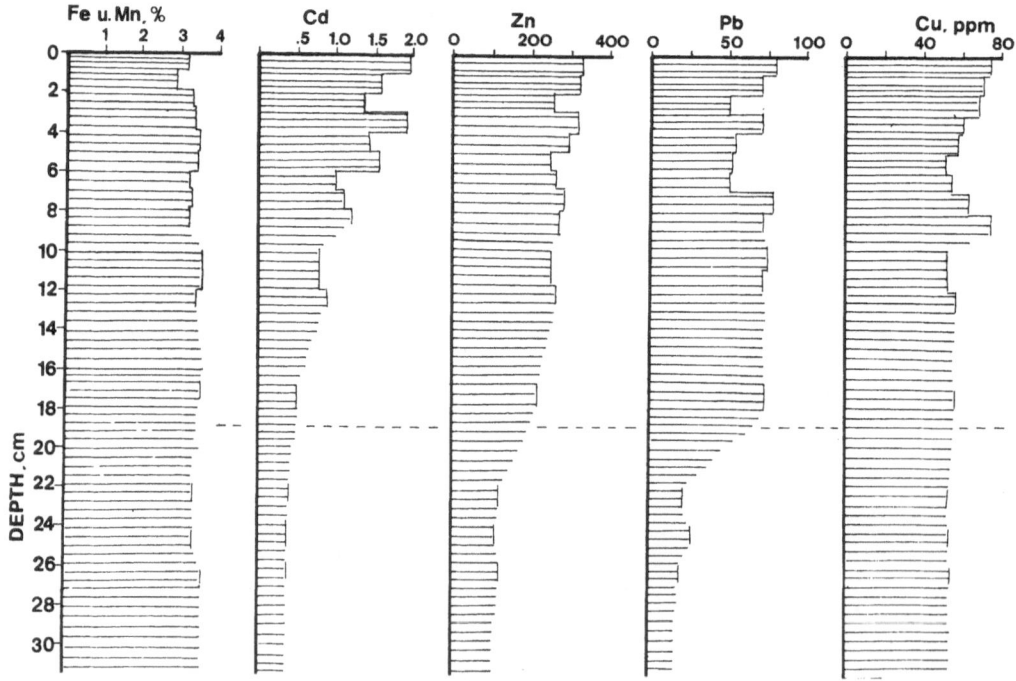

Fig. 4-25: Increasing trace metal concentrations in near-surface sediment layers of the 28 m station, near "Hausgarten". The onset of higher-than-background contents of Cd, Pb, Zn and Cu at 18 cm (dotted line) coincides with increasing amounts of ash and of ^{14}C-inactive fossil carbon; the layer was radiometrically dated at A.D. 1830 ± 20 (after ERLENKEUSER et al. 1974).

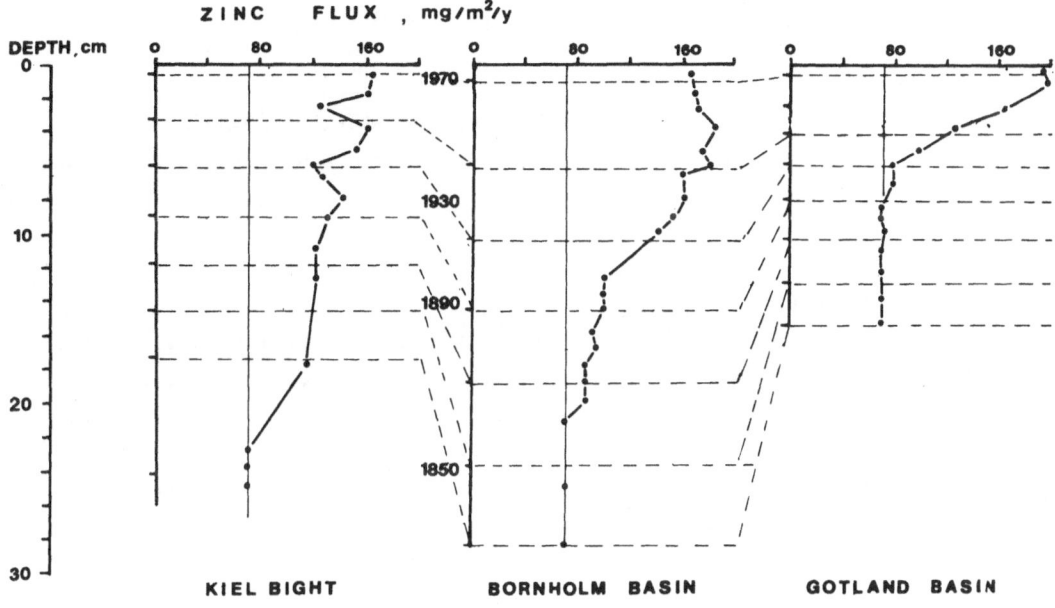

Fig. 4-26: Annual Zn flux to Kiel Bight, Bornholm Basin and Gotland Basin; the position of the layers indicated by broken lines was calculated from ^{14}C-sedi-
mentation rates including data of NIEMISTÖ and VOIPIO (1974). (After:
SUESS and ERLENKEUSER 1975).

Chapter 5:   SEDIMENTARY RECORDS OF BENTHIC PROCESSES

F. WERNER, H. ERLENKEUSER, U.v. GRAFENSTEIN, S. McLEAN,

M. SARNTHEIN, U. SCHAUER, G. UNSÖLD, E. WALGER, and R. WITTSTOCK

5.1   INTRODUCTION

This chapter describes efforts of the SFB 95 in the investigation of processes affecting sediment transport and accumulation in Kiel Bight, an area which offers favorable conditions for studies of this kind. Insight into such processes was applied for the paleo-environmental interpretation of the Holocene sediment column.

Especially promotive of these investigations was the interdisciplinary cooperation with colleagues of the Department of Applied Physics in the development and adaptation of measuring techniques and of methods for data acquisition and processing (E. SCHULZ-DUBOIS, U. PETERSOHN and co-workers).

Studies were focussed on

- determination of bottom shear-stress under the conditions of fluctuating currents typical for Kiel Bight by means of velocity and turbulence profile measurements in the boundary layer in situ (by S. McLEAN, U. SCHAUER) and in the flume (by S. McLEAN);
- conditions of bedform generation in areas of channeled, high energy currents in situ (F. WERNER) and in the flume (G. UNSÖLD);
- sedimentation in the Kiel Bight channels which are controlled by medium energy to lower energy currents (F. WERNER);
- wind-wave induced sediment transport which in the major part of Kiel Bight is a prominent factor of sediment distribution (F. WERNER and U. v. GRAFENSTEIN);
- flume experiments on the critical entrainment conditions of cohesive and non-cohesive sediments (G. UNSÖLD, E. WALGER);
- effects of single, often extreme and rare events on the sedimentary column (F. WERNER);
- effects of bioturbation by erasion of older fabric elements and by creating new ones (F. WERNER, R. DOLD);
- studies on the ecology of molluscs and the interpretation of the shells in the Holocene sediment column for reconstruction of the environmental conditions (M. SARNTHEIN, W. RICHTER).

Especially the necessity to analyze single high energy events and their effects became obvious during these investigations.

The structure of most sediments results from an integration of fluctuating processes; thus, the system only can be fully understood by assessing the impact of single, often extreme and rare events, the importance of which may often surpass that of "every-day events" (EINSELE and SEILACHER 1982).

Fig. 5-1: Bathymetry of Kiel Bight, depth contours in meters (from SEIBOLD et al. 1971). A = investigation area Stoller Grund (oscillation ripples: section 5.3.3); B = investigation area Breitgrund Rinne (section 5.4); C = area of sediment distribution map (Fig. 5-50); V = investigation area Vejsnaes Rinne (section 5.4); S = profile site of shell zonation studies and ^{14}C dating (section 5.5); E = sample profiles for single event studies in Eckernförder Bucht (section 5.3.6); Δ = long-term current-measurement sites (section 5.3.1); O = PMA site (section 5.3.2 - 5.3.3); * = site of long-term wave measurements (section 5.3.4). For a-d see Fig. 5-4.

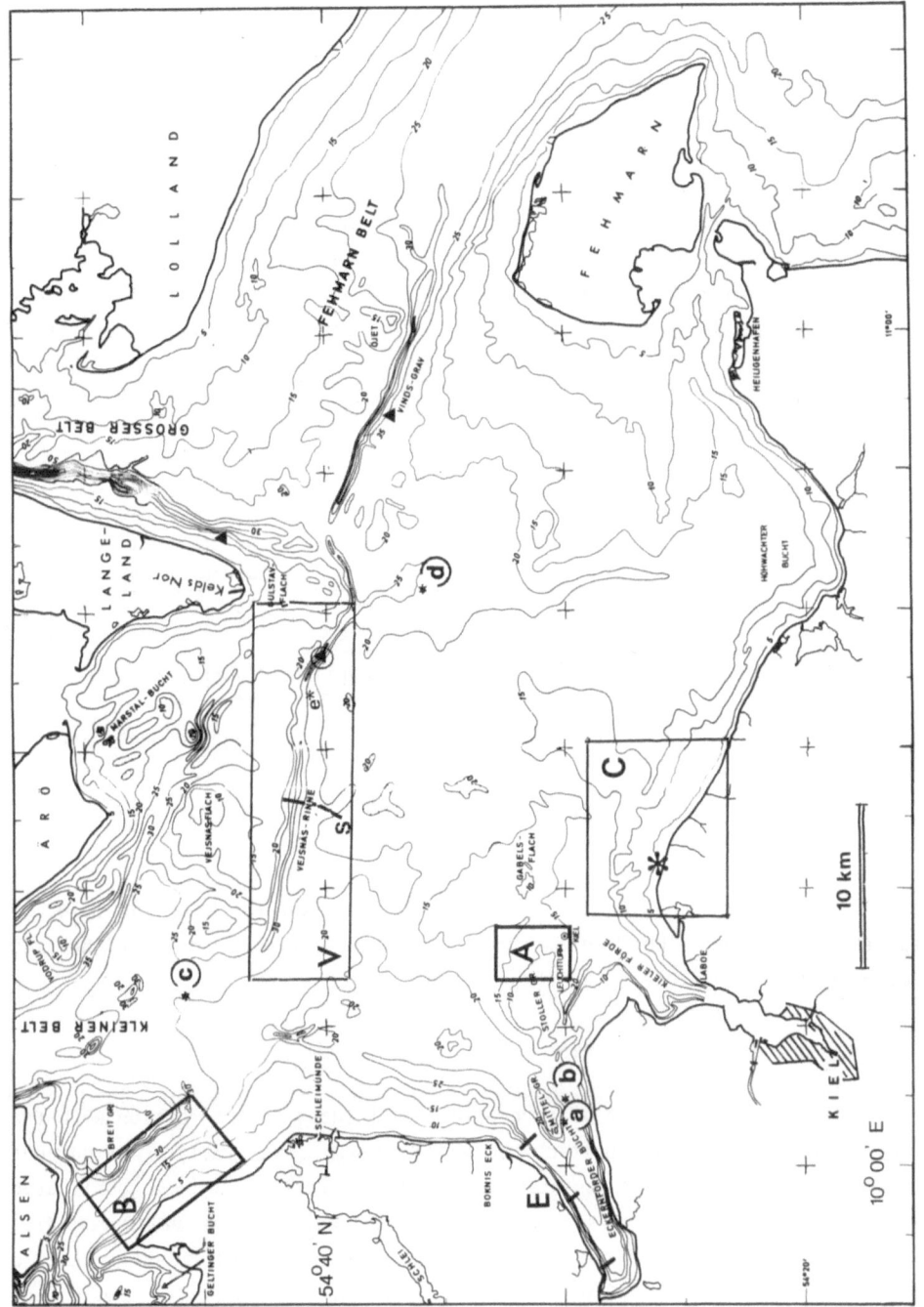

Abb. 5-1

5.2   GENERAL FEATURES OF SEDIMENTARY ENVIRONMENT

5.2.1   Morphology

The main morphologic features of Kiel Bight are a result of the Pleistocene glaciations. Near the coast.wavecut platforms are formed by recent coastal erosion (HEALY and WEFER 1980). Abrasion during the Holocene marine transgression had only the effect of gently smoothing the slopes and plateaus in deeper water without reworking significant amounts of older sediments (WINN et al. 1982). Kiel Bight can be divided into the following morphological units, excluding the littoral zone which is not considered here:

Table 5-1: Morphological units of Kiel Bight

| | | r e l i e f | | |
unit	water depth	large-scale	small-scale	predominant sediment type
abrasion platforms	4-12 m	flat	rough	lag sediment
slopes and plateaus	10-20 m	rough/flat	smooth	coarse to fine sand
shoals	10 m	rough	rough	lag sediment/sand
basins	20 m	flat	smooth	mud
channels	20-35 m	steep	smooth	mud/lag sediment

Below the bathymetric distribution curve (Fig. 5-2) the mean thickness of Holocene marine deposits is plotted. The diagram shows that the maximum sediment thicknesses occur in the mud basins and locally in channel areas.

5.2.2   Hydrodynamic Setting

The water circulation of Kiel Bight is perceptibly influenced by interaction with the adjacent basins of the Western Baltic Sea and Kattegat (see 5.3.1). The current effects on the sea bed depend on this interaction which is controlled by a complicated dynamic system with highly variable temporal fluctuations. This is valid for Kiel Bight as well as for the channel systems of the adjacent belts (Great Belt, Little Belt, Fehmarn Belt, DIETRICH 1951, JACOBSEN 1979). The major dynamic effects of bottom currents on the sea bed are to be expected in places where the flow is channeled, as in the channels crossing Kiel Bight or at narrow sections of the Belt system (Fig. 5-1). In the topographically less confined areas like the Western Kiel Bight,

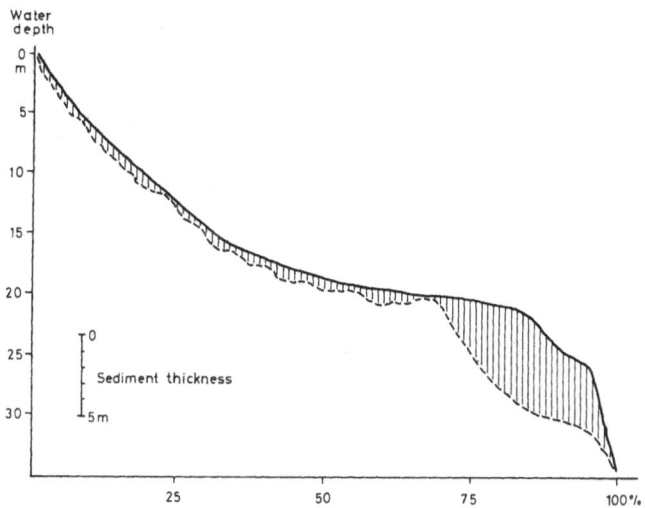

Fig. 5-2: Hypsometric curve of Kiel Bight with indication of sediment thicknesses
(HEALY, 1980).

there is little current activity. Neither do tidal currents have any significance for
sediment motion. The distribution of surface wave energy in Kiel Bight is determined
by the local topographic features and depends only on the local wind field, as wind
induced swell from outer parts is suppressed due to the narrow entrance of Kiel
Bight.

Consequently, the sedimentary record in the shallower parts of Kiel Bight is mainly
controlled by the regime of wind waves, while in the deeper parts bottom currents are
more important. The transition zone lies between 18 and 22 m of water depth (SEIBOLD
et al. 1971).

5.2.3   Sediment Origin

The main sediment sources in Kiel Bight are the retreating cliffs (KANNENBERG 1951,
SEIFERT 1954, SEIBOLD et al. 1971) and the shallow water zones. Sediment input by
rivers and wind is negligible. Contributions from biological sources are quantitatively
unimportant, yet they are all the more important as a source of information widely
used to reconstruct the paleo-environment (WEFER and LUTZE 1978 and chapter 6, this
vol.).

The source material mainly consists of glacial till and intercalated outwash sands and gravel. The outwash sands are of minor importance as submarine sediment source, because they are protected by a widespread till cover of the youngest glacier advance ("Fehmarn-Vorstoß": STEPHAN 1971, PRANGE 1978). For the recent sediment distribution, the following consequences are important:

- due to the wide grain size spectrum of the morainic material even major hydrodynamic events will find their equilibrium grain size range and therefore will have the chance to be recorded on the sea bed (see 5.3.1);

- the source sediment at the coast and on the submarine abrasion platforms (HEALY and WEFER 1980) offers little resistance to erosion in places where a continuous layer of residual sediment (typical gravelly lag sediments, Fig. 5-3) protecting the underlying beds has not yet developed;

- large areas of the abrasion platform are covered with residual sediments (lag sediment layer). Locally they can be overlain by thin moving sand sheets;

- the cliffs retreat by decimeters per year, in places by meters. The dynamic equilibrium is rapidly approximated;

- large amounts of fine-grained material are available from the till cliffs. The sedimentation rates of mud in the basins are correspondingly high (HEALY & WERNER, in press).

5.2.4   Patterns of Sediment Distribution

The general patterns of sediment distribution in Kiel Bight and in the adjacent belts can be explained qualitatively by the hydrodynamic and geological setting as described above. In Kiel Bight, a general sediment zonation parallel to water depth contour lines is obvious, if the sediments are grouped according to grain size characteristics (Fig. 5-3). This generally holds true for the wider belt channels such as Fehmarn Belt and the northern Great Belt. The decrease of wave energy with water depth and the weak water movements in the deeper parts of the basin are reflected by a general increase of the mud content. In the narrow parts of the belts where current intensity is higher, the zonation is less obvious (Fig. 5-4) or - in places - even reversed due to strong bottom currents (WERNER and NEWTON 1975, WINN 1974). This tendency of sediment distribution being controlled more by the topographic features than by water depth only, can be observed in Kiel Bight as well. In the narrow channels well developed there, the sediment is coarser than in comparable water depths of the basins in the same area. Fig. 5-5 shows an isopach map of Holocene mud deposits. The difference in sediment thickness between basins and channels is very high, independent of the sediment type. This shows that the bottom currents in the channels have kept them open up to recent times.

169

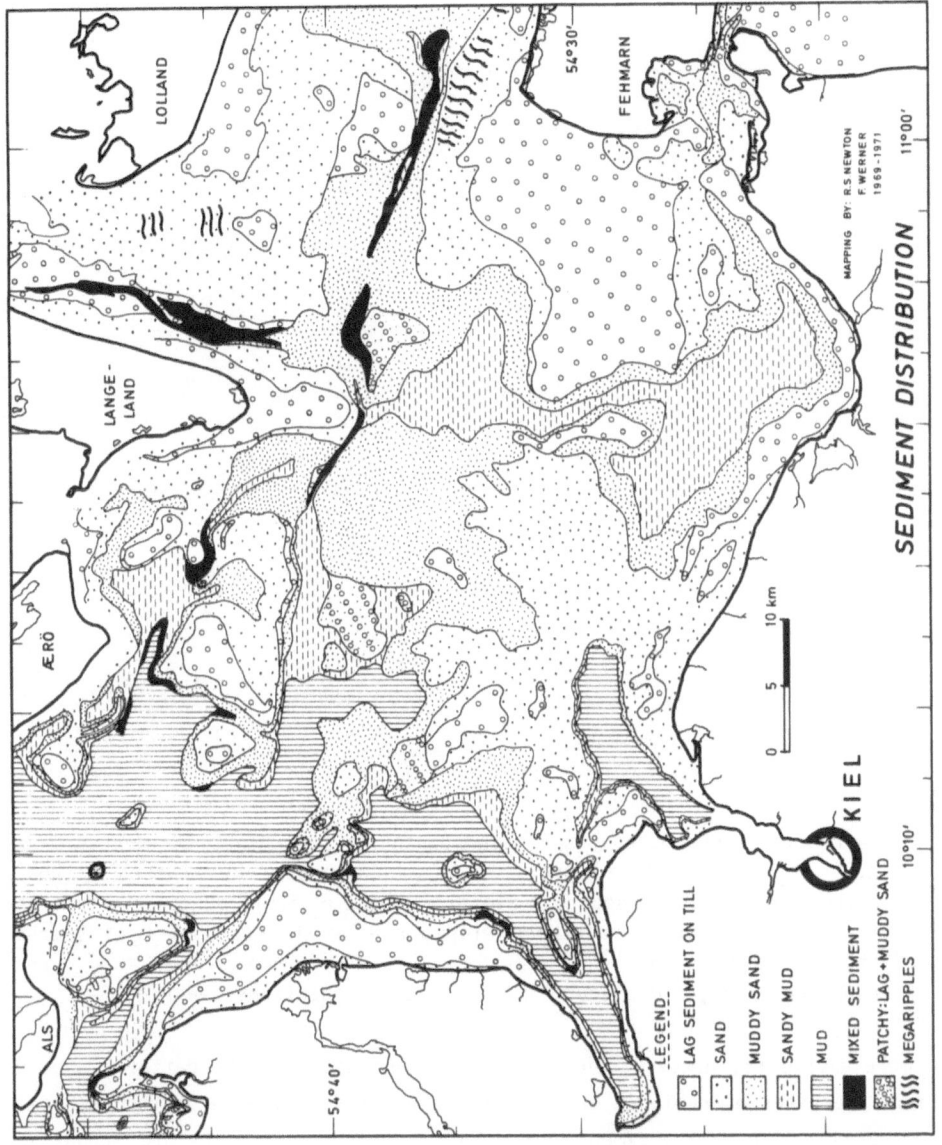

SEDIMENT DISTRIBUTION

Fig. 5-3: Distribution of sediment types in Kiel Bight (from SEIBOLD et al. 1971).

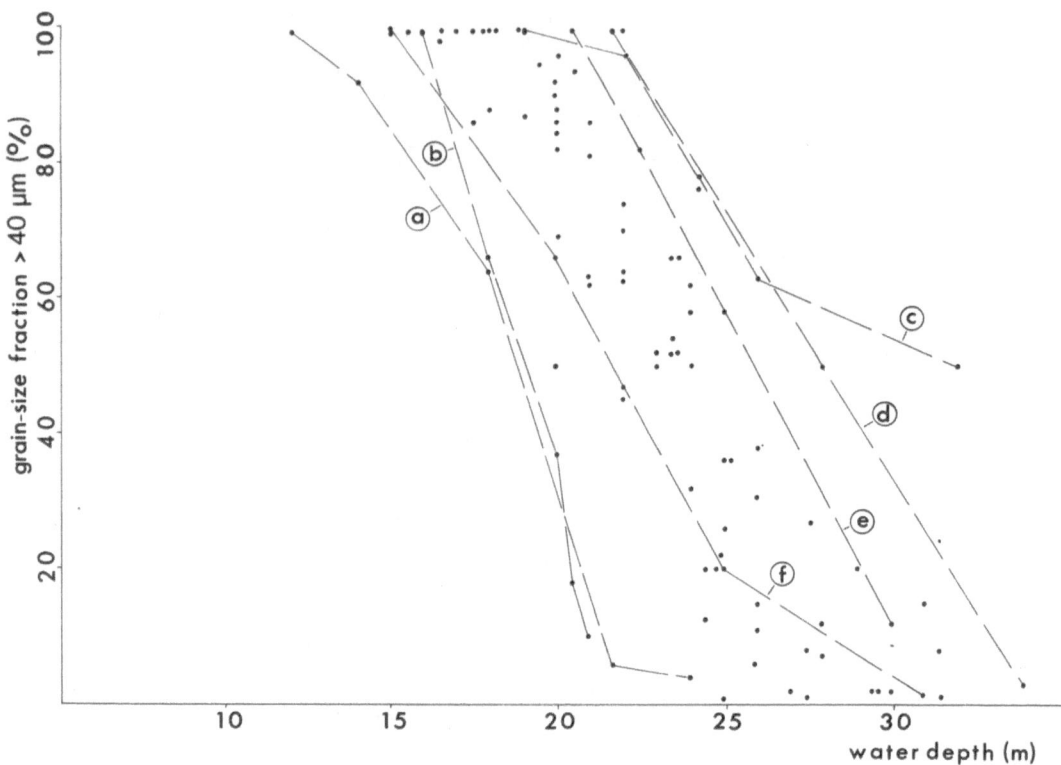

Fig. 5-4: Sand-fraction percentage versus water depth for Kiel Bight sediments. Data points from the slope profiles are connected by dashed lines: a = Eckernförder Bucht, southern slope, b = east of Schleimünde, c = eastern part of Vejsnaes Rinne, northern slope, d = southeastern part of Vodrup Flach, e = western part of Vejsnaes Rinne, northern slope, f = outer Flensburger Förde, slope southeast of Als. For locations see Fig. 5-1.

It is much more difficult, however, to understand quantitatively the patterns caused by the various processes of regional grain size differentiation. Some of the problems related to the corresponding sediment distribution patterns will be discussed in more detail in section 5.4.

171

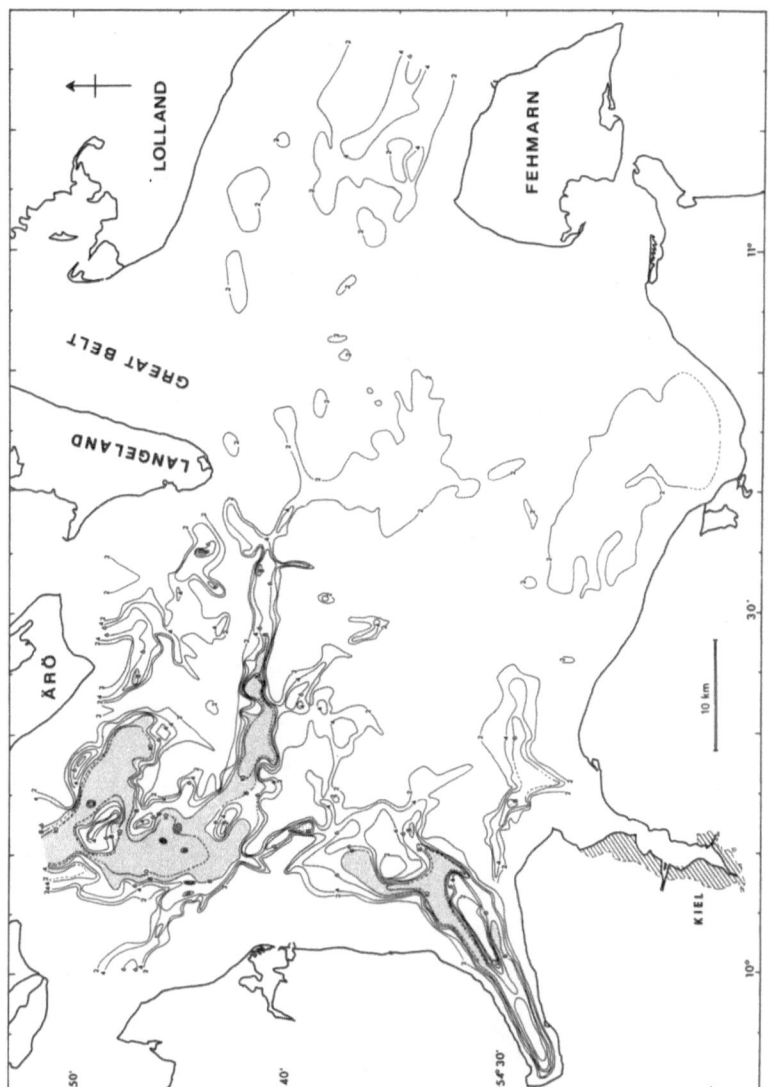

Fig. 5-5: Isopach map of Holocene (mud) sediments, Kiel Bight constructed from 18 kHz sediment echo-sounder data (courtesy F.-C. Kögler, Kiel, unpubl.) by HEALY and WERNER (in press). Contours in meters (2 m interval).

5.2.5   <u>Summary</u>

The main geological conditions for determining the types of Holocene sediments and their areal distribution in Kiel Bight are:

- glacial origin of a basin-, channel- and shoal-physiography and of the source sediment, providing a wide spectrum of grain sizes;
- rapid postglacial sea level rise in the larger part of the area;
- sediment supply by coastal cliff retreat varying in space and time and by submarine wave abrasion;
- lack of fluviatile and eolian input.

Sediment distribution is mainly controlled by wave action in the shallow parts and by bottom currents in the basins and channels.

5.3   EFFECTS OF HYDRODYNAMIC CONDITIONS

5.3.1   The Relation of Wind Stress to Bottom Currents in the Western Baltic Sea

It is well known that a correlation exists between the large-scale wind field and surface currents in the Great Belt (DIETRICH 1951). WEIDEMANN (1950) illustrates this
situation for the Baltic Sea in general:

High air pressure above Scandinavia induces easterly winds (Fig. 5-6). Thereby water
is piled up in the Western Baltic and driven out of the Kattegat. The resulting sea
level difference causes outflow through the Belt Sea. Westerly winds produce the
opposite effect. Due to the resulting sea level difference, water from the Kattegat
flows through the Belt Sea into the central Baltic until a dynamic balance between
sea level difference and currents is attained.

The fluctuation of bottom currents in the Western Baltic Sea also proved to be
related to atmospheric changes. The observational programme for studying the effect
of strong near-bottom currents on the sea bed therefore had to be extended over a
considerable period of time in order to catch a sufficient number of significant
events. Long-term current measurements were carried out during a total of 2.5 years in
the Vejsnaes Rinne (Fig. 5-1). During the same time period long-term current measurements were also carried out in the Danish Straits (JAKOBSEN 1980). The mooring at
30 m water depth was equipped with four Aanderaa current meters installed at 12, 17,
22 and 27 m. The second mooring at 32 m water depth with current meters at 12, 17, 24
and 27 m water depths. For more details and instrument accuracy problems see WITTSTOCK
(1982). The Vejsnaes Rinne was chosen because it offers a favorable site for monitoring
the water flowing from the Great Belt into Kiel Bight (HATJE 1976 and 1977, WITTSTOCK
et al. 1978). For the same reason a case study for channel sedimentation had been
carried out in Kiel Bight previously (see 5.4).

A frequency diagram of velocities and directions from our current measurements
(Fig. 5-7) shows that the highest bottom current velocities (up to 40 cm s[-1]) occur
with inflow events. Inflow frequency, too, is significantly higher at the bottom than
that of outflow.

Therefore, the inflow events must be regarded to have the most important impact on
the sea bed in the Vejsnaes Rinne and channel system of the Kiel Bight. In the
following, the mechanism of their generation in dependence on the wind-stress field
is examined.

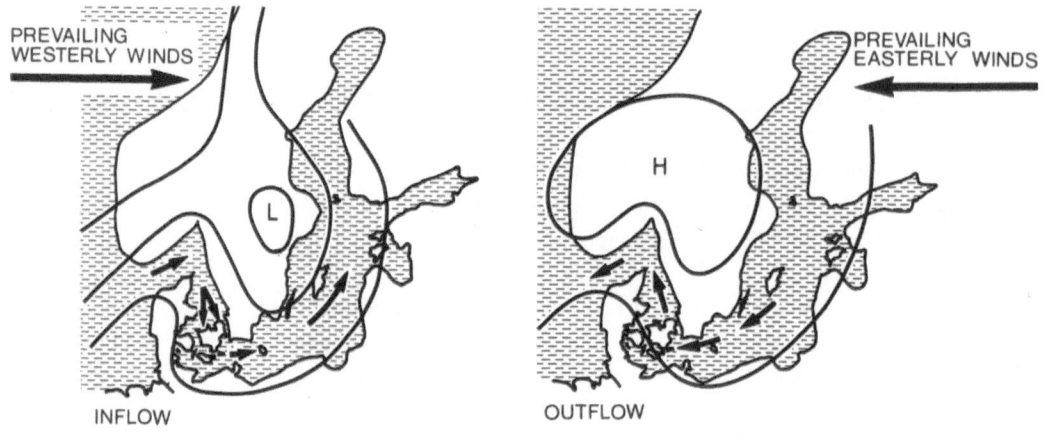

Fig. 5-6: Dependence of inflow and outflow situations on atmospheric circulation (after WEIDEMANN 1950).

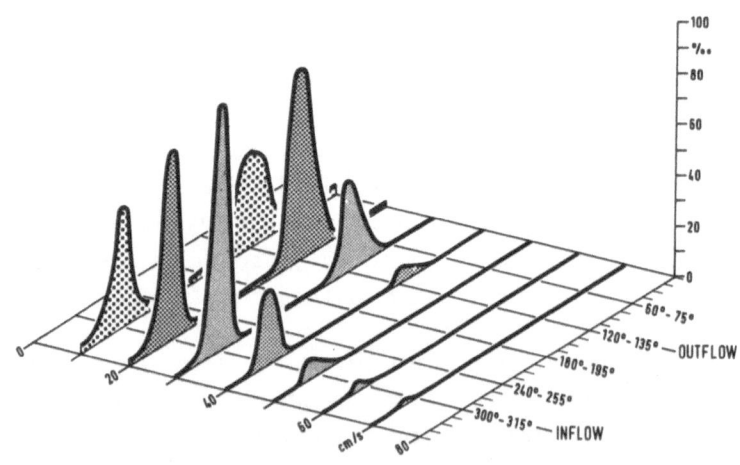

Fig. 5-7: Frequencies of near-bottom currents in Vejsnaes Rinne from long-term measurements (27 m water depth, 104, 403 values). (From WITTSTOCK 1982).

On time scales exceeding 1 day, long-term fluctuations of the near-bottom currents in the Vejsnaes Rinne are well correlated with the wind field, similar to the relations found in the Great Belt by JAKOBSEN (1979). This is well illustrated by a comparison between low-pass filtered time series of the current velocity in 27 m water depth (3 m above bottom) and the zonal wind stress calculated from Danish weather stations (Fig. 5-8). The wind stress represents changes in the large-scale wind field, because local effects on time scales less than 1 day are filtered out. Fig. 5-9 shows the correlation of both variables of this time series in a scatter plot with a correlation coefficient of r = + 0.66.

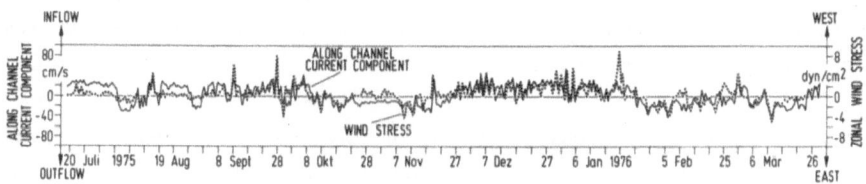

Fig. 5-8: Comparison between along-channel current component (27 m water depth) and the zonal wind-stress fluctuations calculated from Danish weather stations (low-pass filtered 24 hrs.). (From WITTSTOCK 1982).

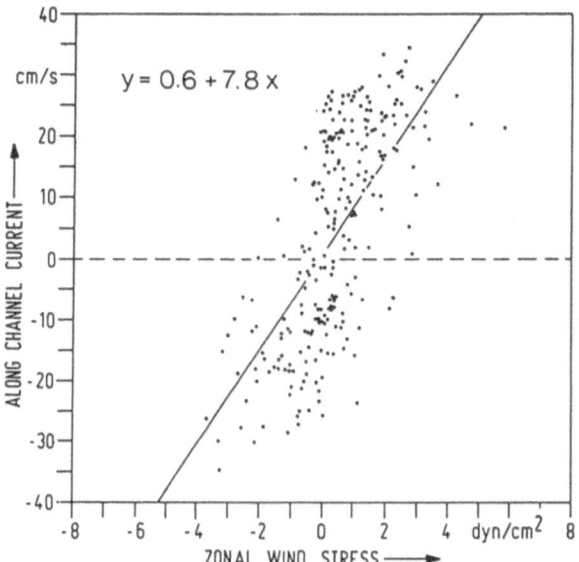

Fig. 5-9: Scatter plot of the along-channel current component and the zonal wind-stress fluctuations on the basis of daily mean values; (Data from time series of Fig. 5-8). (From WITTSTOCK 1982).

A vector diagram of band-pass filtered values (Fig. 5-11) illustrates near-bottom current fluctuations on time scales between 5.5 and 30.0 days in the channel system. The comparison of this diagram with wind-stress vector diagrams suggests a quasi-periodic behavior of both the inflow-outflow fluctuations and the atmospherical changes whose "periods" would be in an order of 11 days on average. A simple resonator model for the Baltic Sea explains this behaviour of the fluctuations as an eigen-period for an oscillation system comprising the open Baltic on the one hand, and the Belts (Fehmarn Belt, Great Belt) which are modelled as branched channels (SVANSSON 1980, WITTSTOCK 1982) on the other. This oscillation is stimulated by sea level changes in the northern Kattegat, generated by atmospheric circulation. The excellent coherence of these systems is documented in Fig. 5-10, using the zonal wind-stress component and sea level fluctuations from the stations at Fredrikshavn and Hirtshals (northern Denmark).

The significance of this resonance system in relation to the absolute values of wind stress as causes for the current velocities in the Vejsnaes Rinne is elucidated by two extreme current events caught in the long-term measurements (Fig. 5-12a and b). In the early December event several velocity peaks are recorded without being connected with stronger westerly winds (Fig. 5-13). In contrast to this, the January event takes place during a severe westerly storm. The comparison between the zonal wind-stress component and the sea level data from the station Strande (Fig. 5-13) gives evidence that Baltic Sea seiches are strongly stimulated in both cases. Kiel Bight is "filled" and "emptied" by these forced oscillations. Results obtained with a linear model for Kiel Bight show that events with higher velocities in the north-eastern part of the channel system are oscillation effects, if a strong forcing of the eigen-period of the Baltic Sea takes place (WITTSTOCK 1982).

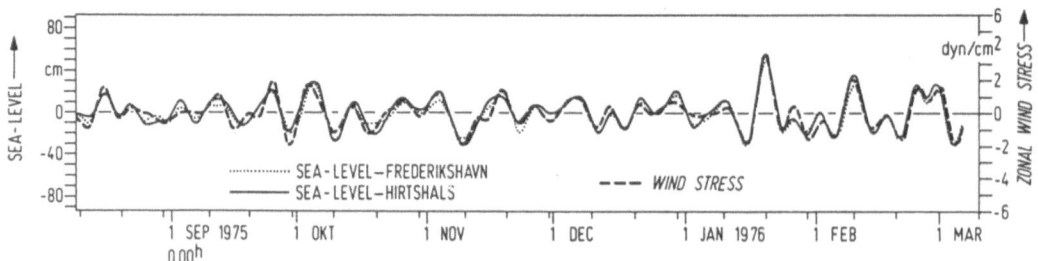

Fig. 5-10: Comparison between the band-pass filtered zonal wind stress and the sea level fluctuations in Frederikshavn (Kattegat) and Hirtshals (Skagerrak). (From WITTSTOCK 1982).

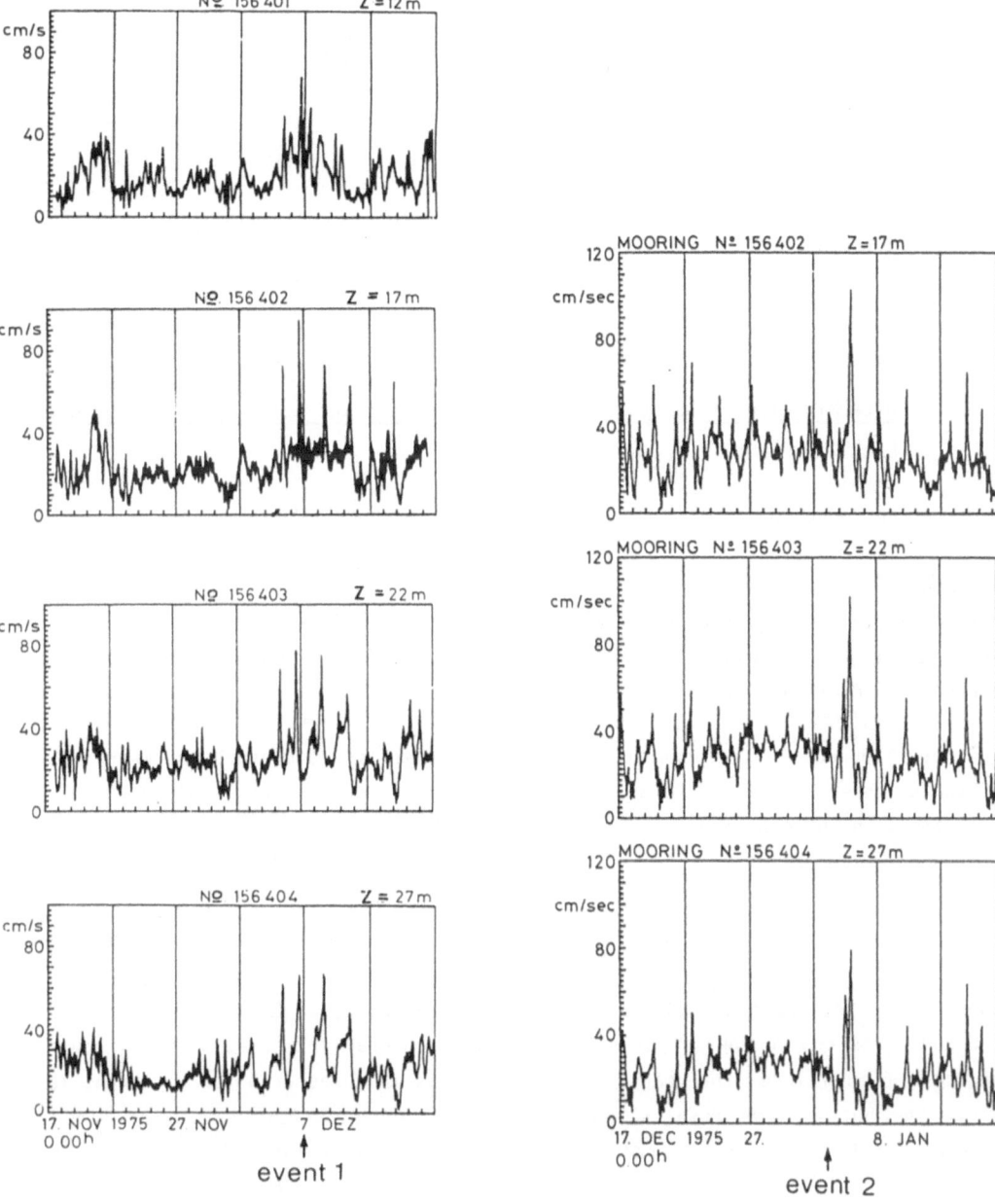

Fig. 5-11: Sections of long-term current measurements in different water depths at the Vejsnaes Rinne site showing event 1 (Dec. 7, 1975) and event 2 (Jan. 4, 1976). (From WITTSTOCK 1982).

178

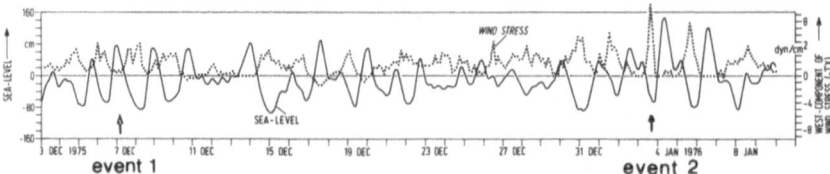

Fig. 5-12: Comparison of zonal wind stress and sea level fluctuations at station
Strande, Kieler Förde showing events 1 and 2 (compare Fig. 5-11). (From
WITTSTOCK 1982).

5.3.2   Determination of Bottom Shear Stress from Current Velocity and Turbulence
Measurements

5.3.2.1 Introduction

Sediment transport involves a complex interaction between the flow and the bottom.
Ultimately an accurate knowledge of the forces acting on the bottom (i.e. the boundary
shear stress) is necessary in order to predict the initiation of sediment movement
and subsequent transport rates.

The normal approach made popular by STERNBERG (1968) is to measure the mean velocity
at two or three levels and to assume a logarithmic velocity profile yielding the mean
velocity with the equation

$$u = \frac{u_*}{k} \ln \frac{z}{z_0} \qquad (1)$$

where u_* is the shear velocity, k is Karman's constant, z is the distance from the
bed and z_0 is the roughness parameter, which in general increases with the roughness
of the bed. Therefore, the mean current speed u is measured at a few different levels
above the bottom, the results are plotted semi-logarithmically and the slope and
intercept determined of a straight line fitted to the points. This determines the
shear velocity and roughness parameter, respectively. This method has been used by
many investigators (e.g. STERNBERG 1968, WEATHERLY 1972, HARVEY and VINCENT 1977,
LESHT 1979) to estimate sediment transport conditions in various locations. The
results of this type of analysis by many authors yield a very large scatter in both

179

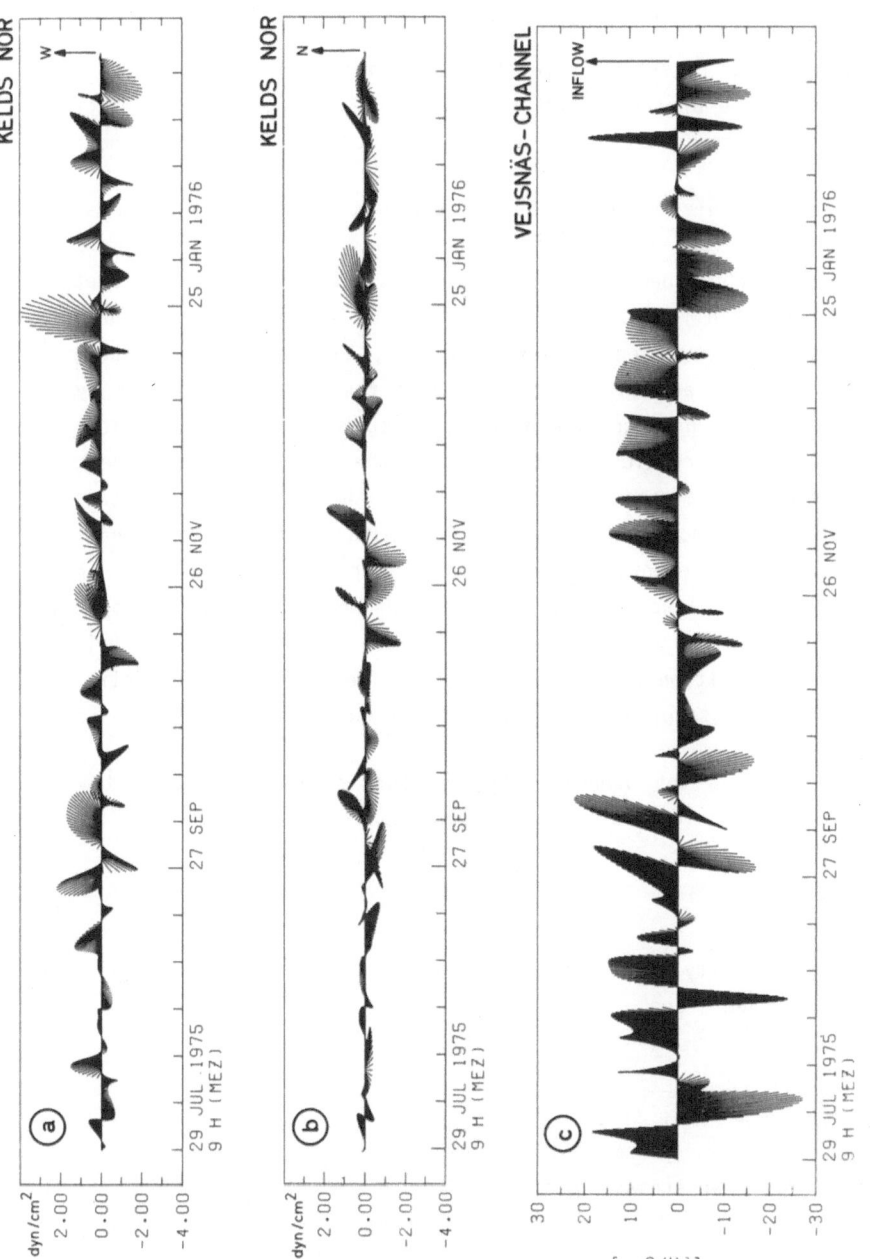

Fig. 5-13: Vector diagram of wind stresses, (a) E-W components, (b) N-S components, and (c) band-pass filtered near-bottom currents (Vejsnaes Rinne, 27 m water depth, 3 m above sea bed). Note that meridional wind stresses are smaller than zonal ones on time scales between 5.5 and 30.0 days. (WITTSTOCK 1982).

$u\ast$ and z_0. This scatter can arise from a number of sources, such as unsteadiness, variable bottom roughness, noise in velocity measurements and sediment movement itself. An additional source of error that will be discussed in more detail later is an inaccurate knowledge of the position of the measurements with respect to the bed.

Another source of error involves the confusion between total boundary shear stress and "skin friction". This skin friction is that part of the total boundary shear stress that acts on a scale of the order of a few tens of sediment particle diameters. This distinction between skin friction and total boundary shear stress is especially important where large topographic scales such as ripples, dunes, and channel curvature are present. In such regions a significant portion of the stress is carried by the form drag which arises from a pressure field that is distributed so as to yield a net downstream force on the topographic feature. Because the changes in pressure occur over lengths much larger than a sand grain, this part of the total boundary stress is ineffective in moving sediment. That is to say, for instance, for open channel flow where ripples or dunes are present, the stress calculated from the product of the depth times the free surface slope would indicate a much larger sediment-transporting potential than would actually be observed. Furthermore, the local skin friction varies considerably with position over such bed forms. The effects of non-uniformity for steady flow have been considered in detail by SMITH and McLEAN (1977a).

Another method of estimating the boundary shear stress is to measure directly the Reynolds stress. If each instantaneous velocity component is defined as

$$u_i = U_i + u'_i \tag{2}$$

where U_i is the mean velocity component and u'_i is the fluctuating component. Then the Reynolds stress components are

$$\tau_{ij} = - \rho \overline{u_i u_j} \tag{3}$$

with τ_{ij} being the bed shear-stress, ρ the density of the fluid, and the overbar indicating the mean value. The term "mean" should properly be an ensemble average; however, in practice a time average is usually taken. SOULSBY (1981) indicates that an averaging time for tidal flows should fall in the 8 to 12 minute range. For times shorter than this the statistical accuracy is degraded and for longer times the non-stationarity becomes important.

The measurement of Reynolds stress requires accurate sampling of the velocity at high frequency (greater than 3 Hz) to avoid significant energy losses especially near the bed. Several investigators have made such measurements in tidal environments (see BOWDEN and HOWE 1963, GORDON and DOHNE 1973, HEATHERSHAW 1976, SOULSBY 1981); however, these were made usually only at one or two levels and ordinarily not in conjunction with a dense vertical sampling of the mean flow. Also, accurate information on the bottom topography was usually not gathered. In that regard it should be noted again that the total boundary shear stress is comprised of form drag and skin friction. Depending on where the Reynolds shear stress measurements are made in relation to any bed forms present, they may or may not include form drag as well as skin friction. For example, very near the bed the Reynolds stress may reflect the skin friction only, while further above the bottom the turbulent stress might include the skin friction plus the form drag due to ripples or mega-ripples. Thus, whereas for channel flow the shear stress tends to be a maximum at the bottom and fall off linearly with distance from the bed, in the presence of bed forms the Reynolds stress can increase, rather dramatically, over relatively short distances near the bed (see SMITH and McLEAN 1977a). Thus much care must be taken in attempting to interpret such measurements. The measurements presented in section 5.3.2.2 are a small step in improving the above situation. A complete sampling scheme would measure the three-dimensional velocity and turbulent stress fields as they vary in time and space; however, this is beyond current technology. What was accomplished was measurement of the full velocity vector and stress tensor at several levels, as well as the streamwise one-dimensional bottom profile. While this perhaps pales next to what is needed, it represents probably the most extensive set of measurements yet attempted in such a system.

The turbulence measurements were accomplished through the use of small mechanical current meters, which were developed by J. Dungan Smith, University of Washington, Seattle. The ducted rotor allows measurement of single components of the velocity vector so that three, orthogonally mounted sensors yield accurate measurements of the full velocity vector. Furthermore the 3 Hz sampling rate of the data acquisition system allows the measurement of the instantaneous velocity (BEHRENDS 1981). Fifteen such sensors typically were mounted on a 2.5 meter high frame that was then set on the bottom. The frame is designed to orient itself so that the mean current comes from head-on. Near the bottom (at approximately 12, 22 and 38 cm) three single current meters were mounted so as to measure the head-on velocity component. The other sensors were mounted in groups of three (triplets) that were at 60, 100, 160, and 214 cm, respectively. These were oriented so that all three measured a sizeable position of the mean speed. For a more detailed description see McLEAN (1983).

5.3.2.2 Sediment Transport Measurements in the Jade Tidal Inlet

Selection of Measurement Sites

It would be best to investigate the unsteady effects first for uniform conditions,
but unfortunately, the inherent instabilities of movable beds make such an approach
almost impossible if one wishes to deal with areas where sediment is being transported.
Nevertheless, a location was sought where the non-uniformity effects could be mini-
mized. The site finally selected was in the Jade, near Wilhelmshaven (see Fig. 5-14
and 5-15). This is a tidal inlet of approximately 20 m depth and tidal amplitude of

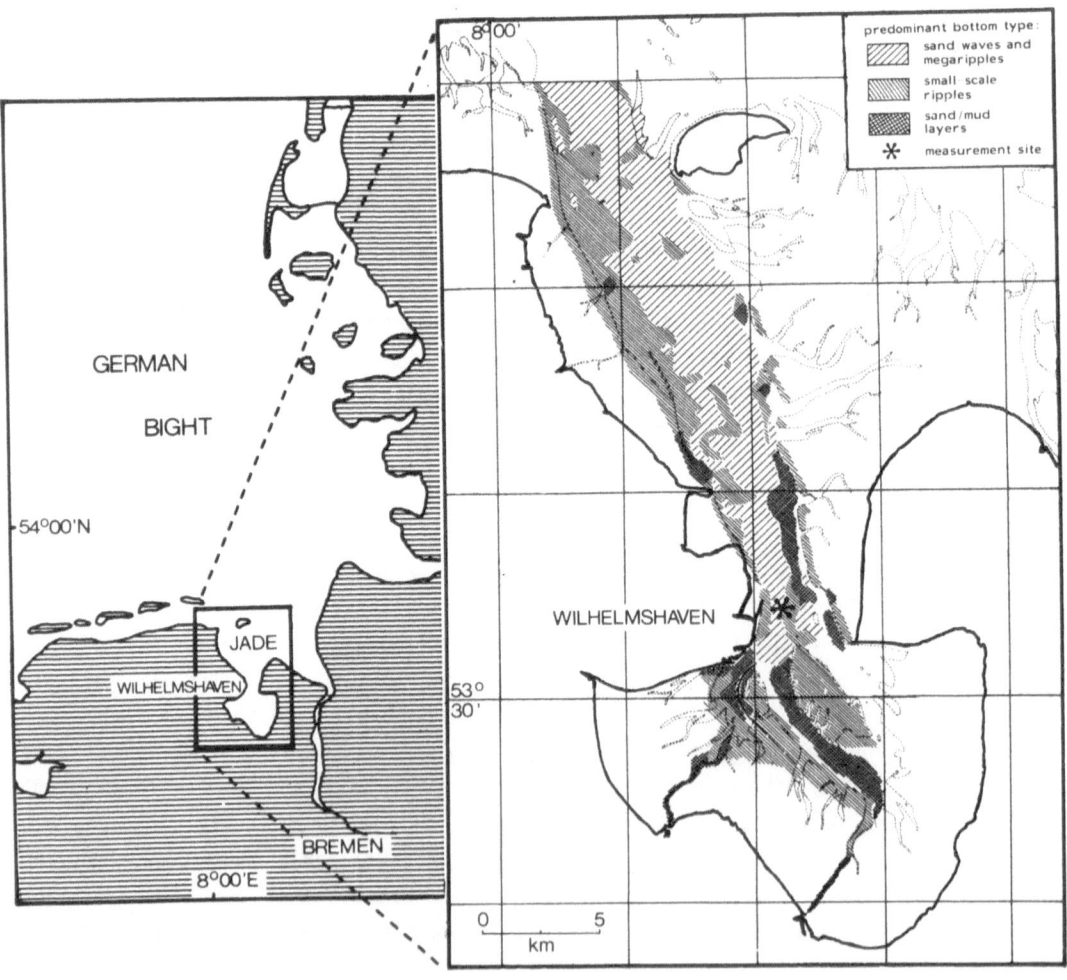

Fig. 5-14: Experiment site (PMA and TMA) in the Jade (German Bight, North Sea).
Sediment distribution map from DÖRJES et al. 1968.

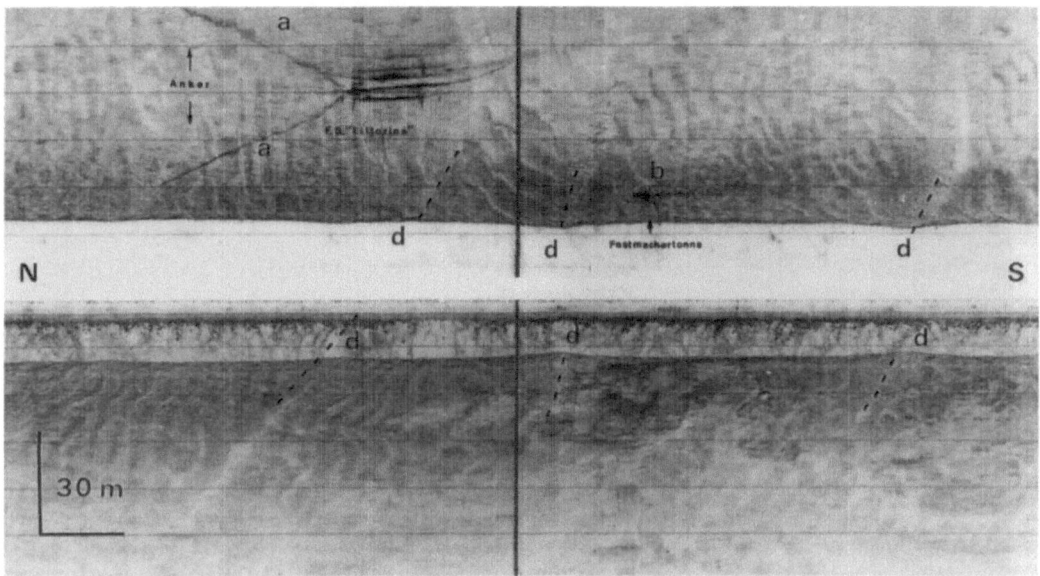

Fig. 5-15: Side-scan sonar record of measurement site. A large sand wave field was
located in the immediate vicinity south of this site. a = anchor chain of
F.S. "Littorina"; b = buoy; d = lee side of sand waves.

about 4 m. Under spring tide conditions surface currents of over 3 knots were
observed. Because the fresh water input to the inlet was negligible and because of an
extensive tidal flat, the water was very well mixed and neutrally stratified.

The site was near a field of large dunes or sand waves, 50 - 100 m long and 1 - 2 m
high but where the measurements were actually made, only smaller mega-ripples, 2 - 4 m
long and 10 - 20 cm high were present (s. Fig. 5-16). Small ripples 1 - 2 cm high and
10 - 20 cm long generally were also present. At these scales, calculations according
to SMITH and McLEAN (1977a) indicated that over most of the vertical sampling region,
the bed forms only would appear as roughness elements (i.e. variations over the wave-
forms would be negligible). Therefore it was assumed that the instrument frame could
be raised and lowered repeatedly without regard to horizontal position. Had large bed
forms been present, accurate positioning, as well as an order of magnitude more measurements would have been necessary to achieve the same results.

The dunes and mega-ripples were comprised of medium sand with a significant fine sand
fraction, but also with grain sizes to over 1 mm. It is possible that the large dunes
were not present at the measurement site because of a relative paucity of sand and
that a more erosion resistant mud bottom was in fact sometimes exposed between mega-
ripple crests. This was difficult to confirm due to the difficulty in obtaining undisturbed and accurately positioned bottom samples. Strong currents and extremely
turbid water made diving very difficult.

184

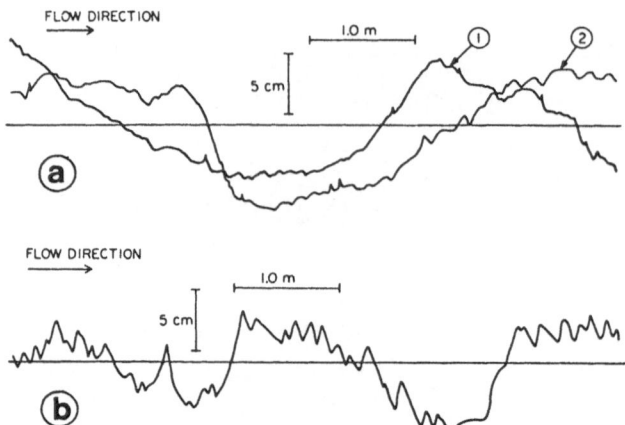

Fig. 5-16: Bottom profiles. (a) Two bottom profiles taken at a time interval of approximately 2 hours. Note the absence of ripples on the mega-ripple in profile 1. This is approximately mid-cycle when current velocities are greatest. Later the ripples reappear as demonstrated in profile 2. (b) Bottom profile at the beginning of the tidal cycle. Note that the steep slopes of both mega-ripples and small ripples face upstream. This causes a relatively high roughness in the early phase of the tide. As the waves re-form, the roughness decreases. If the flow becomes sufficiently large, the ripples even disappear. (From McLEAN 1983).

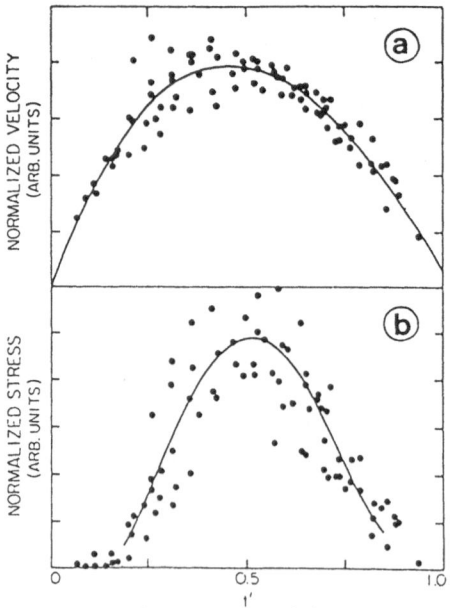

Fig. 5-17: Velocity (a) and Reynolds stress (b) versus non-dimensional time. Here t = 0 and t = 1 represent times of maxima or minima in the water surface elevation. The data have been normalized by the tidal amplitude and the period; so the ordinate is in arbitrary units. Note that the stress (b) lags the mean velocity (a). This is due to the lack of equilibrium between the mean flow and the turbulence field. (From McLEAN 1983).

Results

In the Jade it was found that the tidal cycles were almost symmetrical with variations
due to even small weather disturbances overshadowing any perceptible difference in
ebb and flood. Also, the fact that the large sand wave field was only a few tens of
meters upstream of the measurement site for ebb flow (downstream for flood) did not
seem to cause any noticeable difference in the boundary layer. Much more extensive
measurements would have been necessary to determine reliably the ebb-flood asymmetry.
During each half-cycle the current strength varied with the tidal amplitude, of course;
but the variation with time was similar from cycle to cycle. In order to exploit this
quasi-homogeneity between half-cycles, the measured velocities were normalized by the
tidal amplitude divided by the time between successive depth maxima or minima, and
the stresses were normalized by the square of this value. Fig. 5-17 shows the normalized velocity and Reynolds shear stress at 214 cm as a function of the nondimensional time t', where $t' = 0$ and $t' = 1$ are times of depth maximum or minimum. The
data points here are from the first cruise when the velocity and stress measurements
were most carefully done in the absence of any flow disturbance. The mean maximum
velocity for these 8 half-cycles was about 65 s^{-1} and the mean stress was about
16 dynes cm^{-2}.

The velocity is seen to increase fairly rapidly to a broad maximum that occurs somewhat before the mid-point in the cycle. The stress, on the other hand, has a sharp
maximum that occurs somewhat later than that of the velocity. The lines shown in
Fig. 5-17 are simply least-squares fits in order to identify the maxima more
objectively. The time lag of the maximum stress against the maximum velocity is due
to two processes. First of all, early in the cycle the boundary layer is just
beginning to form. It is thin and does not even extend to 214 cm. This is plainly
seen in Fig. 5-18a which plots the instantaneous values of $u'w'$ as a function of
time for all four triplets. The record from 214 cm is clearly much more tranquil than
the records from below, i.e. the boundary layer, and the turbulence that accompanies
it, has not yet grown to that level. Only occasional eddies pass through in response
to fluctuations in the boundary layer thickness. This allows a much higher shear in
the flow outside the boundary layer where the turbulent transfer of momentum is small.
In Fig. 5-18b, which shows a situation several minutes later, the fluctuations are
still less at 214 cm than below but the turbulence intensities are much more uniform.
Indeed, further getting into the cycle the records become indistinguishable. A second
reason for the lag is that the production of turbulent energy occurs on scales of the
order of the depth (especially towards the middle of the cycle) and as the velocity
increases it simply takes some time for the turbulence (and hence the stress) to
adjust. This is also true of the deceleration phase. Once the turbulence has been
produced, time is required to dissipate it.

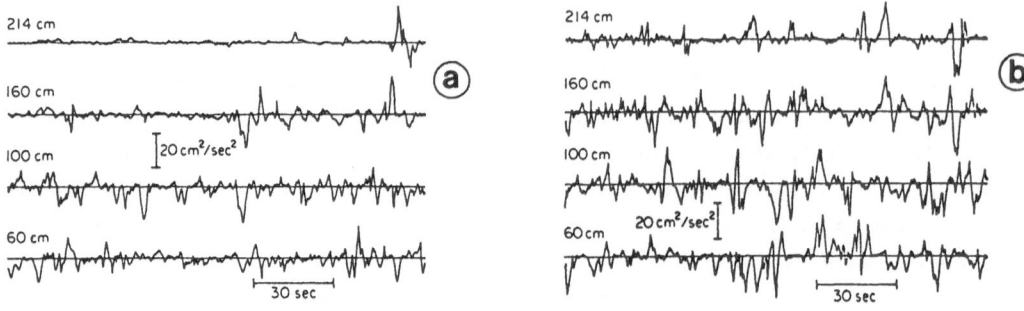

Fig. 5-18: Time series of instantaneous velocity correlation, u'w' at four different levels during two different time periods. The low turbulence intensity at 214 cm level in the first set (a) is due to the fact that the boundary layer has not yet grown to this height yet. The second set (b), a few minutes later in the cycle shows that the boundary layer has grown sufficiently to cause significantly more turbulence at 214 cm. (From McLEAN 1983).

As mentioned above, it is common to estimate the stress from the slope of the velocity profile plotted semi-logarithmically. However, it is evident that an error in the position of the current meters causes a significant error in the friction velocity and roughness parameter determination. For example, with $u* = 4$ cm s^{-1} and $z_0 = 1$ cm, values not unlike those found in the Jade, the shear velocity calculated from current measurements at 25 and 100 cm from the bottom would be in error by 10 % for a vertical positioning error of 5 cm. The error would be more than 30 % if the lower measurement were at 10 cm. It is tempting to say that an error of 5 cm is a bit too high; however, the position of the "mean" bottom relative to the current meters for the last cruise was found to vary over a range of 17 cm. These effects are readily apparent in Fig. 5-19. Here the normalized velocities have been grouped according to phase of the tidal cycle and all the data have been plotted semi-logarithmically and a linear fit has been made. The data points pictured have been corrected using the mean distance from the echo-sounder track to the bottom over the entire length of 6 m. The dashed line is the fit assuming that the bottom was hard and flat. The shear velocities are shown below each profile. Since the tendency is for the frame to settle into the sediment, the uncorrected shear stress (and bottom roughness) would appear much larger than they actually were. The fixed echo-sounder of the earlier experiments accurately reflected the distance at one point but the flow integrates somewhat over the topography so that the velocity profile refers to some spatially averaged surface. In the presence of mega-ripples that were 10 - 20 cm high this mean level could deviate from that measured directly beneath the current meters by up to 10 cm.

In addition, errors in the velocity measurements also cause sizeable errors. For instance, with $u_* = 4$ cm s^{-1}, $z = 1$ cm and current measurements accurately positioned at 25 and 100 cm, an error of 2 cm $s^{-1} \triangleq$ relative error of 4 % produces a relative error in u_* of 30 %. That magnitude of error is quite common due to statistical and instrument error; therefore it is evident that more than two measurements are absolutely imperative. Also, deviations from logarithmic behaviour introduce error, and this can only be detected by a relatively dense vertical array of current meters. With the possibility of instrument failures this means that velocity measurements at 5 or 6 levels at least should be made.

Fig. 5-19 shows a definite tendency for the roughness parameter to reach a minimum near the middle of the tidal cycle. SMITH and McLEAN (1977a) used an energy argument to postulate that the roughness in a sediment transporting flow should increase with increasing boundary shear stress. The sediment extracts energy - and hence momentum - from the fluid, which causes an apparent shift in the velocity profile, thus increasing the intercept value. In essence it is assumed that the higher the boundary shear stress, the more sediment is in motion and the more momentum is extracted. This hypothesis taken by itself would predict high roughness at mid-cycle and low near the slack water times. In general, the individual velocity profiles were quite smooth and fairly logarithmic; however, if one carefully inspects Fig. 5-19 one sees, especially near the beginning of the cycle, a distinct tendency for the shear at the higher levels to be larger than expected. This profile structure is due to the developing boundary layer already discussed and would cause shear velocities and roughness parameters from the simple logarithmic fit to be over-estimated. In order to take this into account the fit was repeated using only the velocities 100 cm and closer to the bed. Although the variation in z_0 was reduced the minimum near the middle of the tidal cycle still was evident.

The explanation of this phenomenon lies likely in the bottom topography itself. At the beginnung of the tidal cycle, the mega-ripples and small ripples are oriented so that their steep faces look upstream (Fig. 5-16). This configuration causes the bottom to appear rougher than when bed forms of the same geometric scale are oriented in the more usual direction (i.e. steep slope facing downstream). This accounts for the large z_0 near the beginning of the cycle. As the sediment begins to move, the shape changes gradually to a more streamlined one (Fig. 5-16a, profile 2). Also, as Fig. 5-16 indicates, the small ripples tend to disappear when the velocities are largest. It is likely that the small ripples are for the most part comprised of fine sand that is carried into suspension under high flows. By about mid-cycle the mega-ripples have reformed into the more common configuration and shortly thereafter the ripples begin to reappear (Fig. 5-16a, profile 2). These ripples cause the roughness to increase as is evidenced in Fig. 5-19.

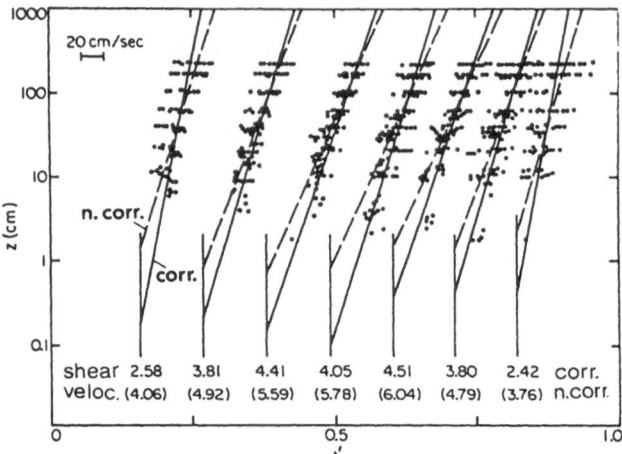

Fig. 5-19: Velocity profiles as a function of time within the tidal cycle. Individual mean velocity profiles, grouped according to phase of the tidal cycle, are plotted semi-logarithmically and displaced so as to make the differences obvious. The solid lines are least-squares fits to the data with the vertical positions corrected using the spatially averaged bottom topography. The dashed lines are the fits without correcting. The numbers below are the shear velocity in cm s^{-1}. It is evident that not correcting the profiles yields much larger stress estimates. (From McLEAN 1983).

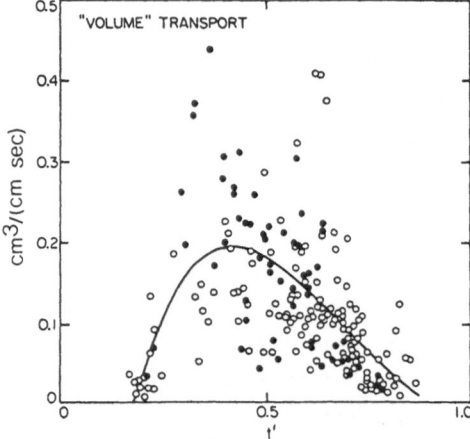

Fig. 5-20: Sediment transport rates versus non-dimensional time. The dots indicate rates estimated by calculating the average of areas of erosion and deposition as indicated by bottom profiles taken at time intervals of the order of an hour. This method measures transport associated with the mega-ripples. The circles on the other hand represent rates calculated by tracking ripple crests. The solid line is a least-squares fit to the data and is meant only to emphasize the fact that the maximum in the transport occurs before mid-cycle, when the velocity, rather than the stress, is maximum. (From McLEAN 1983).

Transport rates as measured by the two different methods (McLEAN 1983) are plotted in Fig. 5-20 versus nondimensional time. Although there is considerable noise in the measurements, the maximum rates clearly occur before the middle of the cycle, at about the time of maximum velocity and significantly before the maximum in the stress. This underlines the importance of differentiating between skin friction and form drag. The Reynolds stress at 214 cm shown in Fig. 5-21 represents the total boundary shear stress arising from skin friction plus the form drag from both the ripples and mega-ripples. This turbulent stress also represents an average over several square meters of the bed. The local skin friction and hence the local bed-load transport rate can vary considerably from this value. When the mega-ripple faces upstream early in the cycle, the steep slope generates a large local skin friction near the crest, much higher than would be present were the sand wave oriented in the usual direction. Thus, though the mean skin friction may be even less than later in the cycle, the maximum transport rates are highest before mid-cycle. It should be noted that this high rate of sediment movement serves only to redistribute the sand (i.e. reshape the bed form) and that when averaged over both ebb and flood the net movement is negligible. This confirms what one would expect; that is, that if significant net transport is to occur it will likely be carried as suspended load.

For a more detailed discussion of these results in relation to boundary layer theories, see McLEAN (1983).

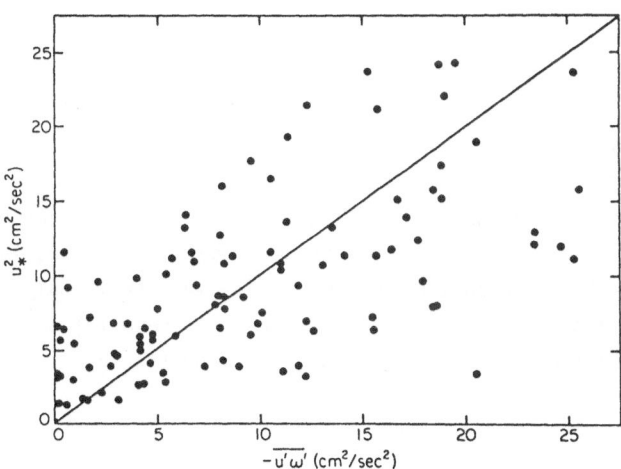

Fig. 5-21: Reynolds shear stress versus the square of the bottom shear velocity as calculated from a least-squares fit to the mean velocity profiles. The large scatter is due to instrument error, inaccurate positioning, and non-logarithmic velocity profiles. This shows that the velocity slope method of estimating stress must be used with caution. (From McLEAN 1983).

Discussion

One of the goals of this study was to investigate the advisability of using the so-called slope method for estimating boundary shear stress in a tidal environment. In Fig. 5-22 the square of the shear velocity (taken from a least-squares fit to the many mean flow profiles) is plotted versus the measured Reynolds shear stress. The scatter here is tremendous, although the trend is certainly obvious. The vertical positions in these data have been corrected using the fixed echo-sounder. Without such information, the scatter would likely be even larger. Of course the conditions under which these measurements were made created a great deal of uncertainty as far as positioning is concerned. If the bottom had been more stable (i.e., where sediment transport is negligible) then variations would be considerably less. On the other hand, these slopes were calculated from velocity measurements at 5 to 7 different levels. This reduces the errors significantly in comparison to measurements made at only two or three levels.

As far as the transport of sediment is concerned, in the Jade there appeared to be relatively little net movement of the sediment in the form of bed load. Usually a mega-ripple migrated less than one wavelength per half cycle and most of the sediment movement caused a change of shape only. The maximum transportation occurred before the middle of the cycle, ahead of the maximum in stress. Also there appeared to be a thin veneer of finer sand that was thrown into suspension during the maximum velocities but returned to the bed during weak flow and subsequently formed ripples that migrated over the large forms.

Suspended sediment can have a marked effect on the velocity profile (see SMITH and McLEAN 1977); however, it is likely that at the concentrations found in the Jade, it was not necessary to include this effect. Nevertheless the velocities were high enough for a significant amount of sediment to be carried over a considerable distance during the span of a high-cycle. Because measurements of suspended sediment were not made with sufficient frequency or accuracy, it is impossible to say whether there was any net flux over an entire tidal cycle. This could only be determined with a much more dense sampling plan, both in space (horizontally) and in time (over both neap and spring tides and with and without storm conditions).

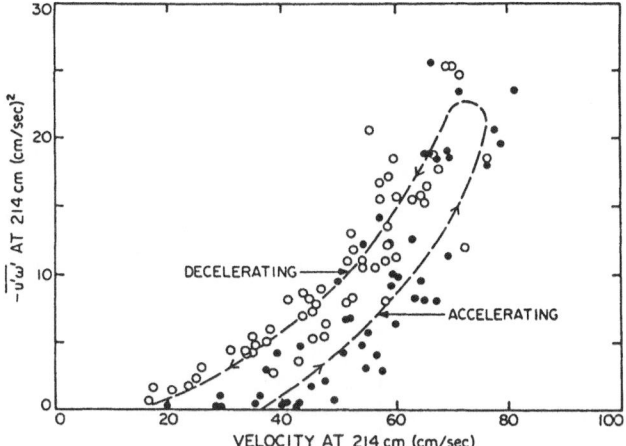

Fig. 5-22: Velocity at 214 cm versus Reynolds stress at the same level. Here the dots are from the accelerating phase of the tide and the circles are from the decelerating phase. There is a clear separation between the two phases suggesting a hysteresis in the mean flow-turbulence system. That is, the turbulence lags the mean flow. (From McLEAN 1983).

5.3.2.3 Long-Term Profile Velocity Measurements

Introduction

In order to estimate the range of the fluctuating bottom currents discussed above (section 5.3.1) and their influence on the sea bed, measurements of the velocity profile in the bottom boundary layer were carried out. For these measurements the profiling device developed for these purposes (PMA) was used. In this device, six Savonius rotors are mounted in the approximately logarithmic bottom distances of 25, 72, 144, 252, and 324 cm on a tripod frame. For more details concerning additional sensors mounted at the frame, calibration and accuracy problems, see SCHAUER (1982). The measuring site in the Vejsnaes Rinne was near to the long-term measurements referred to in section 5.3.1 (Fig. 5-1). As it was considered of major interest to examine the effect of peak current events in respect to sediment erosion, the measurements should comprise at least several weeks. In this section, time series of 14 days in April 1980 and of 10 days in May 1980 are discussed.

Due to technical reasons, it was not possible to use the turbulence measurement system (TMA) at the Vejsnaes Rinne site. For the question of shear stress determination, it was considered to be of general interest, however, to compare simultaneous measurements both on PMA and TMA systems at the same site. This has been carried out in the Jade tidal channel (Fig. 5-15).

192

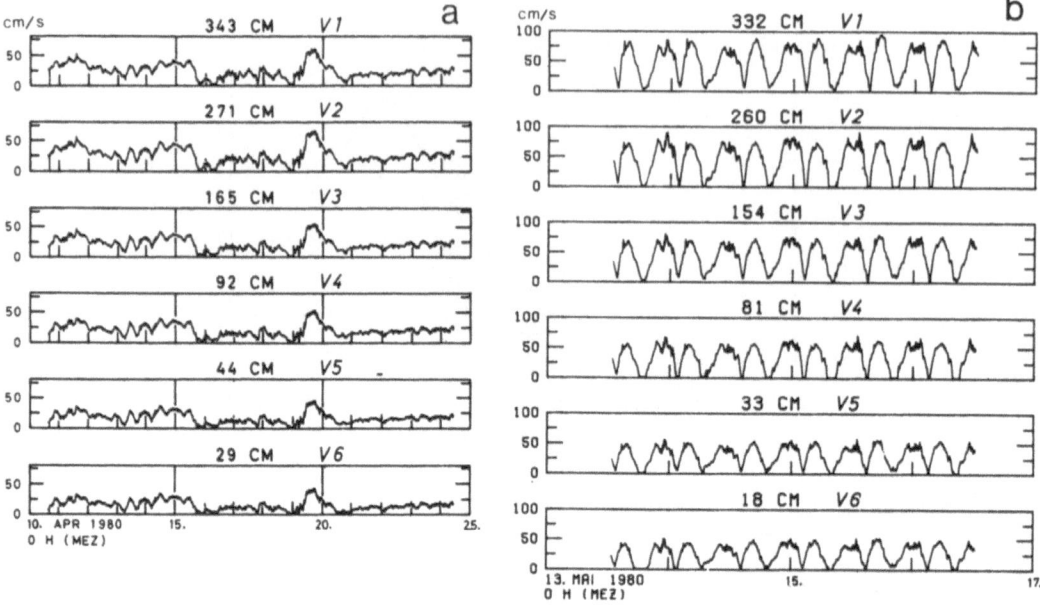

Fig. 5-23: Current measurements with PMA. Section of velocity time series in six
levels in (a) the Vejsnaes Rinne, (b) the Jade. (From SCHAUER 1982).

Results

Fig. 5-23a and b show sections of the velocity-time-series from the Vejsnaes Rinne
and the Jade. From the measured profiles, the friction velocity u* and the roughness
length z_0 have been determined by a fit to equation (1) using the least-squares
method. The resulting time series for these parameters and the correlation coefficient for the fit is shown in Fig. 5-24a and b.   Fig. 5-25 shows sections of
parallel measurements of PMA and TMA.

Applying the usual method of using a "sufficiently" good correlation between velocity
measurements and the logarithmic profile as a criterion for the validity of the law
of the wall (equ. 1) for a single profile, this test would permit here also the
applicability to the majority of our data, as the correlation yields a statistical
significance of 0.95 or even 0.99 (Fig. 5-26a and b). If, however, the entire data
of longer time series are considered with the increased number of degrees of freedom,
the data no longer withstand the criteria of significance. Thus the 10 days-average

193

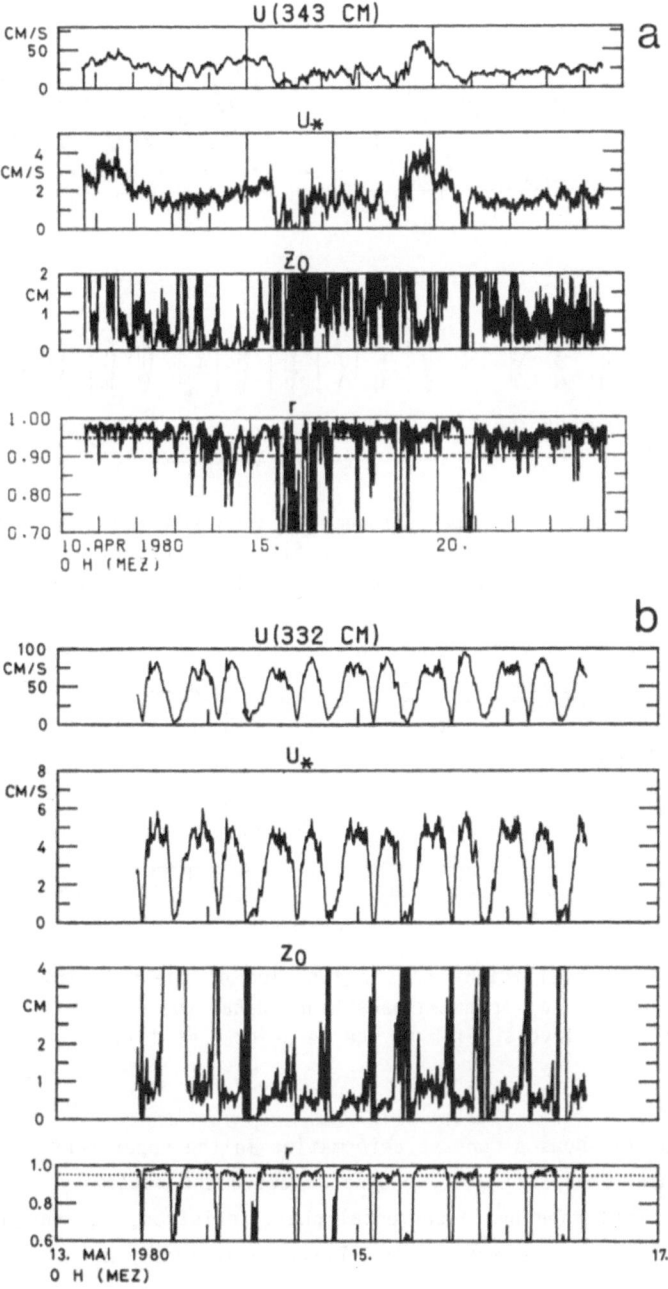

Fig. 5-24: Time series of velocity U, shear velocity u*, roughness length z_0 and the correlation coefficient r for the logarithmical fit according to eq. (1) (a) from data of the Vejsnaes Rinne, (b) the Jade. (From SCHAUER 1982).

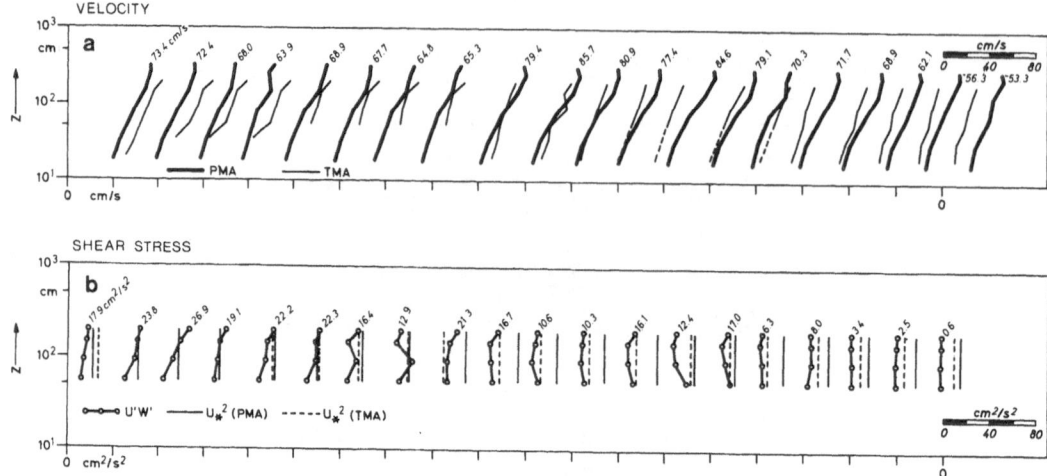

Fig. 5-25: Section of simultaneous measurements from PMA and TMA (a) mean velocity profiles, (b) Reynolds stress profiles from TMA measurements and shear stress values calculated from PMA and TMA measurements using the profile method.

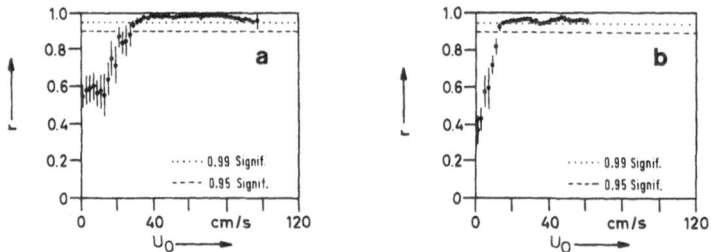

Fig. 5-26: Correlation coefficient r as a function of the velocity in the uppermost PMA-level U_O, (a) from Vejsnaes Rinne data, (b) from Jade data. The dotted significance levels refer to the 64 point 1 ms fit.

profile in Fig. 5-27a shows a concave deformation in the upper part. This corresponds to a shear maximum at approximately one meter above the bottom. A minor part of the deviations may be explained by instrumental characteristics, but the other part must certainly be considered as systematic. These deviations seem to be related to a crucial problem of shear stress determination from velocity profiles. This can be inferred from the enormous scatter of $u*$ and z_O in and between the field data analyses of many investigators as discussed above (see 5.3.2.1). Therefore, it is necessary to give special attention to the deviations found in our data.

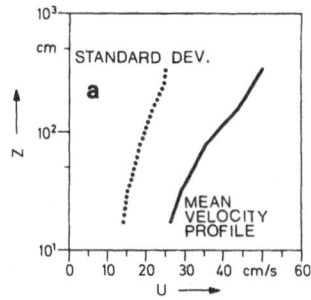 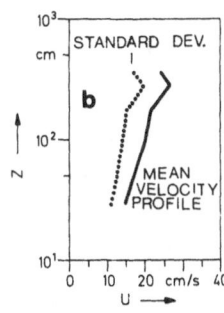

Fig. 5-27: Mean velocity profile and standard deviation from a 14 days average (a) of Vejsnaes Rinne PMA data, (b) 10 days average of Jade PMA data.

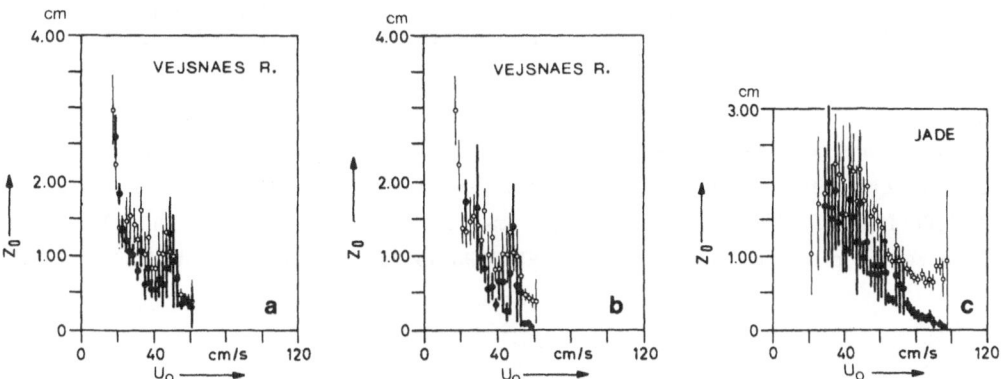

Fig. 5-28: Roughness length z_0 as a function of the velocity in the uppermost PMA level U_0. (a) from Vejsnaes Rinne data, o : z_0 fitted from all 6 levels, ● : z_0 fitted from the lowest 3 levels; (b) from Vejsnaes Rinne data, ● : z_0 fitted from the lowest 2 levels; (c) from Jade data, ● : z_0 fitted from the lowest 3 levels. Bars indicate standard deviation.

In both of our study areas, three types of deviations from the law of the wall and from laboratory results can be identified:

- the roughness length z_0, as determined from assumed logarithmic profile exceeds by several orders of magnitude values expected from the general bottom properties observed in both areas;

- bottom shear velocity u* and roughness length z_0 first increase and then decrease with the height above the bed which is included in the velocity profile used for the determination of u* and z_0. This result signifies the concave character of deviation from the logarithmic profile (Fig. 5-28, Fig. 5-29) in the upper part;

- z_0 decreases with increasing mean velocities by more than one order of magnitude (Fig. 5-28a and c).

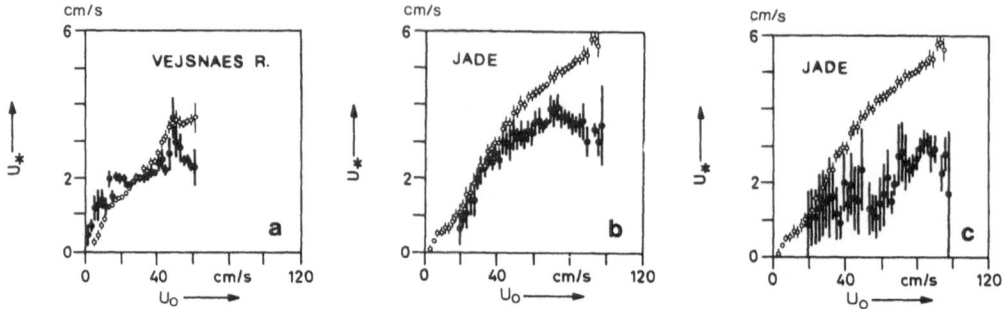

Fig. 5-29: Bottom shear velocity u_* as a function of velocity in the uppermost PMA
level U_0.
(a)-(c),   o: u* fitted from all 6 levels.
(a) from Vejsnaes Rinne data,   •: u* fitted from the lowest 2 levels,
(b) from Jade data,   •: u* fitted from the lowest 3 levels,
(c) from Jade data,   •: u* fitted from the lowest 2 levels.

In fact, z_0 values obtained from ocean floor measurements (STERNBERG 1968) commonly vary
also by several orders of magnitude and do not generally match with Nikuradse roughness
values (SCHLICHTING 1965). Deviations systematically correlated to mean velocities
were observed by other authors as well (STERNBERG 1968, VINCENT and HARVEY 1976).

Discussion

According to experimental and theoretical investigations, three factors are mainly
considered as possible causes of the concavely shaped deviations, namely (1) flow
unsteadiness - according to SOULSBY and DYER (1981) only with decreasing flow velocity - , (2) the hydraulic effects of large-scale bedforms (SMITH 1977), and (3)
density stratification of the fluid due to suspended load (TAYLOR and DYER 1977). In
discussing the effects of these factors, we have the advantage of having a set of
data from different areas which yet showed the same type of deviations. The influencing
factors should therefore be present in both areas. In addition to explaining the
concave profile deformation, they also should be able to account for the variations
observed with different mean velocities. As neither unsteadiness nor bedforms were
observed during the campaign in the Vejsnaes Rinne a significant influence of these
factors must be excluded. Furthermore, a convex deformation as an effect of unsteadiness at accelerating flows postulated by SOULSBY and DYER (1981), could not be
observed with our data.

For testing the influence of a suspension load gradient a model proposed by TAYLOR
and DYER (1977) was adapted. This can explain some of the deviations, but only qualitatively. But even in the case of the Jade observations the concentrations of
suspended matter were not high enough to cause a significant effect.

197

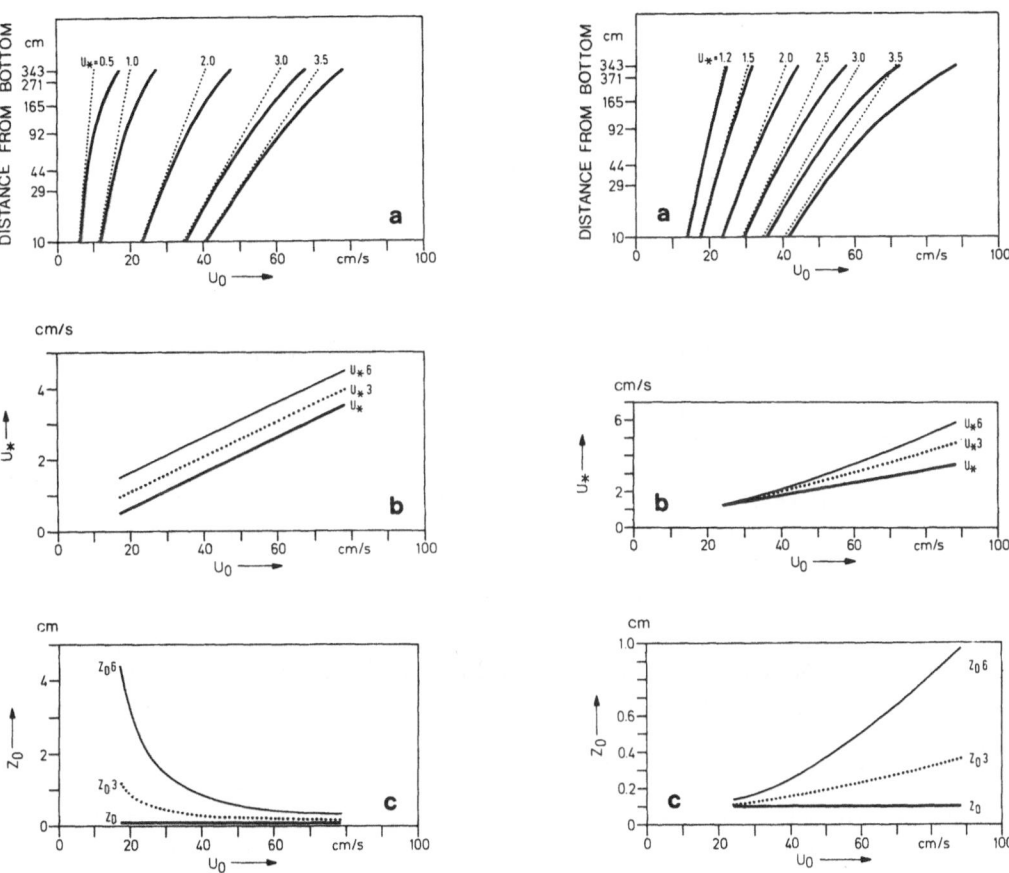

Fig. 5-30:
(a) Development of profile shape with
    velocity, u_0 due to influence of
    suspended matter (dotted: neutral
    profiles).

Fig. 5-31:
(a) Development of profile shape with
    velocity u_0 due to influence of a
    modified mixing concept after
    BLACKADAR (1962).

(b) Bottom shear velocity u* as function of $u_0 \cdot U*6$ and $U*3$ are fitted from 6,
respectively the lowermost 3 levels; u_* = true bottom shear velocity.

(c) Roughness length as function of $u_0 \cdot z_{06}$ and z_{03} are fitted from 6, resp. 3 lowermost levels, z_0 = true roughness length (SCHAUER 1982).

Thus, the factors mentioned do not account for the phenomena observed. Instead, a
comparison with the atmospheric boundary layer offers a promising model. Results from
BLACKADAR (1962) show that even in the case of a neutral, stationary boundary-layer
flow the profile is not really a logarithmic one within a small height above the bottom.
According to this author, instead of a linearly increasing mixing length following the
Prandtl equation ($l = k \cdot z$) the mean turbulence scale will attain a maximum value

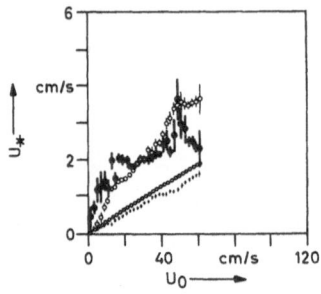

Fig. 5-32: Bottom shear velocity u* as function of velocity u_0 (z = 343 cm) in the
Vejsnaes Rinne. (From SCHAUER 1982).
O : fitted from six levels
● : fitted from the two lowermost levels
o : calculated from the uppermost level (z = 343 cm) for hydrodynamically
smooth flow
• : calculated from the lowermost level (z = 29 cm) for hydrodynamically
smooth flow

already at a distance from the bottom which usually is attributed to the logarithmic
layer (MONIN and YAGLOM 1971). The application of BLACKADAR's (1962) modification of
the turbulent mixing length function to our data could provide an explanation of the
z_0 values decreasing with increasing velocity (Fig. 5-31; SCHAUER 1982).

The concave deformation of the velocity profile observable even in the "undisturbed"
case causes the bottom shear, calculated by the simple application of equation (1),
to depend on the extension of the considered profile. This will lead to considerable
over-estimations.

Comparisons between results obtained by profile adaptation and those calculated from
the velocity at one definite level, assuming a hydraulically smooth flow, show
differences by a factor of 2 for u* (Fig. 5-32) or a factor of 4 for the bottom shear
stress.

Applied to the long-term current measurements discussed in chapter 5.3.1, the different interpretation methods yield a significant discrepancy in the resulting u*
values. Calculated from the profile method over a height of 3.5 m, the shear velocity
attains values up to more than 5 cm s^{-1} (scale A, Fig. 5-33). In this case, the
threshold bottom shear stress for erosion would be exceeded for a considerable grain
size range. However, using the bottom shear velocities calculated for hydraulically
smooth conditions the majority of the observed currents would remain below the
threshold for the characteristic grain sizes in the eastern Vejsnaes Rinne (scale B,

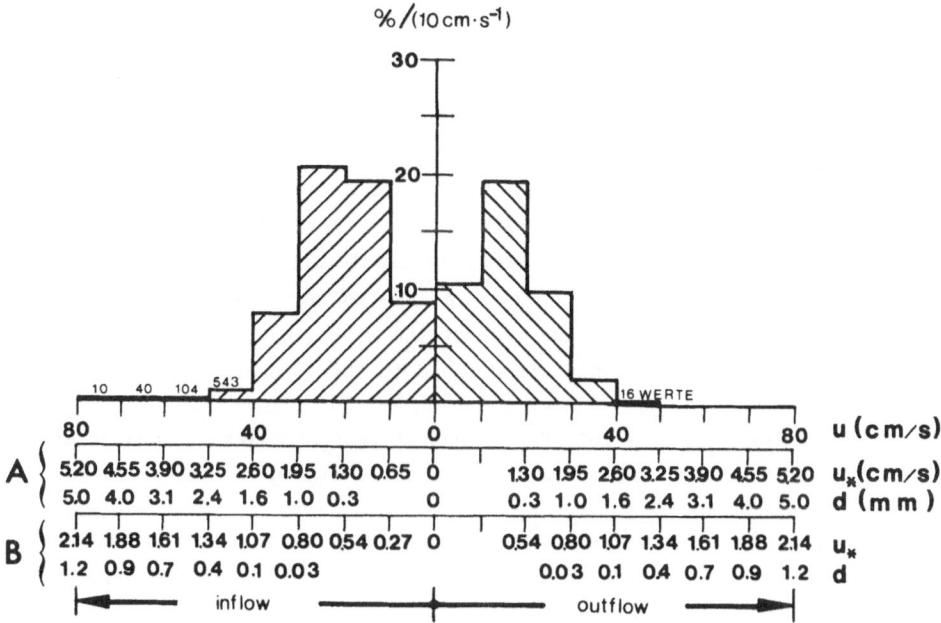

Fig. 5-33: Velocity histogram from the Vejsnaes Rinne long-term measurements (HATJE 1976, 1977) 3 m above the bottom. Scale A: u$_*$ values fitted from six levels, scale B : u* values calculated from the 29 cm-level for smooth flow. Additionally, grain diameters (d) from UNSÖLD (1982) corresponding to the u$_*$ values as critical conditions of grain entrainment are shown. (After SCHAUER 1982).

Fig. 5-33, and Fig. 5-48). It appears reasonable with respect to the geological observations in this area (see also sections 5.3.6 and 5.4) that erosion only occurs together with extraordinary current events as scale B shows in Fig. 5-33.

The improved version of the Shields function provided by UNSÖLD (1982) and the insight into its significance (see also section 5.3.4) provides a sufficiently firm basis for the statement that only the lower values of the bottom shear-velocity are in accordance with the sediment data. This also means that only the sections next to the bottom (Fig. 5-48) can be described in a sufficiently good approximation by the "law of the wall" to warrant the assumption that this function is valid there.

5.3.2.4 Summary

The experiments described above have shown the difficulties encountered in different
environments in determining the bottom shear stress of the dominating flows from both
profile measurements of mean velocities and measurements of high-frequency fluctuations of the flow speed in the bottom boundary layer. The discussion of the
results has shown a selected number of causes for the deviations from the "law of the
wall" at greater distances from the bottom. Certainly there exist other conceivable
causes for these deviations including even a limited validity range of the assumptions underlying the derivation of the "law of the wall", and they easily explain the
wide scattering of drag coefficients often found in the literature.

Thus the proper nature of the velocity and shear-stress profiles at greater distances
from the bottom (e.g. > 1 m) must be considered as remaining an open problem.

On the other hand, it follows from these experiments, particularly of those made in
the Jade, that accurate modelling of sediment transport requires a reliable knowledge
of the boundary shear stress distribution in time and space. This can never be
achieved with fluid dynamical theories alone nor with empirical measurements alone;
rather success could be derived only from the combination of the two. However, the
large scatter in Fig. 5-21 shows clearly that even then great care must be taken in
making measurements and interpreting them within a theoretical framework.

5.3.3   Wind-Wave Induced Sediment Transport

Introduction

As pointed out in section 5.2 the sediment distribution in Kiel Bight and the Western Baltic Sea is largely dependent on wave action. The processes of wave-induced sediment transport are important for different fields of marine sciences involved in the problems of energy transfer to the sea bottom.

In the past, many theoretical attempts have been made to get down to the problem. The result aimed at in a first step was to detect a relationship between wave conditions and the characteristics of oscillation ripples generated thereby. A contribution to the solution of this problem is dealt with below. The further and still more difficult problem to assess the advective sediment transport rate triggered by wave action is excluded here.

All efforts to quantitatively describe the formation conditions of oscillation ripples are based on the Airy linear wave theory (for a short discussion of the admissibility see: DINGLER and INMAN 1977).

Let L = wave length (m),
      H = wave height (m),
      D = water depth (m),
      T = wave periods (s),
  U_{max} = orbital velocity (m s^{-1}),
     d_0 = orbital diameter (m) at depth D (m),
      g = gravity constant (m s^{-2}).

The following three equations result from the model of the linear wave theory:

$$T = \sqrt{2\pi L/g} \cdot \sqrt{\coth(2\pi D/L)} \tag{1}$$

$$d_0 = \frac{H}{\sinh(2\pi D/L)} \tag{2}$$

$$U_{max} = \pi d_0/T \tag{3}$$

Part of the problem of quantifying the formation conditions of oscillation ripples is to determine the conditions of their initiation. These can be characterized by a 'critical grain size' of sediment entrainment as a function of the characteristic parameters of the generating waves. The numerous different efforts to develop such a 'critical-state function' are summarized by MILLER and KOMAR (1980a).

For three different sites in Kiel Bight (lightships Flensburg, Kiel and Fehmarn Belt)

SCHWEIMER (1976) calculated frequencies of bottom sediment mobilizing conditions to be expected from the then available theories in order to test their applicability. The low frequencies found seemed to be unrealistic. Critical review showed that the underlying assumptions about the sediment properties and the mean wave conditions were inadequate. Therefore a new empirical approach was started (v. GRAFENSTEIN 1982, 1984).

Methods

Several previous studies in Kiel Bight (NEWTON and WERNER 1972, FLEMMING and WEFER 1973, WERNER et al. 1976) had shown that large oscillation ripples occur with coarse grain sizes within the patchy lag sediment areas described in section 5.4. Since it appeared possible that these ripples could have been in motion several times a year during storm events repeated observations were carried out to detect likely changes. The changes actually observed were related to the wave parameters of the immediately preceding wave event. These parameters were obtained by evaluation of long-term records of waves measured at a site near the southern coast of Kiel Bight ('*' in Fig. 5-1). The wave measurements were made with a 'reversed echo-sounder' (SCHÄFER 1979).

In the study area at the Stoller Grund, a shoal at the entrance of the Eckernförder Bucht (Fig. 5-1), ripple fields exist on sands of different grain sizes (WERNER et al. 1976). Three sites in different water depths (10.5 m, 12.0 m, and 14.8 m) were repeatedly observed by divers. Additionally, side-scan sonar surveys were carried out to observe the changes of ripple patterns in response to registered wave events.

To get the needed parameter 'grain size' of the newly formed ripples the known corre- lation between ripple spacing and 'grain size' was used (INMAN 1957, KOMAR and MILLER 1973, FLEMMING and WEFER 1973), because ripple spacing can be measured extensively and with sufficient accuracy (crest spacing down to 15 cm) by high-resolution side- scan sonography (Fig. 5-34). Thereby a continuous coverage of different water depths in large areas is possible.

The correlation between the two parameters ripple spacing and 'grain size' valid for the study area was obtained using samples from ripple crests (Fig. 5-36). Accordingly 'grain size' in the sequel means always the phi median diameter of the grain size distributions of ripple crest materials.

During 1980 16 different ripple situations could be related to the wave event generating them, mainly using the crest orientation and age relationships as diagnostic criteria.

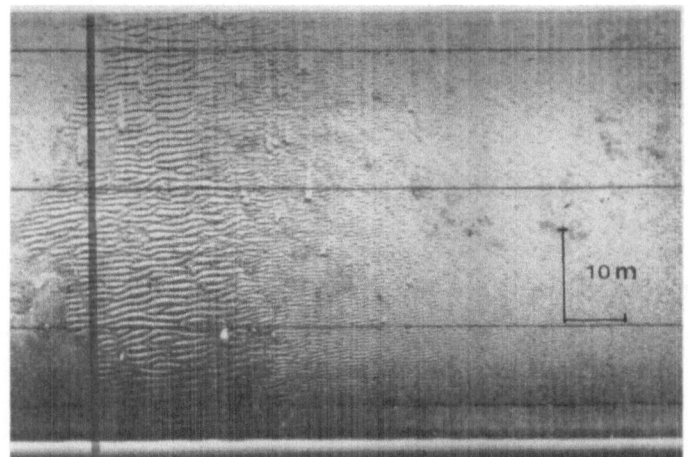

Fig. 5-34: High resolution side-scan sonography showing oscillation-ripple fields of various crest spacing (Stoller Grund, Kiel Bight, from v. GRAFENSTEIN 1982).

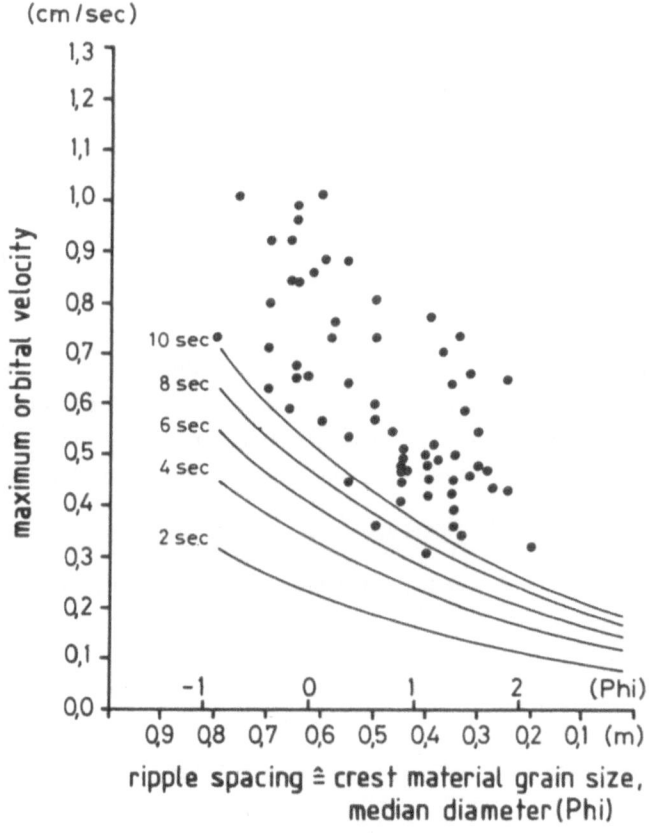

Fig. 5-35: Relation between ripple-crest grain size and ripple spacing of Stoller Grund data. (From v. GRAFENSTEIN 1982).

Results and Discussion

In Fig. 5-35 the maximum orbital velocities observed during a certain wave event are plotted against the grain sizes of the coarsest grained ripples built or rebuilt during this event. For the corresponding range of grain sizes the velocities predicted by an experimentally determined 'critical-state function' are mapped in the same diagram.

As an acceptable approximation to such a function the relation proposed by DINGLER (DINGLER 1974, DINGLER and INMAN 1977, DINGLER 1979) is tentatively used in the following:

Let d = 'critical grain size'   m

           ρ_S = density of the solid   kg m^{-3}

           ρ_F = density of the fluid   kg m^{-3}

           g = gravity constant   m s^{-2}

           $\gamma_S = (\rho_S - \rho_F) \cdot g$ kg m^{-2} s^{-2}

           μ = viscosity   kg m^{-1} s^{-1}

A dimensionless form of the DINGLER function is

$$\frac{\rho_F \cdot U^2_{max}}{\gamma_S \cdot d} = \frac{\pi^2}{c_1} \cdot \left(\frac{d_o}{d}\right)^{2/3} \cdot \left(\frac{\rho_F \cdot \gamma_S}{\mu^2} d^3\right)^{1/g} \qquad (4)$$

with c_1 = 240 (DINGLER and INMAN 1977).

The left-hand side term has the form of a 'Shields parameter' ('relative stress', 'dimensionless shear stress', 'mobility number'). d_o/d is a 'dimensionless orbital diameter', and the last term is called 'dimensionless characteristic of fluid and solid phases' by YALIN (1972), 'dimensionless grain size' or 'sedimentological diameter' by others.

By introducing eq. (3) the DINGLER function may be rewritten in the dimensional form

$$d = c_1^{3/2} \cdot \frac{\rho_F}{\gamma_S} \cdot \left(\frac{\mu \rho_F}{\gamma_S^2}\right)^{1/3} \cdot d_o^2/T^3 \text{ m} \qquad (5)$$

For quartz grains in sea water this equation can be simplified to the form

$$d = c_2 \cdot d_o^2/T^3 \text{ m} \quad \text{with } c_2 = 0.4 \text{ m}^{-1} \text{ s}^3 \qquad (5')$$

Fig. 5-35 shows that all observed values surmount the predicted ones. The same result is obtained with all other 'critical-state functions' proposed in the literature to predict critical orbital velocities under wave action. Certainly reservations as to the validity of all these functions still are appropriate. Nevertheless one can conclude from the diagram that at least in the majority of the observed events ripples with still coarser grains than the observed ones could have been formed, if such grains had been available in sufficient quantities to constitute continuous sediment covers. Conversely, the parameters of the ripples actually built or rebuilt during a wave event in nearly all cases will merely reflect phases of the event's fading away. Only this concept can explain the observation that one and the same wave event at one and the same water depth leaves behind oscillation ripples with different crest spacings on sediments of correspondingly different grain sizes.

In any case, however, the parameters of all ripples actually observed must lie within the limits demarcated by the measured wave parameters and at the same time they must satisfy the conditions specified by a 'critical-state function'. These conditions are checked in Fig. 5-36 and 5-37, using the DINGLER function.

At the outset two assumptions are introduced:
- the observed ripple parameters document critical conditions reached with orbital velocities decreasing from supercritical to subcritical values, pinpointing the conditions of ceasing sediment movement;
- the relation between the crest distances λ and the corresponding orbital diameters d_0 can be described in sufficient approximation by the simple equation $\lambda = d_0$.

Then, using the equations (1), (2), (5') for each set of {L,H,D} of observed wave parameters the corresponding values of the orbital diameter d_0 and of the critical diameter d can be computed. Plotted in the plane d_0 vs. d these pairs of values fill an area representing the range of 'critical conditions' actually possible in the study field. Under the assumption $\lambda = d_0$ the observed pairs of values {λ,d} can also be plotted in the same diagram.

If all of the above assumptions were satisfied, the point cloud {λ,d} must fall into the area within the envelope enclosing the point cloud {d_0,d}.

The results are shown for two different sites:
- the Stoller Grund, a shoal at the entrance of the Eckernförder Bucht (Fig. 5-36),
- the surroundings of the "Forschungsplattform Nordsee" ca. 40 km W of the island of Sylt in the North Sea (Fig. 5-37).

For both diagrams the expectation just expressed is fulfilled in first approximation, indicating that the assumptions made above are justified accordingly. The coincidence

WESTERN BALTIC SEA

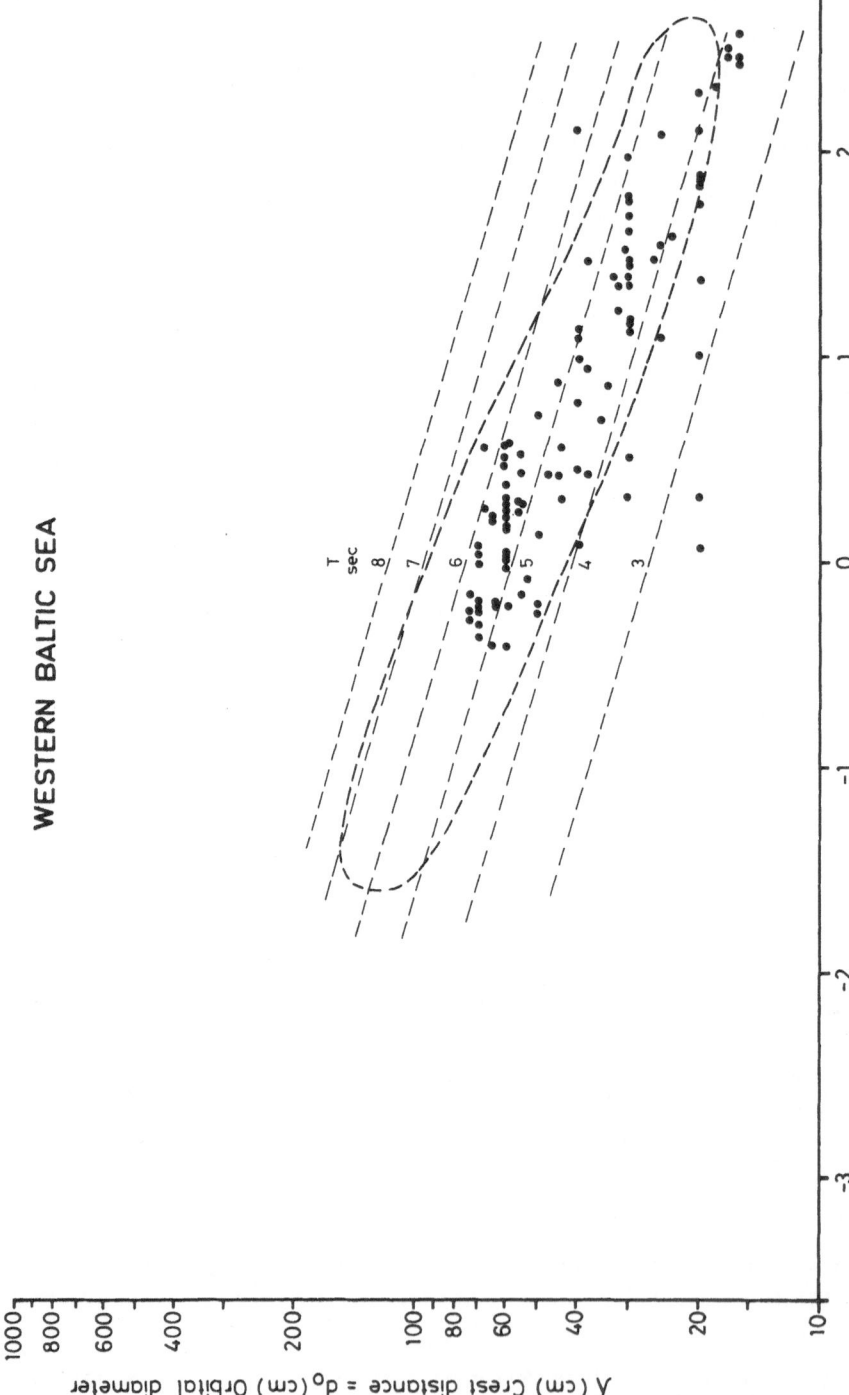

Fig. 5-36: Crest distances λ vs. crest median grain size d of observed oscillation ripples from Kiel Bight. Envelope: Range of observed orbital diameters, which could be critical for grains of diameter d according to the critical state function proposed by DINGLER (1974). Straight lines: Curve family for period T = const. according to the DINGLER function. Note: Optimum coincidence of the point cloud and the range encompassed by the envelope is found for the condition λ ≈ 0.8 d_o.

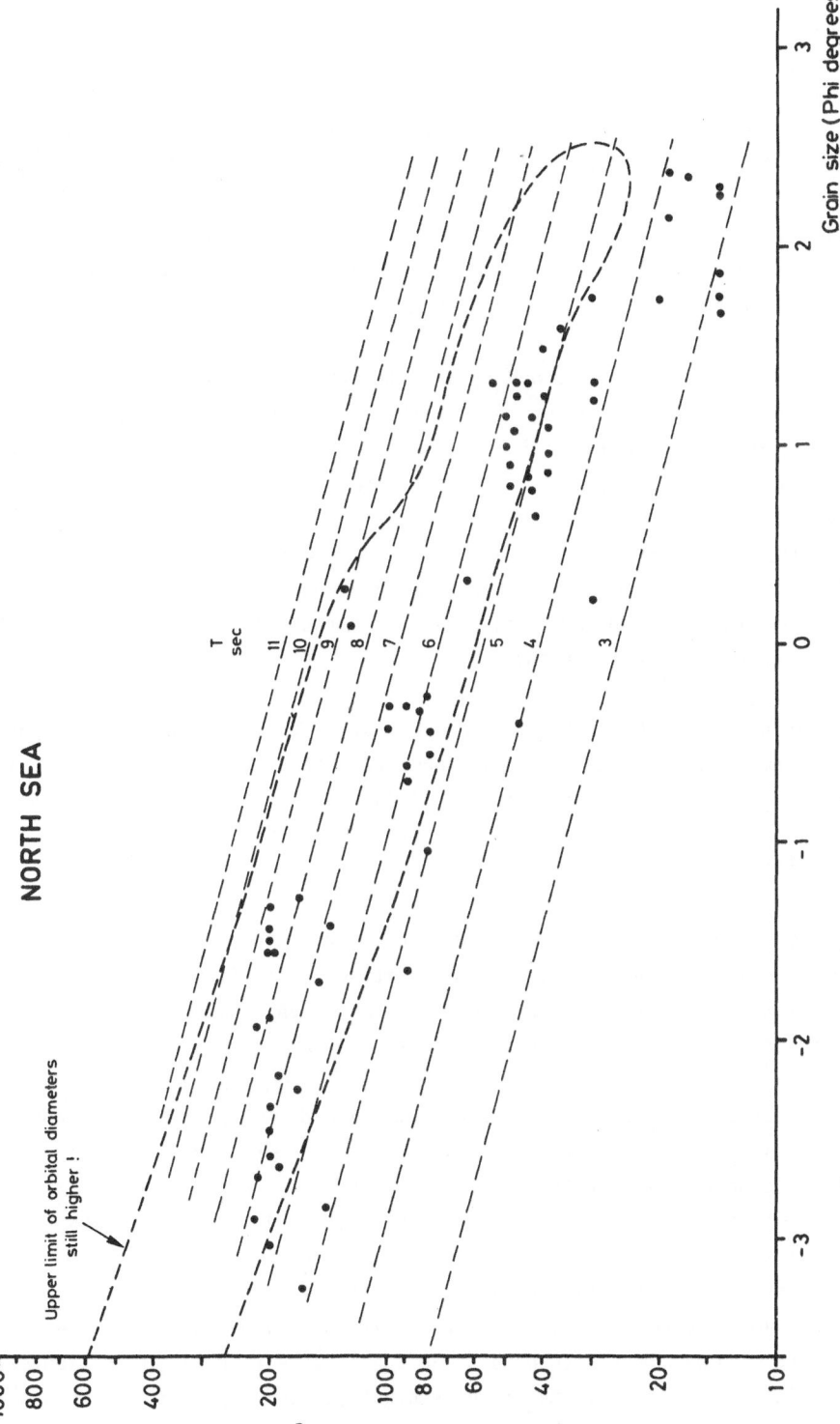

Fig. 5-37: Crest distances λ vs. crest median grain size d of observed oscillation ripples from North Sea (German Bight). Envelope: Range of observed orbital diameters, which could be critical for grains of diameter d according to the critical state function proposed by DINGLER (1974). Straight lines: Curve family for period T = const. according to the DINGLER function. Note: Optimum coincidence of the point cloud and the range encompassed by the envelope is found for the condition λ ≈ 0.8 d_0.

of the two point clouds is better for the coefficient c_1 = 240 given in DINGLER and INMAN (1977) than for the coefficient c_1 = 290 proposed by DINGLER (1979).

In both diagrams the areas corresponding to the strongest wave conditions observed in Kiel Bight are devoid of projection points of observed ripple parameters. This again is an expression of the phenomenon shown above (Fig. 5-35) that the strongest wave events could have moved considerably coarser grains had they been available in sufficient quantity and with adequate sorting to form continuous covers of sufficient extent. By a different approach JAGO and BARUSSEAU (1981) reached a comparable conclusion for the inner shelf of the Roussillon coast (Golfe du Lion).

The limit of the wave conditions recorded by the sediment lies higher in the example from the North Sea (Fig. 5-37) than in the example from Kiel Bight. The grain sizes available in the glacial source material are essentially the same in both areas. Yet the higher level and higher frequency of maximum wave conditions in the North Sea obviously more frequently offer conditions for efficient separation of coarser grains so that they can be assembled into sufficiently continuous sediment covers. v. GRAFENSTEIN (1984) could support this concept by demonstrating that with increasing grain size of the ripple sediment its "lag sediment character" becomes more distinct (indicated by a deficit of fine fraction).

The observed pairs of values $\{\lambda,d\}$ scatter considerably. If they actually correspond to critical entrainment conditions and if at these conditions the relation $\lambda = d_0$ is valid, then from the DINGLER function in the simplified form (5') follows

$$\lambda = c_2^{1/2} \cdot T^{3/2} \cdot d^{1/2} \text{ m} \tag{6}$$

Since the diagrams in Figs. 5-36 and 5-37 are double-logarithmic in the variables λ and d, equation (6) can be represented therein by a family of straight lines with T as parameter. It becomes obvious that most of the scatter in any regression function of λ vs. d is due to the neglection of T as a variable.

In all the relationships discussed above the equation $\lambda = d_0$ seems to fit well enough that for a first approximation no compelling reason can be seen for replacing it by an equation of the type $\lambda = c_0 * d_0$ with $c_0 < 1.0$ as proposed by KOMAR (1974). A review of the field data plots in the plane $(\{\lambda,d_0\})$ (KOMAR 1974, DINGLER and INMAN 1977, MILLER and KOMAR 1980b) shows that the relation in question should be determined as an upper envelope of the point cloud rather than as a trend line. Points to the right of the envelope apparently represent conditions above the threshold of grain movement, associated with grain fluxes higher than those just necessary to form mature ripples. This concept is in accordance with the results displayed in Fig. 8 of DINGLER and

INMAN (1977) and in Fig. 2 of DINGLER (1979). If so, the relation $\lambda = d_0$ should characterize the threshold conditions for the formation of oscillation ripples. Closer inspection of Figs. 5-36 and 5-37 shows a distinct tendency of the point clouds representing the observed pairs of values $\{\lambda, d\}$ to be displaced a bit downwards out of the range occupied by the pairs of values $\{d_0, d\}$ as predicted from the observed wave parameters by the DINGLER function. By shifting the corresponding envelope downwards one finds in both examples a maximum of coincidence for the relation $\lambda = 0.8 * d_0$. Critical consideration of the methods used in measuring the crest distances λ suggests that $c_0 = 0.8$ should be taken as a minimum estimate and c_0 thus is to be expected lying somewhere in the range $0.8 < c_0 < 1.0$. The relation $\lambda = c_0 * d$ with c_0 a little smaller than 1.0 in a second approximation fits well into the model described above.

Conclusions

From the above findings follows the geologically relevant insight that fossil oscillation ripples nearly never record the maximum level of wave conditions, but some level below it. How much below, will depend upon the availability of correspondingly coarse grains and of the level of the most frequent wave conditions efficient in grain size differentiation.

The DINGLER function performs sufficiently well to warrant its tentative use in evaluating fossil ripple data for the reconstruction of ancient wave conditions. The relation $d_0 = \lambda$ seems to provide sufficiently good estimates of d_0 from ripple crest spacings. By inserting it together with the corresponding grain size values into the DINGLER function estimates for the corresponding wave periods T are obtained. Since only the two equations (1) and (2) are available for the determination of the three unknown variables $\{D,L,H\}$ an estimate of one of them must be procured from other sources of evidence. Plausible values of wave length L or wave height H may be assessed combining oceanographical experience and paleogeographical considerations; acceptable values of the water depth D may be derived from paleobiological observations. Since all such guesses are uncertain, the results derived there from must be checked against one another for internal consistency (KOMAR 1974).

5.3.4   Critical Entrainment Conditions of Sediment Transport

Introduction

Since sediment cannot be transported unless mobilized at all, the study of 'critical entrainment conditions' is basic to all problems concerning near bottom advective transport of sedimentary material.

This all the more, as sediment carried by a current well above the critical conditions must pass once more the narrow range of near-critical conditions, if it is to come to rest again and form sediment in the proper sense of the word. Therefore it should be these conditions which are fundamentally determining sediment texture. Moreover, we are forced again and again to evaluate the functional relation between sediment texture and the corresponding critical conditions to assess regional current conditions: direct current measurements generally are neither feasible with sufficient spatial and temporal resolution, nor do they yield at the present state of the art unambiguous assessments of bottom shear stresses, as the foregoing discussions have shown.

This is particularly true for the record of high-energy events which are both difficult to observe because of their rarity and because they are risky for the measurement equipment.

Thus, a better knowledge about the critical sediment entrainment conditions is without doubt a necessary though not sufficient prerequisite for an improved understanding of sediment transport processes.

These considerations led us to the decision to work in this field. Our experiments were carried out in a recirculating tilting flume with a length of 18 m and a width of 0.57 m. The measuring section was 13 to 15 m downstream of the entrance.

Critical Entrainment Conditions for Clayey Cohesive Material

The motivating facts for attacking this realm of intricate problems were:
- sediments of such kind cover the deeper parts of Kiel Bight and therefore pertinent questions arose from many observations in different contexts as described in this volume;
- the knowledge about the erosion resistance of such materials is still much poorer than that about non-cohesive, coarser grained material.

As also intended by other workers in the field, the objective was to study relationships between the minimum hydraulic shear stress ('critical shear stress'), necessary to mobilize particles of the clay material as response variable, and some soil mechanical parameters used to characterize its state of consolidation as predictor variables.

However, in contrast to former investigations the main idea was to examine the influence that the fabric of the clay sediment was expected to exert on these relationships. Therefore the sediment samples were prepared and consolidated under controlled conditions in the laboratory and then very carefully transferred into the flume to avoid mechanical disturbances as far as possible. These experiments were described and discussed in detail by EINSELE et al. (1974) and OVERBECK (1979). In the following, only a few points of general importance are reviewed.

In preparing samples at least free of perceivable inhomogeneities of fabric it proved to be necessary to generate them by letting settle the clay material out of a highly concentrated suspension (8 percent by weight of Ca-Mg-Kaolin in tap water). Thereby the clay formed a coherent network of flocs already at the very beginning of the settling process. When a suspension of only slightly lower initial concentration (5 percent by weight) was used with the same settling method, the formation of perceptibly graded sediments was inevitable.

This effect was even stronger with the alternatively used method of pouring a clay suspension drop by drop into a settling tank, even though pouring it at low rates and as evenly distributed as possible over its whole cross section. In applying this method it was intended to prepare samples under autoconsolidation conditions like they can be reached in a sediment column of up to 1 m thickness. However, the combined effects of the inevitable gradation processes, of disturbances of the sedimentation by the enclosing walls and of the transfer into the flume made it impossible to obtain sufficiently consistent results by this method.

The samples prepared by settling out of highly concentrated suspensions subsequently were brought to higher consolidation states by compressing them under gas pressure. Thereby the relative importance of still present inhomogeneities of the fabric was further diminished and thus in this range of consolidation states sufficiently consistent results could be obtained.

The main conclusions on the basis of these experiments can be summarized in the following statements:

- as was expected, the hydraulic critical shear stress increases with increasing consolidation which was characterized by decreasing void ratio or by increasing vane shear strength, and more strongly so with horizontal than with vertical vane shear strength;

- as expected, there is clear evidence that the fabric is extraordinarily important
for the erosion resistance of cohesive material. Considering in turn that numerous
factors determine the fabric of clayey sediments, the broad range of results
obtained by different authors using different clay materials, different methods in
sample preparation, different current conditions and different definitions of the
'critical erosion state' cannot be surprising;

- of general importance is the undisputable finding, that for a clay sediment of a
given state there exists a definite range of really 'subcritical' hydraulic shear
stresses, at which - after a small initial and soon vanishing erosion - the sediment surface remains stable at least for periods surpassing the maximum exposure
time of up to 3 days used in the experiments. The cycle (increased hydraulic shear
stress, faint erosion vanishing again, stable bed) can be passed through repeatedly
until the beginning of a slight but then continuous erosion.

A careful description of the experiments and observations, a lot of methodical considerations, hints and a thorough discussion of the results comparing it with those
of many other workers in the field is given by OVERBECK (1979).

Critical Entrainement Conditions for Very Fine Grained, Well-Sorted, Non-cohesive
Material

Considering that the knowledge of the critical shear stresses for correspondingly
fine grained but non-cohesive material must be the basis for assessing the influence
of cohesivity on the erosion resistance, we attacked the problem of extending the
range of grain sizes studied by SHIELDS (1936) downward to much smaller ones. At that
time we had no notice of the work of WHITE (1970) and MANTZ (1977), concerning the
same question. Thus our determination offers an independent affirmation.

From a review of the literature we concluded that the usual procedure of visual
assessment of the 'beginning of bed movement' was very subjective and therefore unsatisfactory. In order to obtain an undisputable result the visual assessment therefore
had to be replaced by systematic measurements of the low flat-bed transport rates,
even though this is a very time consuming procedure.

Ten well sorted grades of pure quartz material with median diameters between 0.26 mm
and 0.003 mm were investigated.

The low bottom shear stresses typical for such experiments were calculated from depth
and slope of the carefully adjusted uniform open channel flow and additionally
checked by measuring vertical velocity profiles using both a Prandtl tube and LDA
(Laser Doppler Anemometry). This method proved to be of decisive help in obtaining
reliable values of very small shear stresses.

The results are summarized in Fig. 5-38, incorporated into a unified picture developed by UNSÖLD 1984 where all published data available, giving sufficient experimental details, were reevaluated.

The parameter of the curves is the dimensionless transport rate

$$Q_S^+ = \frac{Q_S}{\rho_S \cdot g \cdot d \cdot u_*}$$

where Q_S = transport rate as weight per unit width per second

u_* = bottom shear velocity

d = grain size diameter

ρ_S = density of the grain

g = gravity constant

These curves are contour lines of the lowermost part of the surface

$$Q_S^+ = f(X,Y)$$

where X = grain size Reynolds number

Y = Shields parameter ('mobility number', 'dimensionless shear stress')

This surface is schematically depicted by YALIN (1972, Fig. 5.3, p. 115). His diagram still contains the 45° - extrapolation originally proposed by SHIELDS (1936) which now is definitely shown to be invalid: instead of remaining constant the bottom shear stresses corresponding to constant transport rates decrease further with decreasing grain diameter. The diagram Fig. 5-38 given by UNSÖLD (1982) is corrected for this fact.

However, this diagram also shows that for the system 'quartz-water' the tendency of the critical shear stresses to decrease with decreasing grain size holds true only down to grain diameters of about 0.015 mm.

The increased erosional stability actually found for the two finest materials studied (d = 0.009 mm and d = 0.003 mm) can be definitely proven to be caused by interparticle forces resulting from the drastically increased specific surface area generating cohesivity (UNSÖLD 1982).

The long debated question, whether the phenomenon of a finite 'critical shear stress' actually exists can be clarified, if one considers the sequence of events encountered in passing through the 'near critical region' of current conditions in sufficiently small increments of the applied shear stress:

214

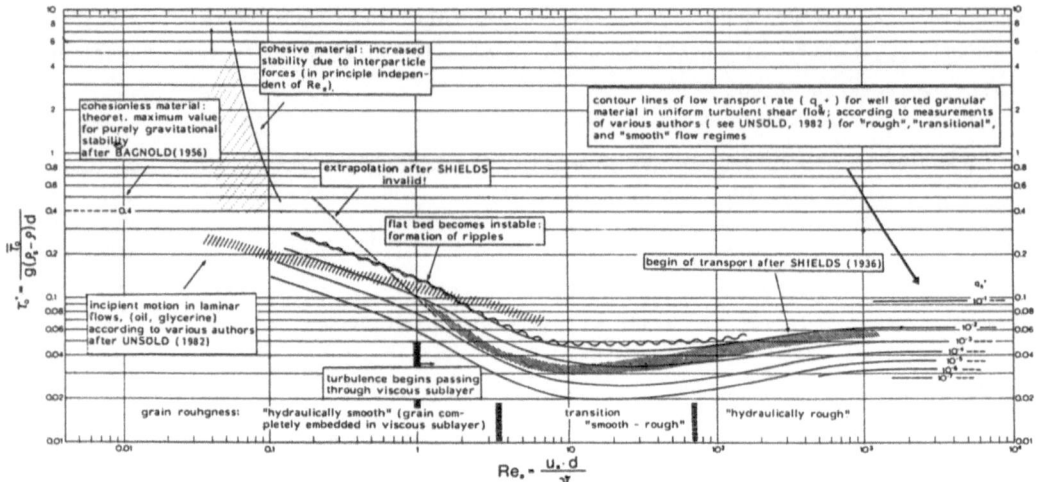

Fig. 5-38: Improved and extended Shields diagram, based on determinations of transport rates. (After UNSÖLD 1982).

Immediately after the preparation of the sediment sample surface even a very low current velocity will mobilize some few grains, but very soon they come to rest again and then the surface remains completely stable under the corresponding shear stress. The cycle (slightly increased shear stress, small initial transport rate, soon vanishing again, sediment surface remaining stable) can be passed through repeatedly until a small-rate transport sets in which remains constant and is not decreasing again.

This behaviour of the sediment surface in the region of near-critical current conditions is the same as observed in the experiments with the cohesive material, which were described above.

It can be explained with the following model, derived from ideas expressed by GRASS (1970) and described schematically in Fig. 5-39. The underlying assumptions are:
- the frequency distribution of the instantaneous local velocities has a finite upper limit;
- the instantaneous local velocity which must hit a certain grain so that it is pushed out of its actual position in the bed surface - its 'individual critical velocity' - is not a unique function of its grain size and therefore neither a conservative property which the grain posesses once and for all: it changes with any jump of the grain, depending on the position at which it will come to rest again, and it likewise changes with any motion of grains in its momentary neighbourhood;
- the individual critical velocities of a certain grain will vary within a certain range essentially depending upon the shape of the grain size distribution to which it belongs; nevertheless one might expect that on average they will have a definable frequency distribution;
- the individual critical velocities of all the grains form a joint frequency distribution with finite upper and lower limits.

Under these assumptions the observable phenomena can be explained as follows:
With increasing mean shear velocity and therefore with increasing mean local velocity, the maximum instantaneous local velocity scans over the range of the individual critical velocities of the bed particles. All grains which happen to be extremely exposed at that moment and thus have an extraordinarily low individual critical velocity will be mobilized. But if the increase in shear velocity is performed stepwise and if the step durations are sufficiently long, they soon will find a more stable position with an individual critical velocity just or well above the actual upper limit of the instantaneous local velocities: the initial weak transport will vanish again. When repeating such steps the lower flank of the frequency distribution of the individual critical velocities becomes steeper and steeper, and its dispersion becomes smaller and smaller. If finally a condition is reached where the mobilized grains

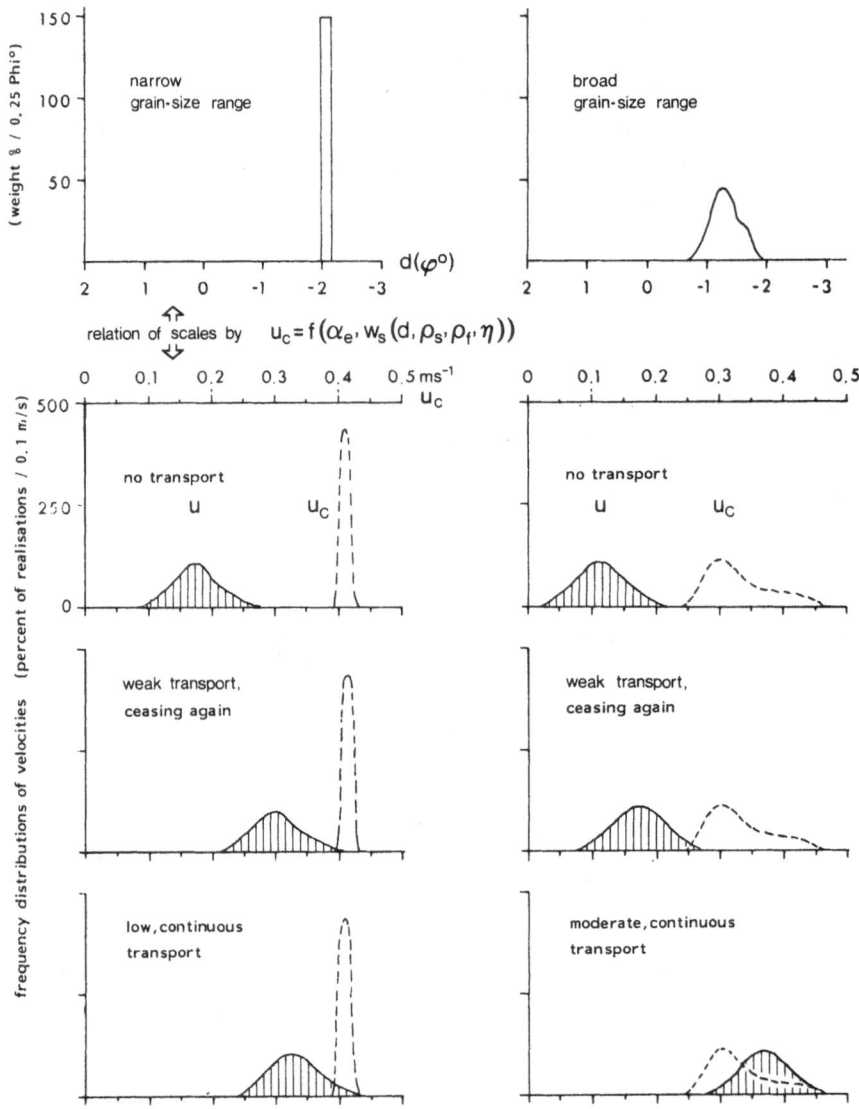

Fig. 5-39: Tentative scheme of critical entrainment conditions. Frequency distribution
of the 'instantaneous local velocity' ($u = \bar{u} + u'$) based on LDA measurements at the 'grain level'. ------ Frequency distribution of the
'momentary critical velocity' of the grains u_c forming the surface of the
bed (schematic). α_e = exposition parameter of the single grain (stochastic
variable); W_s = settling velocity; d = grain diameter; ρ = density of
sediment and fluid, resp.; η = viscosity.

cannot find a sufficiently stable position any longer, then at once a low-rate transport begins which does not vanish any more. This will be the case, if the upper limit of the instantaneous local velocities is still below, but somewhere near the upper limit of the joint frequency distribution of the individual critical velocities of the bed particles. If the increase in mean shear velocity is made continuously instead of stepwise one should expect from this model that a weak but continuous transport begins already at the lowermost threshold of any grain movement at all.

Thus, the 'critical phenomena' spread over a certain range of mean shear velocities, but within this range two limits are discernible:

A lower limit, below which indeed no motion at all is observable, and an upper limit, above which continuous transport is inevitable. Between these two limits self-stabilisation of the bed surface is possible. The lower limit cannot be defined precisely, because it is highly dependent upon the previous history of the bed surface. The upper limit is accentuated more precisely: it is this limit which, in a first approximation, is described by the Shields function.

The model described above leads among others to the conclusion that the formation of residual sediments ('armouring') is a direct consequence of the supposed condition that the frequency distribution of the instantaneous local velocities has a finite upper limit. This can be checked by corresponding experiments, as they are in progress now.

Critical Entrainement Phenomena for Sediments with a Very Broad Grain Size Spectrum Under Conditions of Residual Sediment Formation

If the frequency distribution of the instantaneous local velocities has a finite upper limit, then the grain size range of the mobilised material must also have a finite upper limit which should be distinctly below the upper limit of grain sizes available in the bed. Indeed this proved to be so.

In order to check the expectation, the process of residual sediment formation ('armouring') is studied in flume experiments using sediments with a broad grain size spectrum (width up to 8 Phi degrees).

At the present state of our knowledge it can be stated that the upper limit of the erodible grain sizes is predictable with acceptable approximation by the Shields function, as can be expected if the model described above is adequate.

In this context the following should be noted: the statement that the frequency
distribution of the instantaneous local velocities has a finite upper limit implies
that this distribution is not 'Gaussian' in the strict sense, even if it happens to
appear symmetrical!

Experiments to Measure Forces Acting on a Single Grain of the Sediment Surface

In connection with the experiments to investigate the process of 'armouring' the en-
deavour to measure directly the dynamic hydraulic forces acting on a single grain has
to be seen as well. It poses one of the basic problems of the sediment transport mechanics as evident, for example, from the bed load functions of EINSTEIN (1950) and
YALIN (1963).

Since the still often cited study of EINSTEIN and EL SAMNI (1949), fully new experimental techniques were developed, and therefore we decided to attack the problem
anew.

For such experiments two prerequisites are necessary:
- velocity measurements with sufficient resolution in time and space in the immediate
  surrounding of the grain;
- a force transducer with sufficient sensitivity.

For the velocity measurements the techniques of the Laser Doppler Anemometry (LDA)
were used.

They can be viewed as a special case of the wider class of Photon Correlation Methods, as they are carried out in the Institute for Applied Physics by SCHULZ-DUBOIS
(1983) and his coworkers. Inter alia they were used there studying certain problems
in fluid dynamics such as measurement of weak secondary laminar flows (HILLE et al.
1983) or hydrodynamic instability (PFISTER et al. 1983). In the course of such studies
they were improved in different respects (e.g. SCHAETZEL 1983, SCHULZ-DUBOIS and
REHBERG 1981, PFISTER et al. 1983).

A force transducer of sufficient sensitivity was developed by HILLE et al. (1983).
They succeeded by utilizing the new semiconductor strain gauge technique, which made
it possible to reach a sensitivity of about two orders of magnitude higher than
attainable with the classical strain gauge technique and still about one order of
magnitude higher than can be achieved using piezo-electric sensors.

Thereby it became feasible to devise a lift-force sensor which can measure forces varying in the range down to 10 micro Newton and up to 500 HZ at a signal to noise ratio of 20 dB.

This sensitivity is in the range needed to measure the hydrodynamic lift forces acting on a single captive grain of 12 mm diameter in a turbulent channel flow at boundary shear velocities in the range between about 1 and 3 cm/s. For details of construction and testing procedures see HILLE (1984).

From the measurements made with this device so far, the following preliminary results were obtained:
- On a steel sphere with 12 mm diameter whose center was situated 3.5 mm above the bottom, a dynamic lift-force coefficient C_L as defined by EINSTEIN (1950) was found with a value of C = 0.047. This is in the same order of magnitude as the value C_L = 0.065 reported by EINSTEIN and EL SAMNI (1949).
- No direct correlation is detectable between time series of the instantaneous velocities in the immediate vicinity of the grain and synchronous time series of the lift forces;
- Detectable are cross correlations with short time lags between velocities and forces. They suggest that the forces may be correlated to the velocity field only at instances when turbulence structures, whose magnitude is comparable with that of the grain, are hitting the grain.

For a detailed discussion of the measuring technique, the results attained so far, and remaining problems see HILLE (1984).

5.3.5   Bedform Response to Hydrodynamic Conditions

5.3.5.1 Introduction

The distribution of bedforms is generally considered as a good indicator of current dynamics in shelf seas. The possibility to map the sea floor in plane view which is provided by modern acoustical survey techniques, in particular by side-scan sonar, permits analysing the bedform response to hydrodynamic effects with high spatial resolution. Such information on current and sediment dynamics normally includes three advantages in comparison to current measurement:
a) the bedforms record events of major significance of sediment transport;
b) the recorded information represents a temporal integration of fluctuating dynamics;
c) the information is two-dimensional and can completely cover large areas.

On the other hand, much attention has to be paid to distinguishing between bedforms which are active in the geological presence, those being relic forms and those being a mixture of both. In Kiel Bight and the Belt Sea, bedforms have been mapped using side-scan sonar by WERNER et al. (1974), WINN (1974), WERNER and NEWTON (1975), WERNER et al. (1976), KUIJPERS (1980) and WERNER et al. (1980).

The distribution of bedforms reflects the main current systems of the Western Baltic Sea (Fig. 5-40). High-energy bedforms, i.e. sand waves, dunes or mega-ripples, sand ribbons and larger comet marks (WERNER et al. 1980) are nearly exclusively confined to the Belt Channels, whereas in Kiel Bight only transitional or 'undeveloped' large-scale bedforms occur.

5.3.5.2 Regional distribution of bedforms

The bedforms in the belts display a wide variety of forms and distribution patterns. Large-scale sand waves, i.e. forms with 50 or more meters of wave length, were found only at a few locations: the Fehmarn Belt, the southern part of the Great Belt (Fig. 5-40) and at some positions in the Little Belt and the Sound. Even in these areas, sand wave fields occur in restricted zones only. These restrictions are not due to the current patterns, but to other factors which have to be seen mainly in connection of the sand transport system to a sediment source and in a local current acceleration due to topographical conditions (WERNER et al. 1974, WERNER and NEWTON 1975). Both of the sites in the Langeland Belt (Fig. 5-40) are linked with an extended shallow offshore platform in the upstream areas of these currents (WERNER and NEWTON 1975). This holds true for both currents, the outflow from the south and the inflow from the north (see 5.3.1). The shape of these sand waves is concave and nearly symmetrical in cross section, as is typical for alternating stream directions

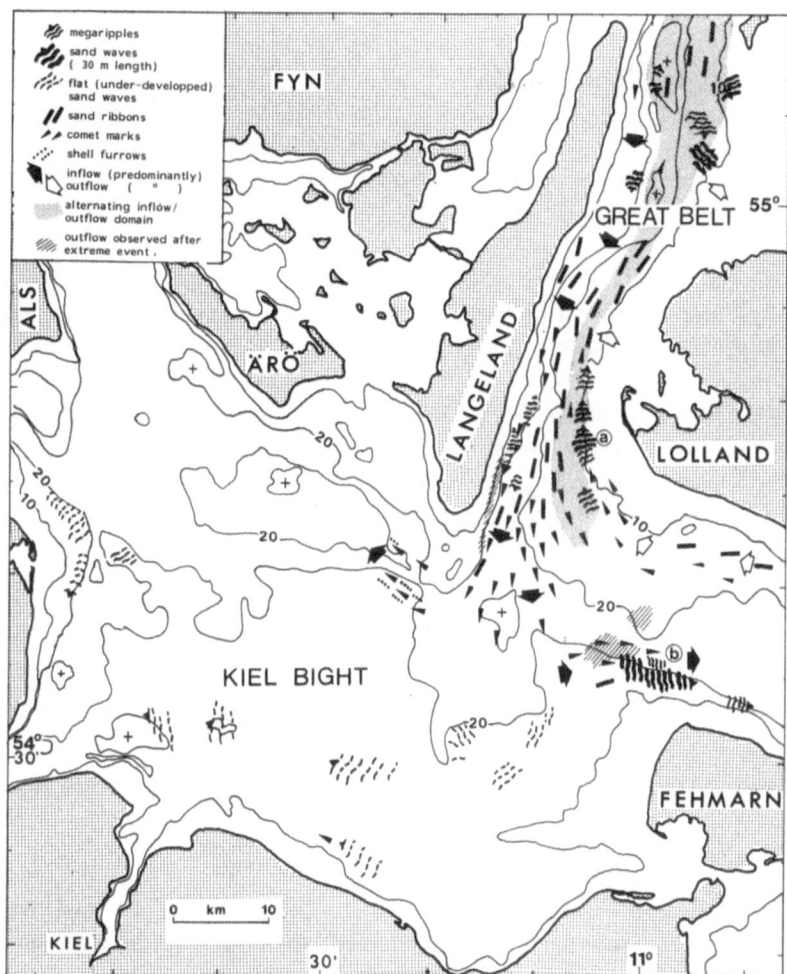

Fig. 5-40: Distribution of large-scale bed form types in Kiel Bight and the northern Great Belt. (Modified after WERNER et al. 1980).

(ALLEN 1980). In contrast to this, the Fehmarn Belt sand waves have a convex stoss side and are strongly asymmetric with steep leeside angles (Fig. 5-41). This form has been attributed to the action of unidirectional currents by ALLEN (1981) and is in agreement with the dominance of inflow currents (from W to E) in this area (WERNER et al. 1974).

At one locality (flanks of eastern Vejsnaes Rinne) 'shell furrows' have been found (WERNER et al. 1980). They are very flat, longitudinal 'furrows' whose bottoms are sprinkled with large shells of <u>Arctica islandica</u>. These furrows are cut into cohesive muddy sand and may be of similar origin as forms from tidal channels described by FLOOD (1981).

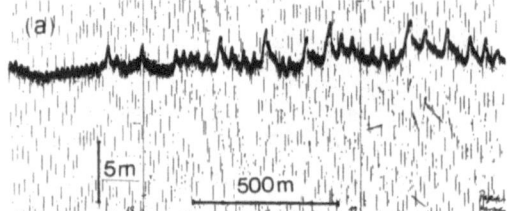 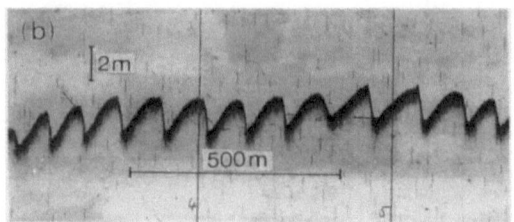

Fig. 5-41: Echo-sounder profiles of a sand-wave field in (a) the southern Great Belt
(from WERNER and NEWTON 1975); (b) the southern Fehmarn Belt. (a and b in
the map of Fig. 5-40.)

The main features shown in the bedform distribution map (Fig. 5-40) are inferred
mainly from longitudinal current marks, in particular from sand ribbons and comet
marks which indicate current directions more precisely than transversal forms (WERNER
and NEWTON 1975).

They are characterized by
- a fan-shaped spreading in the inflow direction at the southern end of the Great Belt;
- slight dominance of the inflow direction at the western and central part of the
  Great Belt (Langeland Belt) and strong dominance at the southern side of the
  Fehmarn Belt;
- dominance of the outflow direction at the eastern side of the Great Belt and
  northern part of the Fehmarn Belt;
- constant inflow direction in the eastern Vejsnaes Rinne area.

This directional pattern neither represents one definite situation nor must
necessarily be taken as an average. These observations are in reasonable agreement
with the regional pattern of the known current systems (WYRTKI 1953, JACOBSEN 1980,
and section 5.3.1). The pattern reflects that in the upper part of the water mass,
outflow currents are more effective and more frequent and that the sea bottom at the
right-hand sides is more affected due to the Coriolis force (Fig. 5-40).

Repeated sonographic surveys have shown that there are zones which are exclusively
dominated either by inflow or by outflow currents, respectively, while in others the
dominating flow frequently changes with opposite direction, as documented by bipolar
comet marks (WERNER et al. 1980). The remaining areas are dominated either by inflow
or outflow, but occasionally invaded by opposite flows.

Two further features of the pattern may be mentioned. One is the northeast directed bedform field of the eastern Vejsnaes Rinne consisting of comet marks and shell furrows (WERNER et al. 1980, KUIJPERS 1985). Tracing the inflow streamlines at the southern end of Great Belt, one encounters the difficulty of obtaining an acute-angled corner which is not in agreement with the dynamic behavior of a current system. But current measurements in the Vejsnaes Rinne (WITTSTOCK 1982, SCHAUER 1982, HATJE et al. 1976) have shown that all significant inflow events in the Vejsnaes Rinne are connected with Great Belt inflow events. Thus the existence of a right-hand eddy system in this area must be expected.

The second feature is the west to east direction in the area of the Fehmarn Belt sand waves. That direction also is not in a streamlined continuation with the Great Belt inflow system. The bedform pattern, consisting mainly of comet marks, is acute-angled to the bathymetric contours in the Fehmarn Belt area and has a continuous connection to the southern Kiel Bight. Again this pattern is not in accordance with a direct flow from the Great Belt through the Fehmarn Belt, but may be explained as a consequence of currents postulated by the resonator model of WITTSTOCK (see 5.3.1), according to which current-pulses may be linked to a backflow from Kiel Bight.

5.3.5.3 Interpretation of bedform dynamics

Many attempts have been made to define the hydrodynamic conditions of bedform generation in order to infer flow characteristics. Fields of hydrodynamic stability have been determined by laboratory experiments for ripples and sand waves (ALLEN 1981) and from natural observations (DALRYMPLE et al. 1978). The generation of longitudinal bedforms which are very important in our area, is currently extensively discussed, mainly with respect to the wide distribution of various types found in the deep sea (e.g. FLOOD 1982, LONSDALE 1982). Many aspects of the generation of such bedforms can be studied more easily in the shallow water and in the laboratory flume than in the deep-sea environment. This aspect enhanced the motivation to carry out the following investigations (WERNER et al. 1980, McLEAN 1981).

For application on our case, the major problems may be grouped as follows:

(a) In order to estimate at which current conditions the bedforms were in equilibrium with the flow, the bottom shear-stress should be measured. However, the corresponding experiments failed because of general difficulties (section 5.3.2);

(b) As the major part of the study area is covered with longitudinal bedforms, stability conditions referring to these different morphological types should be known. For these reasons, experimental studies on comet marks and sand ribbons have been carried out (WERNER et al. 1980, McLEAN 1981);

(c) The currents in the study area have a highly variable, intermittent characteristic and significant current events are active during short periods only. Thus, all bedforms have an explicitly historical aspect. The bedforms observed at present were formed by past events, and the age of different forms may vary considerably. An approach we used to solve the problem was to repeatedly observe bedform patterns in certain time intervals and to try to identify the current events responsible for the observed changes. However, the historical aspect also implies that generation and subsequent maintenance of bedforms are due to different current events.

Laboratory flume experiments on the generation of sand ribbons (McLEAN 1981) have shown that strip-like patterns of different bed roughness are able to initiate a three-dimensional current system which develops and maintains this roughness pattern. This is due to inhomogeneities in the field of boundary shear stresses which are created by the non-uniform roughness distribution and which induce a secondary flow. Although the cross-section component is as small as about 2 % of the downstream velocity, the flow system can be interpreted as the much-cited 'helical flow' where the cross-stream component helps to sweep sand from the rough strips (consisting of gravel) towards the smoother sand ribbons. An important factor in the development of sand ribbons is the relation between flow depth and ribbon spacing. A theoretical model developed by McLEAN (1981) on the basis of his flume experiments shows a clear dependence of the normalized cross-stream shear stress perturbation on the spacing where the maximum secondary circulation occurs when the ratio of spacing to depth is about four. This ratio was actually measured in the best developed sand ribbon fields of the Great Belt (WERNER and NEWTON 1975 and Fig. 5-42). However, in many other places with less developed sand ribbons, the spacing depends largely upon the statistical distribution of roughness elements of the bed and is, therefore, rather irregular.

Due to the inherent feed-back mechanism between roughness distribution and sand transport, the secondary flow system causes a high degree of maintenance of the once formed sand ribbons.

The feed-back mechanism is effective also in the generation of sand ribbons in nature. The development of longitudinal bedforms is favored by a reduced sediment supply leading to thin sand layers covering an older, coarse relic bed which in places may be exposed to the bottom flow and then offer its higher roughness property to the flow. This frequently happens behind the randomly distributed boulders where shear stress fields similar to that of the sand ribbons are developed by the wake flow (WERNER et al. 1980). The tendency towards a preferred spacing thus would cause the comet mark circulation cells to adjust to those circulation cells corresponding to the sand ribbon geometry maintained by the feed-back mechanism between sand transport and shear stress distribution.

Sediment transport is reduced on the sand ribbons, the bottom shear stress being about 50 % of the mean. Although the directional pattern (Fig. 5-40) may suggest the existence of a considerable sediment transport towards the south, examinations with sub-bottom profiling failed to show any significant fan-like sand deposition in that area since the Holocene marine transgression (WINN 1974).

5.3.5.4 Response of Bedforms to Current Events

The observations made on the temporal variation of bedform patterns in the belt area (WERNER and NEWTON 1975, KUIJPERS 1980 and unpublished data) can be summarized as follows:

(1) Sand waves. The two main sand wave fields in the area shown in Fig. 5-40 have very different mobilities. The Fehmarn Belt sand waves are found to be stable for decades, which is mainly indicated by the fact that their crests are densely populated by the long-living bivalves like Arctica islandica and Mya arenaria (WERNER and ARNTZ 1974). Observed small-scale ripples on the sand waves indicate that sand migrates at least during major events, but without displacing the forms as a whole. In contrast to this, the bipolar shape of the sand-waves in the eastern Langeland Belt (Fig. 5-42) indicates responses to present-day conditions.

(2) Sand ribbons. All sand ribbons checked from time to time were found to be stable in their location.

(3) Comet marks. There are indications of rapid formation of comet marks, based on the presence of bipolar forms (WERNER et al. 1980) and on the observation that between two different surveys the direction of the wake erosion was changed to the opposite. During winter 1979 extraordinarily strong easterly storms caused an extreme outflow situation, resulting in erosional tails up to 15 m long in the western part of Fehmarn Belt channel, an area where for years only inflow-oriented comet marks had been found. This change was evidenced by two surveys the time interval between the two side-scan surveys being four months.

(4) Mega-ripples. Mega-ripple fields may be reformed or newly generated from plane sand bottom during a single current event. This has been evidenced by two side-scan surveys before and after a strong, but not extreme inflow event (KUIJPERS 1980). As on the site of this field of 5 m mega-ripples on several surveys in the preceding years, plane bed (or small-scale ripples not seen in the sonography) was found, thus it may be assumed that the mega-ripples will be levelled down during events of an order affecting the sea bed several times a year.

During surveys in the preceding years a plane bed (or small-scale ripples not detectable with side-scan sonar) had been mapped instead of the 5 m mega-ripples seen at the last survey; combining the regional distribution of bedform groups in the Belt area

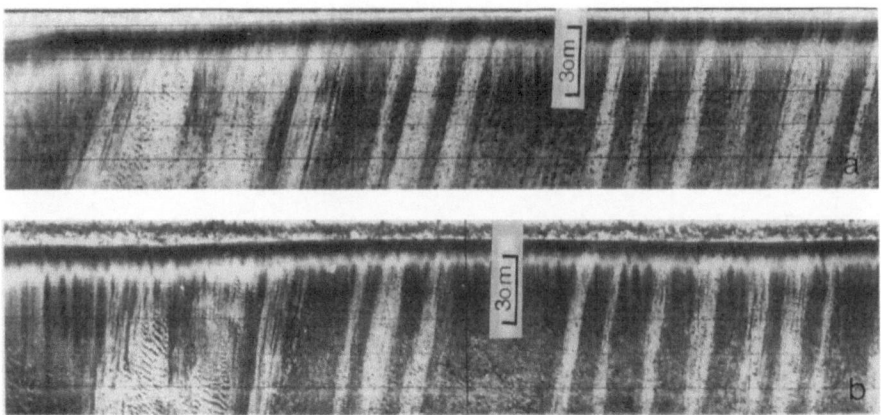

Fig. 5-42: Sand-ribbon field in the southern Great Belt in two side-scan sonographies, (a) from July 1972, (b) from September 1973, showing identity of
ribbon patterns. Minor differences are mainly due to change of form and
polarity of comet marks and to mega-ripple developments. Length of scales:
30 m. (WERNER and NEWTON 1975).

(Fig. 5-40) with their response to current events allowed an assessment of the dynamic
range of bedforms (Fig. 5-43). WERNER and NEWTON (1975) used the regional grouping of
the tail lengths of comet marks to relate them to the corresponding current conditions
and evaluated their positions within fields of different transverse bedforms. Flume
experiments on the hydrodynamic conditions of generation of comet marks (WERNER et al.
1980) actually indicated a positive correlation between current velocity and length
of comet marks within the dynamic range of the stability field of ripples and dunes
(Fig. 5-43). With increasing current velocity the intensity of the helical flow pattern developed in the wake of the obstacle increases.

The correspondingly increasing length of the comet marks could be explained by
assuming that the intensity of the cross stream components in the turbulence pattern
of the surrounding does not increase to the same extent. As this turbulence pattern
is determined by the vortex fields of small-scale ripples this is obvious. A sudden
increase of the comet length was noted when the small-scale ripples disappeared at
the upper limit of their stability. In this case, dunes would be formed on the sand
bed. However, under the natural conditions where comet marks are found, dunes (or
mega-ripples) are rarely observed because of the reduced sand supply. Due to the
absence of dunes as disturbing elements, the comet marks could develop to very long
forms similar to 'narrow' sand ribbons or 'sand streaks'. Both the development of
sand ribbons from randomly distributed comet marks and the regional distribution of
the bedforms (WERNER and NEWTON 1975) suggest that the maximum flow velocity is
related to the formation of sand ribbons.

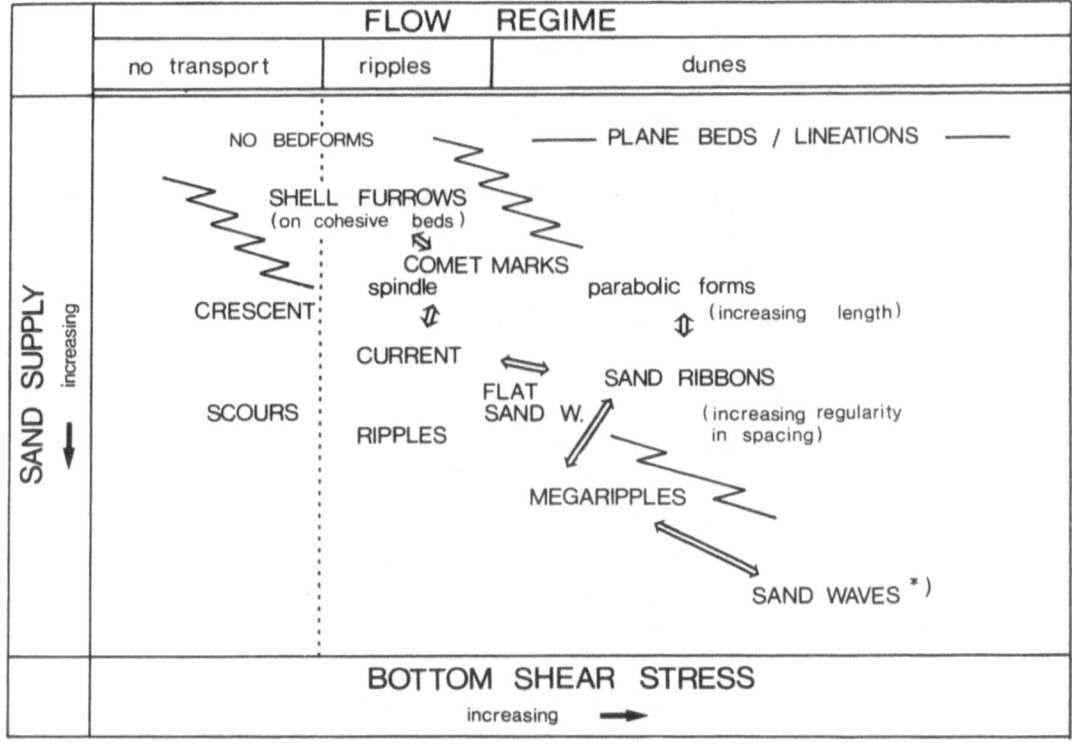

Fig. 5-43: Schematic diagram shows the main bedform types of Kiel Bight and the Belt areas in relation to flow conditions. Different co-existing forms are connected by double streaks. Transitions between bedform fields are separated by zigzag lines. For conditions of formation of crescent scours, shell furrows and comet mark types see WERNER et al. (1980).

5.3.6   The Impact of Single Events

5.3.6.1 Introduction

Recently, the characteristics and significance of storm-influenced sediments of ancient and modern sequences have been extensively discussed, in particular with respect to an approach to an "event stratigraphy" in mesozoic epicontinental seas of Mid-Europe (SEILACHER 1982, EINSELE and SEILACHER 1982, NELSON and NIO 1982, AIGNER and REINECK 1982). Certain characteristics of internal structures of depositional units, including an erosive base layer, shell assemblages, ripple lamination, etc., have been used to define corresponding sequences as "tempestites" (SEILACHER 1982). Modern examples were reported from the North Sea (GADOW and REINECK 1969, AIGNER and REINECK 1982), the Gulf of Mexico (MORTON 1981), and from the Bering Sea (NELSON and NIO 1982).

Using trends in these and other sedimentological parameters normal to coast lines, proximal and distal storm sequences were defined. In the Baltic Sea, WERNER (1967, 1968) has interpreted trends and frequency of sand and silt laminae in Kiel Bight as indicators of storm-induced deposition. The papers cited refer to the interpretation of sequences where the related meteorological events are unknown, with exception of that of MORTON (1981) who studied deposits of hurricanes immediately after the event. In our present study we tried to follow up in the sediment the impact of single known events to sediment accumulation. In two examples storm-generated sediment layers were investigated, one concerning the mud basin of Eckernförder Bucht, the other one the Vejsnaes Rinne.

5.3.6.2 Storm Layers in Eckernförder Bucht

In winter 1979, Kiel Bight was affected by three extraordinarily strong easterly storms. Such events cause maximum wave energies, high water levels and violent coastal erosion in the western Kiel Bight at the same time high current speeds were measured at the bottom of Eckernförder Bucht (e.g. GEYER 1965).

During the following year, box samples were taken from the bottom of Eckernförder Bucht (Fig. 5-1) in order to investigate possible effects of these storm periods on the structure of the sediment surface (KHANDRICHE 1984). In most of the cases a distinct surface layer of 10 to 25 mm thickness was found, in contrast to the findings in many samples taken from this area during previous studies (e.g. WERNER 1967, 1968). In the deep of the basin, this surface layer has a definitely higher sand content than the section below (Fig. 5-44). Its structural characteristics vary (Fig. 5-45). In the outer part of Eckernförder Bucht, ripple fabric occurs (Fig. 5-46). The fabric shows that during such events bed-load sand transport occurs as an exceptional process in the mud basin.

The storm layers also provide some information on the magnitude of maximum sedimentation rates in the study area.

Average and maximum sedimentation rates measured there are discussed in chapter 5.2.3. According to these results, average sedimentation rates of 1.4 mm y^{-1} for the deeper part of the outer Eckernförder Bucht roughly agree with bulk values derived from sediment thickness and age of the Littorina transgression. The extreme value of 8.4 mm y^{-1} cited in 5.2.3 is derived from a short core interval, but the sedimentary structure of this interval is not known. WERNER (1968) describes several sections without bioturbation up to 11 cm thickness from sediment cores of the outer Eckernförder Bucht. He interpreted these layers as having been generated by storm-

229

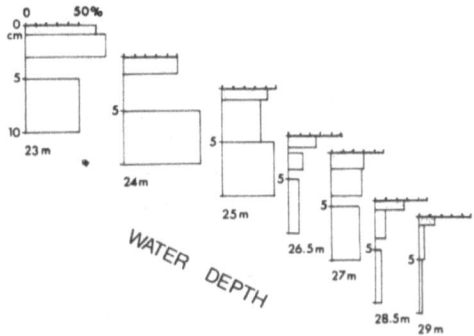

Fig. 5-44: Sand content (% ≥ 63 μm) of a storm layer (stippled) on top of surface sediments on the northern slope of Eckernförder Bucht (after KHANDRICHE 1984).
For location see Fig. 5-1, eastern most profile.

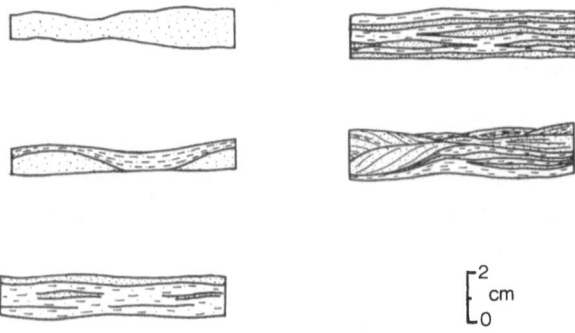

Fig. 5-45: Structures of storm layers observed in box cores of Eckernförder Bucht
sediments after storm events of winter 1979 (after KHANDRICHE 1984).

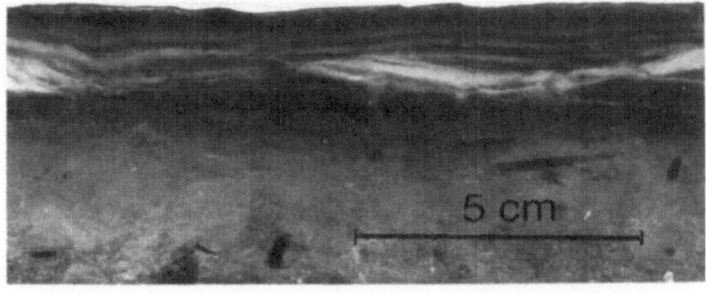

Fig. 5-46: X-ray radiograph (negative) showing typical structure of surface sediment
(Eckernförder Bucht) with top layer interpreted as storm deposits
(a) 25.5 m water depth, (b) 28 m water depth (after KHANDRICHE 1984).

induced water movement. The cores were taken from the narrow part of the Eckernförder Bucht channel off Boknis Eck. No similar sections of ripple bedding of millimeter scale and parallel lamination could be found in other cores from the wider Eckernförder Bucht. Therefore this structure pointed to an exceptional case of topographically intensified current regime. In contrast to these exceptional conditions, the storm layer found at the sediment surface represents a regional effect which can be considered as a regular contribution to the sediment column. The question however is up to what proportion such storm layers may contribute to the total sediment column. The importance of this contribution may be assessed from the time series of highly fluctuating sedimentation rates measured with sediment traps moored in Eckernförder Bucht (SMETACEK 1980b and chapter 2, this vol.). During the storm events sedimentation rates suddenly increased due to admixture of minerogenic material. The evidence from the sediment cores shows that such storm layers, due to the mixing process of bioturbation, will have only little chance to be preserved. The probability to be preserved increases with increased water depth and from the outer, better ventilated areas of Kiel Bight to the sheltered inner parts with its reduced benthic life.

Quantitatively, one has primarily to take into account that dynamic events of the order of magnitude in question will also cause a certain erosion of the sea bed. Indications for erosion in our area are high peak values of sediments collected in the cited sediment trap experiments (SMETACEK 1980b).

As it is not known to which amount these values are due to local sea bed erosion, they cannot be used for estimating the relation between 'normal' sedimentation rates and those due to storm events. If one concedes, however, that one half of the layers is due to local erosion, and the rest imported from higher slope levels and the heavily affected cliff coast during such events, about 10 to 15 mm of wet sediment would be approximately the maximum amount of storm-event deposits in the near-coast basins.

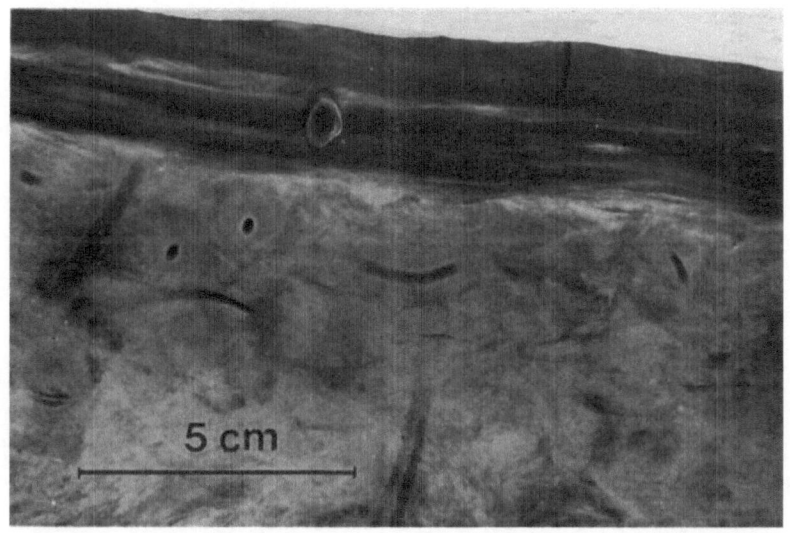

Fig. 5-47: X-ray radiograph (negative) of eastern Vejsnaes Rinne surface sediment (near PMA current measurement site, see Fig. 5-1), showing top layer of laminated mud and fine sand, interpreted as storm event deposit. (Courtesy G. UNSÖLD).

5.3.6.3 Storm Layers in Vejsnaes Rinne

The sediments in the eastern, narrow part of the Vejsnaes Rinne consist of sandy mud. One might expect that extreme current events as recorded in long term measurements (section 5.3.1 and 5.3.2) cause sedimentation or erosion processes observable in the sedimentary structure. But the sediments below the top layer are totally bioturbated (Fig. 5-47) showing that the maximum thickness of single event layers does not exceed the depth range of bioturbation activity. The fossil result would be similar to that obtained from the Eckernförder Bucht sediments. Fig. 5-47 shows a section of a box sample which was taken shortly after the strong inflow event of November 24, 1981. Apparently as an effect of this event, a top layer of ca. 3 cm thickness had developed. Since in this part of the channel little Holocene sediment has accumulated (section 5.4.2), it must be concluded that this sediment layer and corresponding deposits from similar events do not contribute to the final sediment column; it occupies its position only until submitted to later erosion. Apparently each strong current event starts with an erosion phase carrying away enough sediment for the budget to be balanced (section 5.4.2).

5.3.7   Summary

On the basis of a <u>resonator model</u> with Kattegat and Belt Sea representing the channel and the Baltic Sea representing the reservoir it is possible to predict current fluctuations from certain meteorological situations. The model makes clear that the resulting current velocities can attain high values, which may be registered as 'current events'. Yet it is still hardly possible to predict the erosional effects of such currents on the sea bottom, unless the near bottom velocity profile (up to about 1 m distance) has been determined by at least three points.

Measurements of <u>near bottom velocity profiles</u> in the Vejsnaes Rinne and in the Jade tidal current environment showed significant deviations from the logarithmic velocity profile. Several models which could explain such deviations are discussed. The atmosphere model proposed by BLACKADAR (1962) offers an analogy which merits further consideration. Reynolds shear-stress calculations from high-frequency velocity measurements in the boundary layers threw light on the basic difficulties in determining the boundary shear-stress distribution of natural flows.

The <u>efficacy of surface waves</u> on the sea bed was studied by an investigation on the formation of oscillation ripples. Ripple fields in different water depths and on sediments of different grain size were surveyed by high resolution side-scan sonar each time after a wave event, which could be expected to have rebuilt the ripples. The corresponding wave parameters were determined from long term wave records. The data set gathered was evaluated using the 'critical-state function' proposed by DINGLER (1979). It fits so well that its tentative application seems to be warranted. The results showed that oscillation ripples never record the maximum level of wave events, but some stages of their declining phases, dependent upon the recording capability and availability of sediments with corresponding grain sizes.

Experimental studies on <u>critical conditions of sediment entrainment</u> in a laboratory flume yielded the following results: the erosion resistance of cohesive sediments strongly depends upon their fabric, notwithstanding other factors. A correlation between critical shear-stress and vane shear-stress is perceptible, but not of predictive quality.

In studying the critical entrainment conditions of non-cohesive material the visual assessment of the critical state was replaced by measurement of the small transport rates. Thus it became possible to develop a Shields' type diagram into a family of curves with transport rate as parameter. For the range of very fine grained non-cohesive material the results of WHITE (1970) and MANTZ (1977) could be confirmed: in contrast to the extrapolation by SHIELDS (1936) the critical shear-stress decreases with decreasing grain size down to a limit where cohesivity begins, caused merely by the

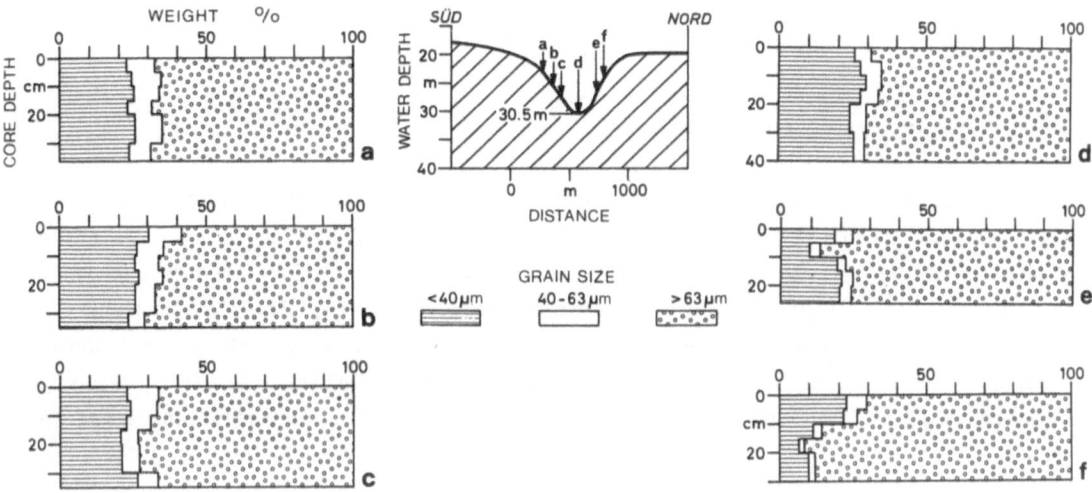

Fig. 5-48: Grain size composition of surface sediments of the eastern Vejsnaes Rinne
(PMA site, see Fig. 5-1 ⬤). (From SCHAUER 1982).

increase of specific surface. For quartz this limit lies near a diameter of 16 μm.

Studying the process of 'armouring' (formation of lag sediment) offered some insight
into the single grain processes of the entrainment mechanism, which helped to develop
a unified idea of this phenomenon.

Sonographic mapping of the regional distribution of current bedforms revealed that
in most parts of Kiel Bight no large-scale bedforms occur. In the Belt channels, however, such forms are widely distributed and can be used as current indicators. They
delineate zones of different influences of in- and outflow.

The influence of storm events on sedimentation in the deeper parts of Kiel Bight has
been studied in two cases. During an extreme easterly storm period 2 cm of laminated
sand and clay were deposited on top of sediments in Eckernförder Bucht. This is a
maximum sedimentation rate. In the Vejsnaes Rinne, a strong westerly storm produced a
top layer of comparable thickness. Most of the specific fabric of such layers is ex-
tinguished by subsequent bioturbation in basins or by later erosion in channels.

5.4   SPECIAL PATTERNS OF SEDIMENT DISTRIBUTION IN KIEL BIGHT

5.4.1   Patchiness of Lag Sediments

In water depths down to 12 m, gravelly-pebbly residual sediments ('lag sediments' in
Fig. 5-3) are frequently found with random surface sampling, although their relative
abundance may be quite variable in different regions. This is indicated only roughly
in Fig. 5-3. They are particularly abundant offshore retreating cliffs or on submarine
hills, but also there they are frequently interspersed with patchy sand areas
(Fig. 5-49 and 5-50, BRESSAU 1952, HINTZ 1958, FLEMMING and WEFER 1974). As revealed by
side-scan mapping, lag sediment patches occur also in sandy and muddy sand areas in
greater water depths (Fig. 5-49).

This patchiness can be due to a combination of different causes. Off retreating
cliffs, the lag sediments resting on the abrasion platform are residuals of the
eroded sediments. They may as well reflect the morphology of the youngest till cover
(Fig. 5-51). In other cases inhomogeneities of the underlying sediment sources, e.g.
alternations of till and outwash sands may be the cause of patchiness (WINN et al.
1982).

The patchiness implies that sediments of very different grain sizes (pebbles, gravel,
coarse sand and medium to fine sands) are simultaneously exposed to wave and current
action at definite depths. The finer sediments show a close relation between grain
size and water depth (Fig. 5-4, WERNER 1967, SEIBOLD et al. 1971). This relation
may reflect mainly the decrease of wave energy with water depth, as found also in
other comparable basins (JAGO and BARRUSEAU 1981). However the question remains why
coarse sediments in corresponding depths are not covered with finer sands though
these are in equilibrium with the prevailing hydrodynamic conditions effective there.
Two answers are possible:
- even coarse sand is moved occasionally in water depths down to about 16 m, as shown
  by the occurrence of oscillation ripples on such sediments (see 5.3.3; NEWTON and
  WERNER 1972);
- in wide areas, fine sand is not available in sufficient amounts.

The mobile fine sands are often organized in large-scale streaks alternating with
coarse sediment (Fig. 5-51, WERNER et al. 1976). The streaks are considered to represent an initial stage of mega-ripples or sand waves. They are thin veneers of fine
sand having almost no relief. A sequence of samples across one of the western sharp
boundaries of a streak visible on the side-scan sonograph (Fig. 5-51) proves a strict
separation of two grain size populations (Fig. 5-52). From the gradual transition
visible in the same side-scan sonograph it can be inferred that the two grain size

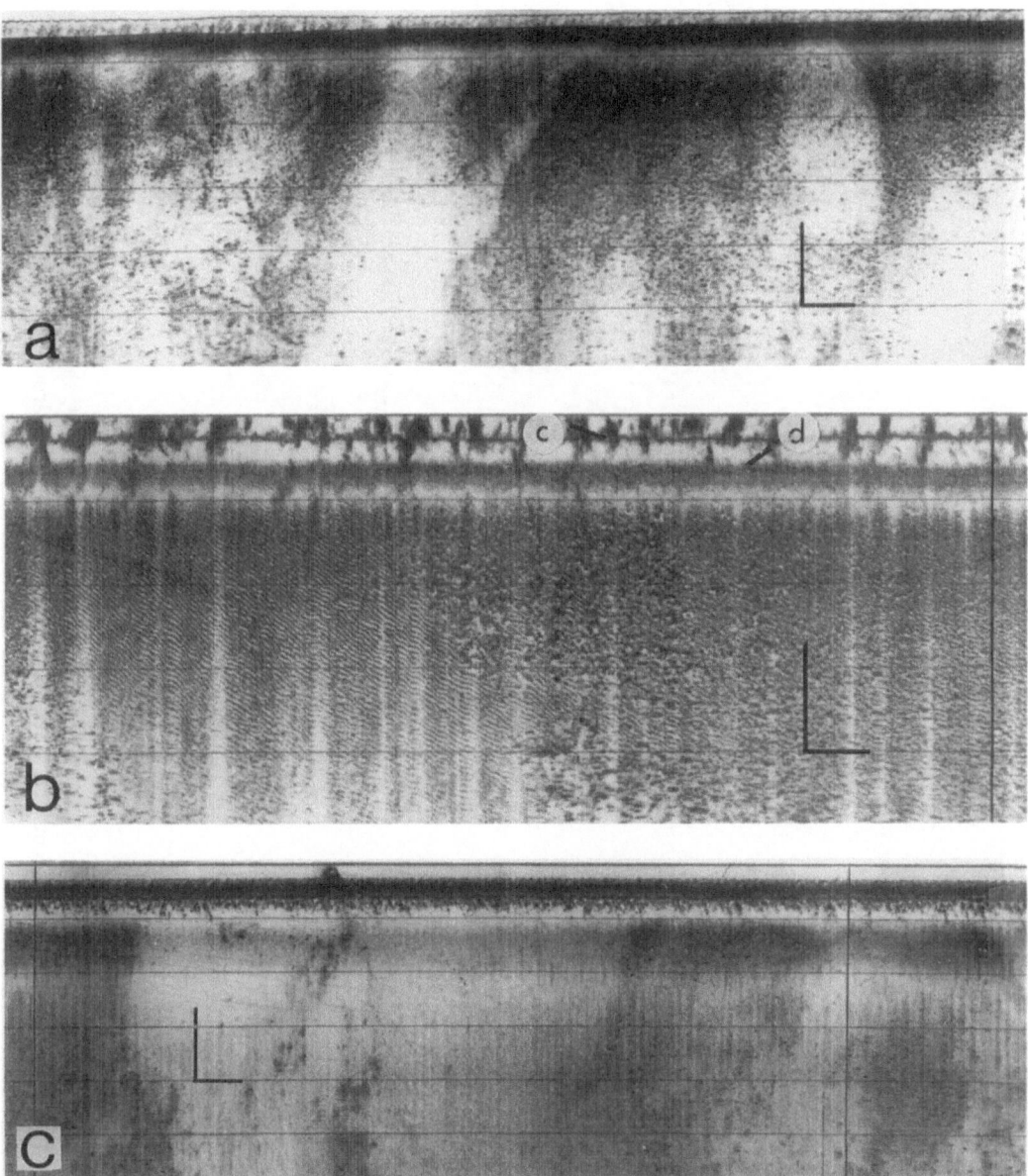

Fig. 5-49: Side-scan sonographies showing different sediment distribution patterns typical for lag sediment areas. (a): Patchy distribution of coarse residual sediments and sand, 12-13 m water depth. (b): Alternating areas of well-sorted coarse sand with ripples and gravel lag sediments with pebbles and boulders, 13 m water depth. (c): Coarse sediment areas in silty fine sand, 19 m water depth. Length of bars in all figures: 20 m. For location see Fig. 5-1.

236

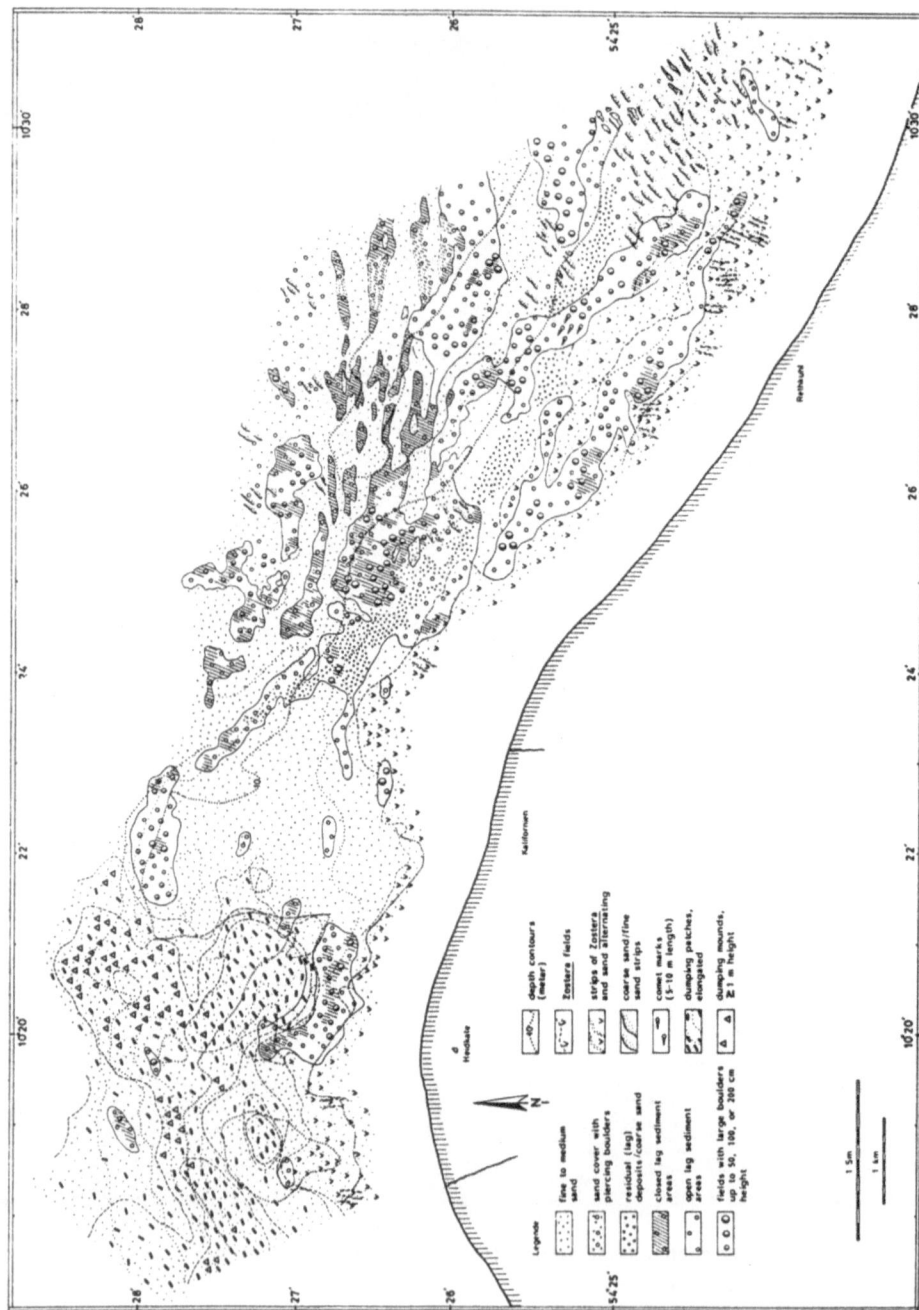

Fig. 5-50: Sediment distribution after side-scan sonar surveys in an offshore lag-
sediment zone of the southern Kiel Bight (WERNER 1980). Area 'C' in
Fig. 5-1.

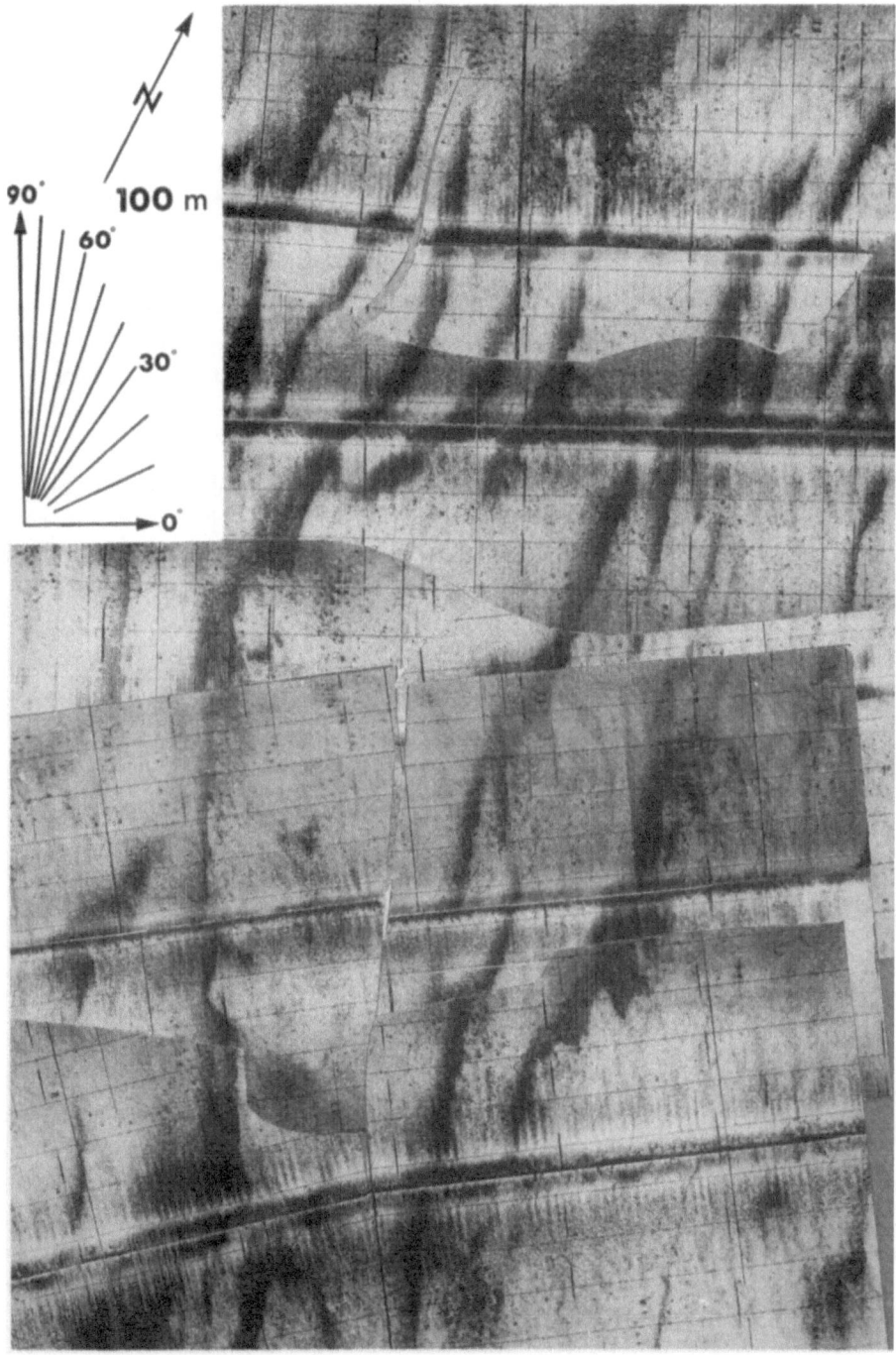

Fig. 5-51: Lag-sediment (dark) and thin veneers of migrating fine grained sand on Stoller Grund ('A' in Fig. 5-1). Sonographic mosaic (WERNER et al. 1976).

populations gradually merge into each other at the eastern boundaries of the streaks. Similar asymmetric streaks apparently are very common features on continental shelves, as shown by many sonographic surveys (NEWTON et al. 1973, McKINNEY et al. 1974, TABAT 1980, KNEBEL et al. 1982).

A general theory on their formation does not exist as yet. The following considerations may give a tentative explanation.

Given some patchy sediment distribution of irregular geometry in the area, the highly different bed roughness are thought to cause in average a preferred trend of sediment transport towards the sand areas in analogy of the process of sand ribbon formation described in section 5.3.5 (McLEAN 1981).

The basic mechanism of the processes involved can be seen in the statistically higher chance of any sand grain in motion for settling down on a bed of lower roughness (sandy areas) than on one of higher roughness (gravel areas). From the experiments cited above, an excess bottom shear stress of about 50 % was derived on the rough beds (see also RUBIN and McCULLOCH 1980). Differences of this order of magnitude should be sufficient to cause a general tendency to stabilize the sediment patterns, whose geometrics will be controlled by the local current systems.

In areas with prevailing unidirectional flow the maximum elevation of these transport bodies is located downstream of the center of the streaks.

An expression of this fact is that the downstream border of a streak is sharp and the upstream border appears to be blurred by the underlying roughness more or less penetrating the thin sand cover. Maintenance of the sharpness of the downstream border of a streak is supported by the reversed direction which the cleaning effect must have there.

The studies on the wave effectivity (see 5.3.3) have shown that also the coarse grain size population can be correlated with water depth, if relative maxima within a depth class are considered. It has to be mentioned, however, that not all of the lag sediment is dynamically active in this way. Beyond a certain grain diameter no further movement will occur. Exact knowledge about the size of the critical diameter does not exist. At least, however, stones of block size, seem to be immobile. They can be observed everywhere on lag sediment areas interspersed within ripple patterns (Fig. 5-49b).

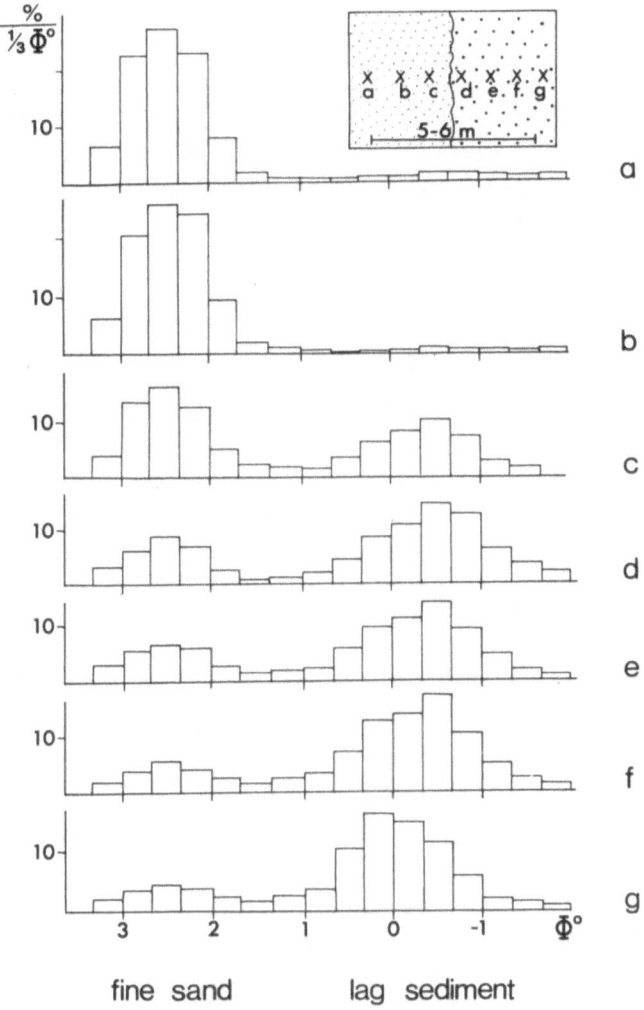

Fig. 5-52: Grain size variation across the boundary of an asymmetric sand strip of Stoller Grund shown in Fig. 5-51 (WERNER et al. 1976).

5.4.2   Channel Sedimentation

5.4.2.1 Introduction

As the deep water circulation in Kiel Bight uses the channels as paths of propagation, one should expect that the recent sediment distribution as well as the sedimentary filling during the Holocene is affected by this circulation. In the following, such current effects in Kiel Bight channels will be discussed.

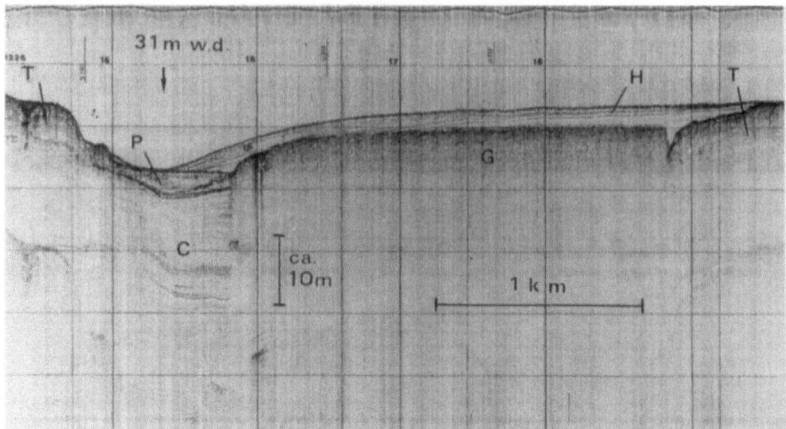

Fig. 5-53: High-resolution seismic section (Uniboom) across the central Vejsnaes Rinne, showing convex-layered Holocene sediment (H) with acoustic turbidity (G) due to gas content, older lacustrine postglacial sediments (P) and glacial barrier deposits (varved clays) in the sub-glacial valley (C). Glacial till (T) on the channel flanks.

The main features observed in the main channels which may be attributed to such effects are:

- reduced sedimentation towards the channel axis;

- a coarsening of sediments towards the channel axis;

- a marked asymmetry in sediment filling is noted in most cases, as seen in sediment echo-soundings.

5.4.2.2 Vejsnaes Rinne

Sub-bottom profiling revealed the morphology of the younger sediments lying on glacial till or, in the center of the channel, on older glacial or postglacial sediments (EXON 1972, WINN et al. 1982, Fig. 5-53). A series of parallel 30 kHz-echogram cross sections shows a varying degree of sediment filling along the channel axis (Fig. 5-54). From east to west, the volume of the marine mud sediments increases but an open part of the channel can still be observed at the western channel end. As this feature seemed to be significant with respect to current dynamics, the variation of relative sediment filling along the channel has been evaluated from that series of echosounder profiles. In order to define a measure for this, the following limiting conditions are considered (Fig. 5-55). If the potential maximum sedimentation ($\triangleq$ 100 % degree of filling) in the channel were only controlled by the average regional sedimentation rate, one would obtain the sediment surface level by adding the regional

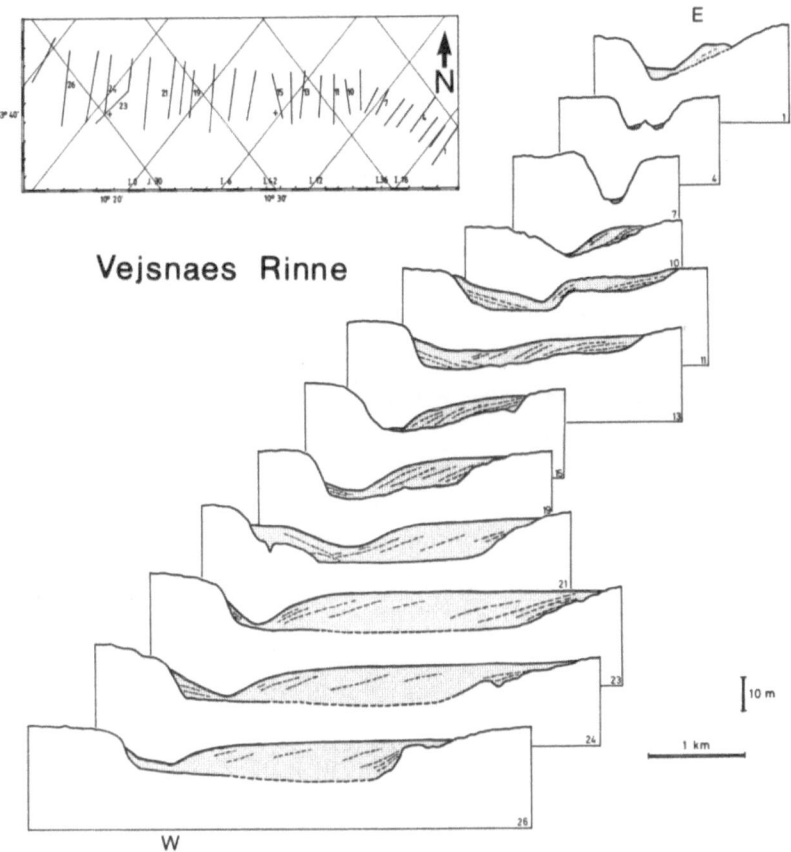

Fig. 5-54: Sediment echogram sections across Vejsnaes Rinne, based on a 30 kHz-echo-
sounder survey. Area 'V' in Fig. 5-1.

sediment thickness to the pre-marine basis level. But muddy sediment cannot grow up
to a higher level than is determined by the 'wave base'. This level is regionally
marked roughly by the transition from 'mud' to 'sandy mud' facies (see Fig. 5-3).
In places - like in the vicinity of the Vejsnaes Rinne - it can also be indicated by
very thin sand layers which overlap the boundary between Holocene mud accumulations
and adjacent Pleistocene plateaus. Such a situation can be seen on the right-hand
side of Fig. 5-53. The corresponding depth marks the local level of the wave base.
Knowing the regional average sedimentation rate and the level of the local wave base,
a theoretical '100 % line' can be constructed for each of the cross sections as indicated in Fig. 5-55. The area of a cross section not filled by sediments between the
basis of marine mud and the 100 % line is defined as 'sediment deficit'. In Fig. 5-56,
the 'sediment deficit' is calculated for the sub-bottom profiles and plotted against
distance along the Vejsnaes Rinne. On the same horizontal scale, the absolute values

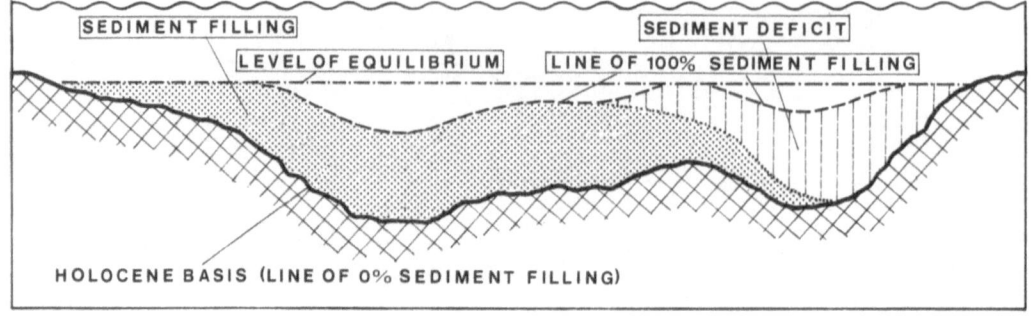

Fig. 5-55: Scheme illustrating the definition of 'sediment deficit' in channel
sections.

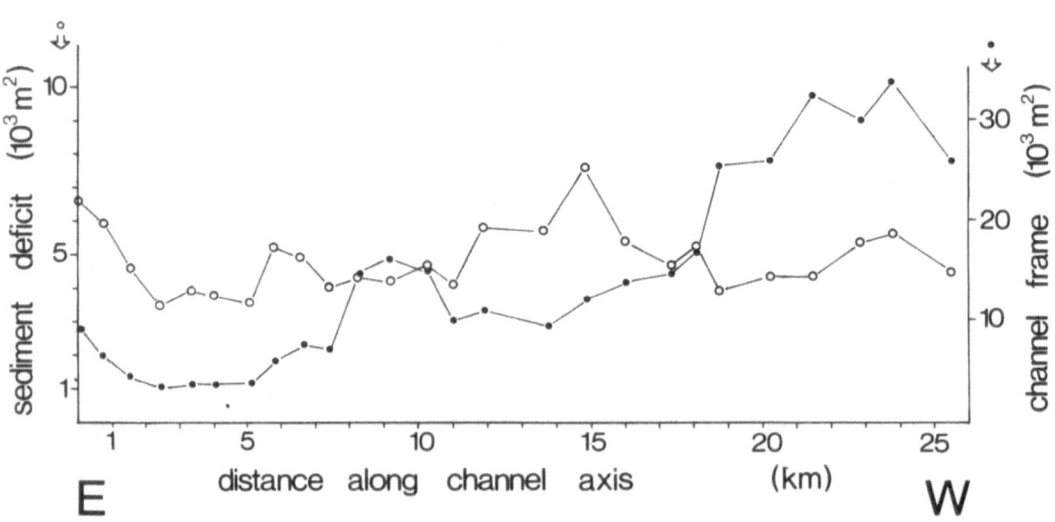

Fig. 5-56: Variation of 'sediment deficit' as illustrated in Fig. 5-55 along Vejsnaes
Rinne (Fig. 5-54) according to evaluation of echo-sounder sections (o).
Cross-section area between basis line to level of equilibrium (.)
(= channel frame).

of the open cross-section areas are plotted. Two results can be recognized:

- a fairly continuous decrease of the 'sediment deficit' from east to west;
- the open cross-section volume remains approximately constant.

These results strongly suggest an interaction between currents and sedimentation within the channel. Whereas in the eastern part practically no sediment has been accumulated during the Holocene, in the western part the overall sedimentation rate at the southern channel side equals that of the adjacent mud basin (Fig. 5-3) wherein the Vejsnaes Rinne ends. It can be assumed that the forces which kept the eastern part open and those which impeded the complete filling of the western part are the same. Therefore, the responsible currents can be supposed to prevent a future sedimentation to plug the western part as they have prevented in the past the sedimentation in the eastern part. These relationships indicate that at present conditions of equilibrium exist between sedimentation and erosion throughout the entire Vejsnaes Rinne ('zero net sedimentation'), controlled merely by the condition of continuity.

The asymmetric sediment filling also can be explained in terms of current dynamics: the position of the recent open channel at the northern side of the ancient inflow channel cross-section may be due to a deflection by currents by Coriolis force.

Also a number of other observations can be explained by the above model:

- evidence of lateral accretion of sediment;
- varying sedimentation rates in lateral and longitudinal direction;
- sediment coarsening towards the channel axis;
- evidence of stronger action of inflow currents than of outflow currents (see 5.3.2).

These observations are discussed as follows:

Lateral accretion. Throughout the channel the sediment echograms show an oblique internal stratification in the mud cushions of the southern flank (Fig. 5-54). The strata typically show a convex shape.

Sedimentation rates. In the western part of the Vejsnaes Rinne, one section containing extended mud deposits on the southern flank was cored with a number of piston and box cores. The results of ^{14}C-dating from six of these cores are summarized in Fig. 5-57. Three features are of particular interest:

- the sedimentation rates are considerably varying during the Holocene;
- an accentuated maximum of 13 mm y^{-1} occurs at the channel slope in a relatively advanced stage of development;
- the uppermost layer is characterized by minimum sedimentation rates throughout the cross-section.

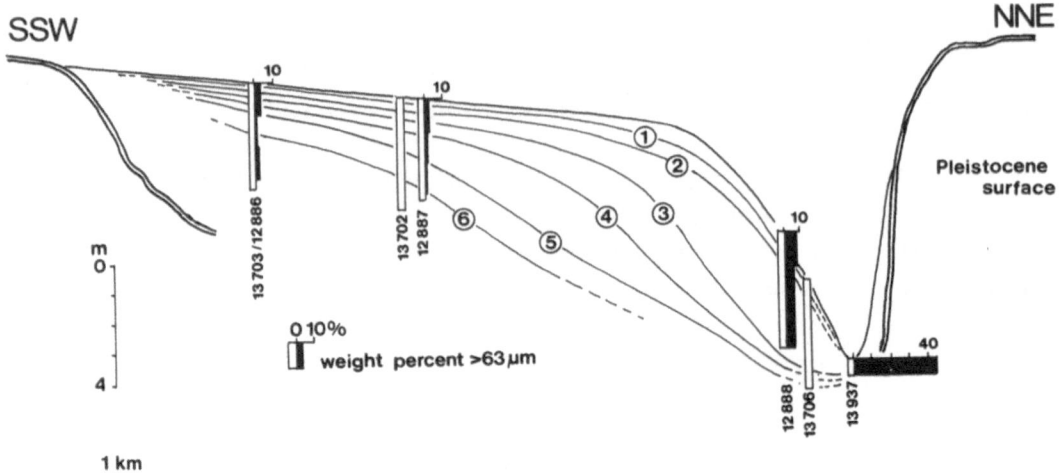

Fig. 5-57: Cross-section of Vejsnaes Rinne ('S' in area 'V', Fig. 5-1) showing grain size variation and isochrones according to numerous ^{14}C-determinations (ERLENKEUSER unpubl.) in indicated sediment cores. Numbers in circles indicate k-years B.P. (SFB-Bericht 77-79, S. 35).

All these results are in agreement with the model of a channel sedimentation tending towards dynamic equilibrium. The low values of sedimentation rates at present support the assumption that a quasi-equilibrium is approached. The maximum rate of up to 13 mm y^{-1} occurring between 2 and 3·10³ y.B.P. suggests corresponding high erosion rates in more 'upstream' parts of the channel.

The sedimentation in the Kattegat under competent influence of two current systems leading to high sedimentation rates east of the Skagerrak inflow and to very low rates near the Swedish coast (FÄLT 1982), are probably analogous processes.

Sediment coarsening. The results from sediment cores clearly show that the coarsening towards the channel axis is accentuated in the younger past (Fig. 5-57). Also this phenomenon supports the assumption of an increasing erosion frequency in the channel towards recent times.

Inflow-outflow relation. The asymmetry of the sediment fillings in other channels in Kiel Bight can consistently be explained by the assumption that the efficacy of inflow exceeds the efficacy of outflow and that the currents are deflected by the Coriolis effect. This is in good agreement with long-term current measurements (Fig. 5-7 and 5-33) and discussed in sections 5.3.1 and 5.3.2.

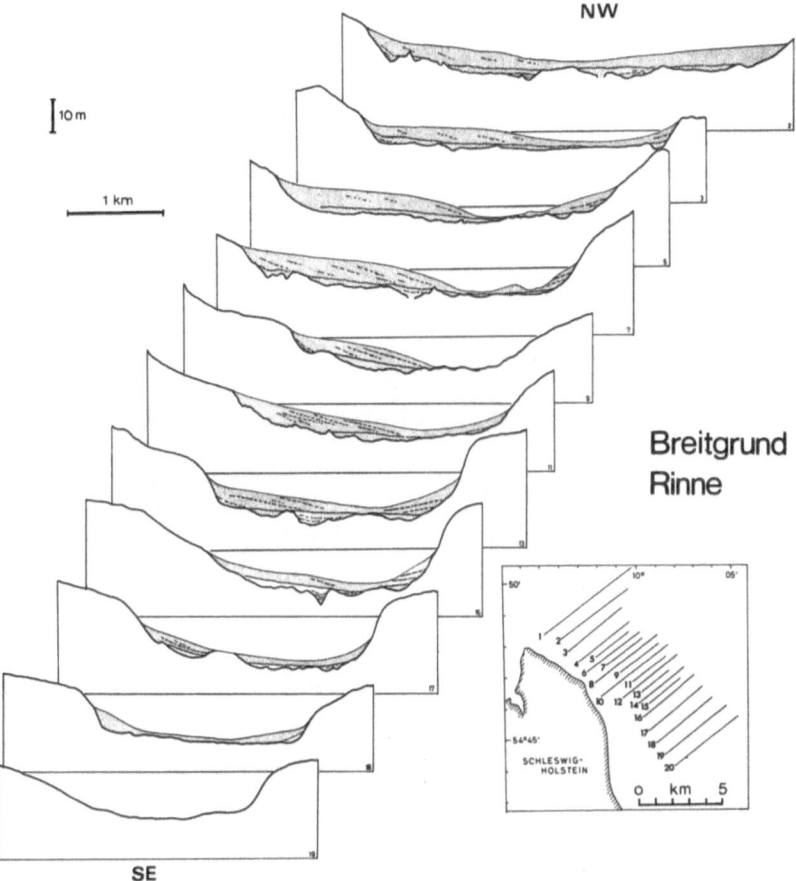

Fig. 5-58: Sediment echogram sections across Breitgrund Rinne, based on 30 kHz-echo-
        sounder survey. Area 'B' in Fig. 5-1.

5.4.2.3 Breitgrund Rinne

Sediment deficits detected by subbottom profiling occur in other channels of Kiel
Bight also. The Breitgrund Rinne connects the Geltinger Bucht with the basin in
northern Kiel Bight (see area "B" in Fig. 5-1). The echosounder sections across the
channel resulted in the diagram shown in Fig. 5-58. The characteristic features
appearing in that diagram are similar to those of the Vejsnaes Rinne:
- asymmetric sediment filling;
- negative correlation between sediment deficit and pre-marine channel cross-section (Fig. 5-59);
- lateral accretion of sediment filling;
- a slight increase of sand content towards the channel axis.

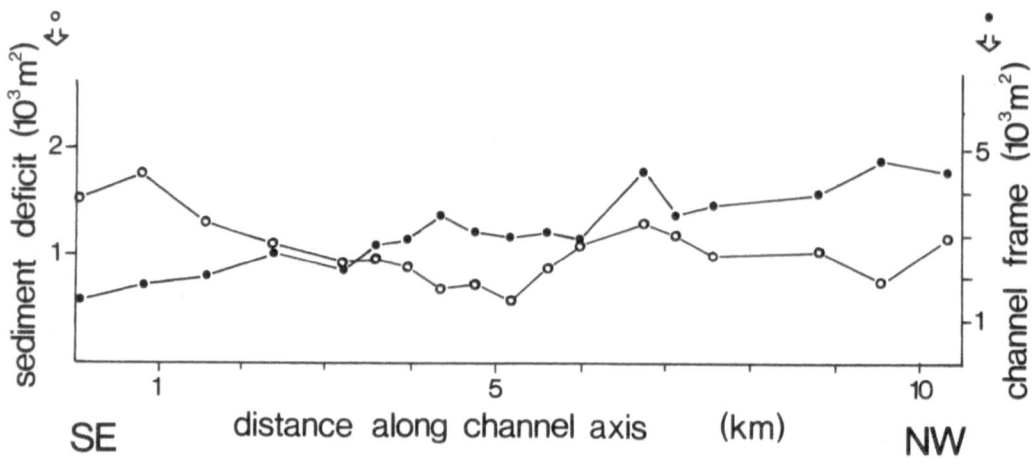

Fig. 5-59: Variation of 'sediment deficit' as illustrated in Fig. 5-55 along Breitgrund Rinne (Fig. 5-58) according to evaluation of echo-sounder sections. Scales as in Fig. 5-56.

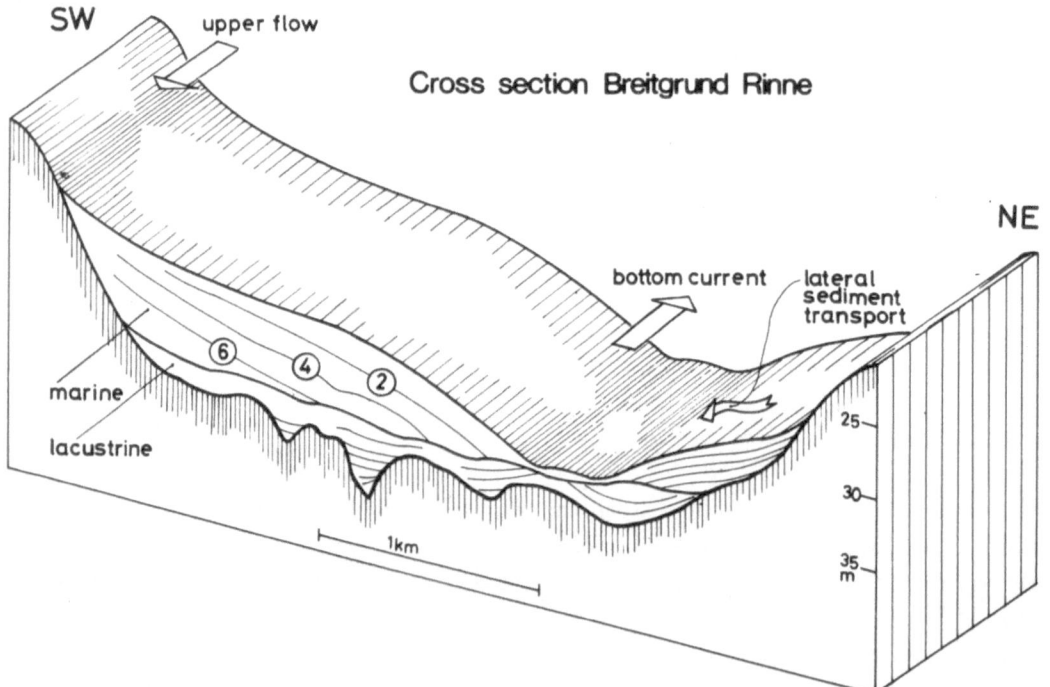

Fig. 5-60: Model of sedimentation in Breitgrund Rinne with isochrones according to ^{14}C-data. Numbers in circles indicate k-years B.P.

In general, the 'channel effects' are less marked than in the Vejsnaes Rinne (compare Figs. 5-56 with 5-59 and 5-54 with 5-58, respectively).

Fig. 5-60 summarizes the main results in a sedimentation model. The indicated isochrones in the sediment filling are determined from ^{14}C measurements of three sediment cores from a channel cross section.

5.4.3 Mixed Sediments

In places, channels and mud areas in Kiel Bight are bordered by a narrow zone of so-called mixed sediments (Fig. 5-3; SEIBOLD et al. 1971). It consists of abundant fine-grained material which is thoroughly mixed with coarse material of grain sizes up to gravel and rubble size. The grain size distribution is therefore similar to that of till (Fig. 5-61). A marked peak in the fine-sand fraction is typical.

It is obvious that the gravel and rubble fraction within the mixed sediments are comparable to the lag sediments of the shallower zones and can also be explained as relics from the underlying moraines. Occasional sampling of the underlying till base as well as evidence from seismic profiles shows that this basis occurs in very shallow depth below the surface, at the most in a few decimeters. The fact that the mixed sediments occupy the zone where two other sediment types adjoin, the mud and the sandy sediment types, shows the modern origin of the finer components.

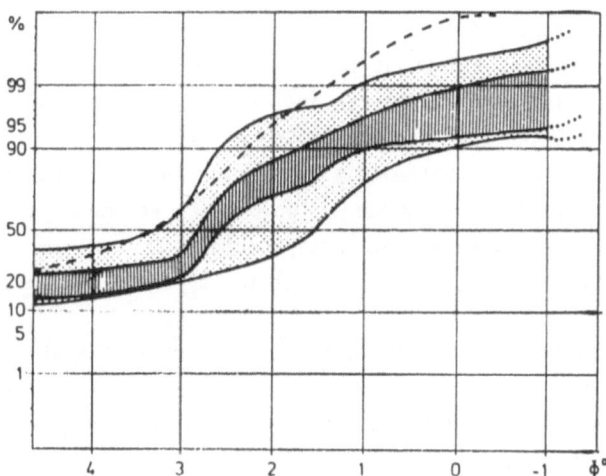

Fig. 5-61: Grain size distribution of typical mixed sediments according to DOLD (1980). Hatched: samples of one locality (DOLD's station T40). Dotted: samples from 4 different localities. Dashed line: average grain size distribution of muddy sand sediment of Kiel Bight.

The trends of mud and grain size fractions are crossing each other in the mixed sediment: clay and silt contents decrease in the sand matrix towards the shallower areas while sand contents rapidly decrease towards the mud zone within little distance from their borders (section 5.2). These sediments have therefore a marked palimpsest character according to the definition of SWIFT (1973) and may be compared with similar sediment types occurring in the outer zones of many open shelf areas where the sedimentation rate is near zero. Correspondingly, the mixed sediments in our area are best developed where the bathymetric gradients are low while in areas of steep gradients the zone of sand sedimentation passes directly into the mud zone.

The mixing process itself is a phenomenon of bioturbation. DOLD (1980), who studied the burrowing activities of the macrobenthos of Kiel Bight at a limited number of reference stations, found specific abundances of a few species of polychaetes and bivalves on those of his stations which had the character of mixed sediments (in particular: Pherusa plumosa, Rhodine gracilior, Astarte spp. (mainly A. elliptica), Arctica islandica, Macoma calcarea).

As a further characteristic of mixed sediments, the occurrence of ferro-manganese nodules of the limnic-brackish type (WINTERHALTER 1966, DJAFARI 1977) with substantial, disk-like accretion crusts is exclusively restricted to these zones in Kiel Bight.

According to HARTMANN (1964), DJAFARI (1977), GHIORSE (1980), SUESS and DJAFARI (1977), and HEUSER (pers. comm.), the following conditions control the formation of the nodules (see also chapter 4 this vol.):
(1) Mud basins with sediment thickness of at least several meters, where the reduction zone allows solution of manganese and iron and the diffusion of ions through the sediment-water interface into the bottom water;
(2) Temporary oxygen deficiency below 40 % saturation in the bottom waters;
(3) Bottom areas adjacent to the mud basins which bear sediment particles of shingle size on their surfaces as nuclei of precipitation;
(4) Conditions of little or no sedimentation at the sites of precipitation.

Combinations of all these factors occur only at some localities at the mixed sediment zones bordering the mud basin of northern Kiel Bight while other mixed sediments are free from ferro-manganese nodules. The places with nodules are at water depths slightly deeper than most of the other mixed sediment occurrences which is well in agreement with the described conditions and with the necessary oxygen fluctuations.

5.4.4   Summary

In this section, some special features of the sediment distribution in Kiel Bight are
discussed which are considered to have a more general significance.

The patchy character of sediment distribution in the lag sediment zone on abrasion
surfaces is a general feature of all environments where material with a sufficiently
broad grain size spectrum is available. Inhomogeneity of the underlying glacial
sediments, pre-marine relief features and sediment reworking processes control the
grain size differentiation generating the patchy sediment distribution.

Since the marine transgression into Kiel Bight originally subglacial channels were
reformed by a sediment filling, regularizing the originally widely varying cross
section areas to smaller and rather equal dimensions with a quasi-equilibrium
between sedimentation and erosion.

Where the slope is gentle, the mud basins of Kiel Bight are rimmed in places by a
zone of so-called mixed sediments which represent a palimpsest facies. The thorough
mixture of younger fine sediment with coarse relic material is an effect of bioturbation. Mixed sediments are the habitat of ferro-manganese concretions in Kiel Bight.

5.5   RECORD OF ENVIRONMENTAL CONDITIONS

Considering the different constituents forming the Holocene sediment column with
respect to their significance as environmental signals, mollusc shells and biogenic
structures were selected for special investigations in the SFB 95 project.

5.5.1   Shell Zonation

One of the central objectives of the SFB 95 project has been to investigate the
interrelationships between the hydrographic regime and the biological production. It
is obvious that only a small part of the biological parameters can be traced into the
geological past as indicators of possible fluctuations in the 'Kiel Bight model'.
Benthic calcareous skeleton remains are rare in most parts of the sediments because
they are largely subjected to dissolution, including both macrobenthic shells and
microorganisms (RESIG 1965, LEWY 1975, WEFER 1976a). Thus the upper part of the
sediment column in Kiel Bight, except for the surface layer, is mostly free from
calcareous foraminifera, and mollusc shells are only found scarcely interspersed or
in occasionally occurring shell layers. On the other hand, both foraminifera and
molluscs show a clear depth zonation in the living state, being dependent upon the
stratified hydrographic structure of Kiel Bight (LUTZE 1965, EXON 1972, LUTZE 1974,

WEFER 1976a, ARNTZ et al. 1976). Particularly ARNTZ et al. have investigated the shell-bearing molluscs with respect to the hydrographic structure and its fluctuations and have found significant correlations between composition and frequency of faunal patterns and the average depth position of the thermo- and halocline. Moreover, the ecological conditions of mollusc development have been studied by population experiments using moored sediment platforms in different heights above the bottom (RICHTER and SARNTHEIN 1977, see also chapter 6, this vol.) and artificial sediments on the sea bottom (RUMOHR 1980). Diatoms are often abundant throughout the profile of mud sediment cores, but they have not been investigated except for an older study carried out in coastal sediments (DAHM 1956). Moreover, the very small diatoms are much more subjected to transport than the other organisms and therefore may not clearly reflect the hydrographic zonation. Accordingly, the shell remains of the mollusc fauna yet appeared to be the most promising tool to trace the Kiel Bight hydrography in the Holocene record.

The shell contents of core profiles from the western Vejsnaes Rinne, where sediment thickness is high (see 5.4.2) were analysed in 5 cm sampling intervals with respect to the species composition. The Vejsnaes Rinne sediments were considered as parti- cularly suitable for such studies because they are located in a hydrographic key po- sition where the fauna can be expected to react sensitively to hydrographic fluctuations (ARNTZ et al. 1976). In addition, grain sizes and organic carbon content were also determined, and ^{14}C ages were measured from the organic carbon fraction. These data and the distribution of the more significant shell species in the core profiles are presented in Fig. 5-62. The patterns show marked differences in vertical and horizontal distribution of the different species. Generally, the sediments of the open channel parts with their lower sedimentation rates show a smaller faunal diver- sity than the sediment blanket on its southern side. This is interpreted as an effect of the relative oxygen abundance which is lower in the deeper parts than near the pycno- cline at about 22 to 25 m water depth where the optimum living conditions of molluscs were found (ARNTZ et al. 1976). The thin-walled Abra alba shells having their maximum frequency in the deeper open channel indicate that this heterogeneous distribution may not be influenced substantially by local differences in carbonate dissolution.

Table 5-2 summarizes the results of the species distribution patterns from the core profiles as compared to the modern ecological characteristics of the species described by ARNTZ et al. (1976). According to this interpretation, the evolution of the marine Holocene of Kiel Bight can be divided into the following four stages:

(1) The oldest stage occurs immediately after the transgression until approximately 7000 y.B.P. It is characterized by shallow water forms: Mytilus edulis, Cerasto- derma edule, Hydrobia ulvae, Macoma baltica. The controlling environment is in- terpreted as shallow and brackish.

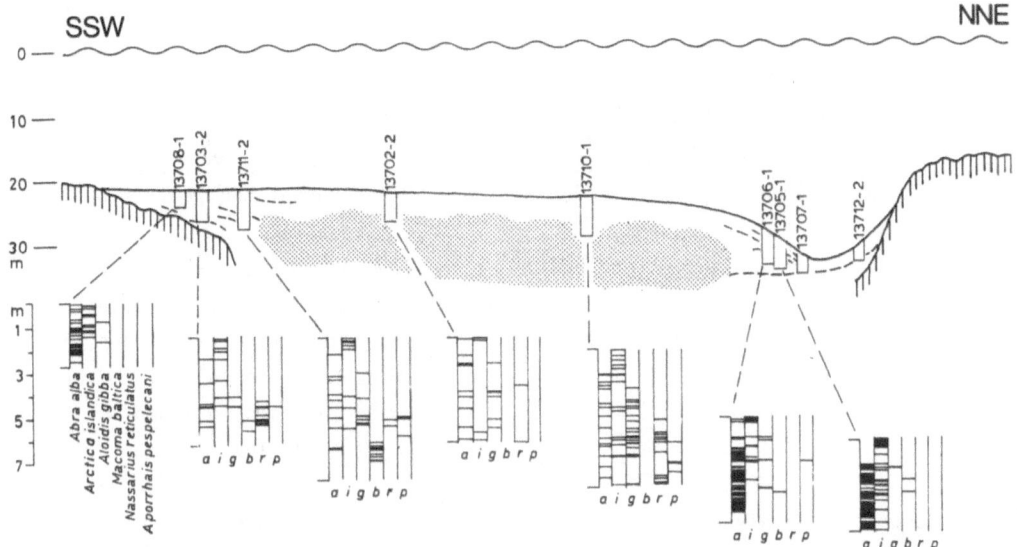

Fig. 5-62: Cross-section of Vejsnaes Rinne ('S' in area 'V', Fig. 5-1) showing distribution of mollusc species in Holocene sediments based on evaluation of shell content in sediment cores. Shadowed: gas-containing sediments. (After RICHTER and SARNTHEIN unpubl.).

(2) The second stage follows between 6500 and 5500 y.B.P. It is characterized by a relative abundance of gastropods (Apporrhais pespelicani, Nassarius reticulatus, Tryphonopsis truncatus) which live at present in fully marine environments or at least in the Baltic inflow areas. Also, the pelecypod Aloidis gibba indicates by its recent distribution little tolerance to low salinity (JAECKEL 1952), showing maxima of occurrence in well-aerated zones near the pycnocline and below the strong salinity fluctuations of the shallow water. During this stage Arctica islandica also appears for the first time. This period can be considered as a maximum of marine conditions in Kiel Bight in accordance with EXON (1972).

(3) The third stage between 5500 y.B.P. and about 3000 - 2500 y.B.P. is documented by a mollusc fauna consisting mainly of bivalves, including Abra alba and locally Mya truncata, in the early beginning Aloidis gibba and partly Arctica islandica. Moreover, Balanus sp. shows a maximum abundance. Abra alba and the Balanus maximum are interpreted as indicators of oxygen saturation in the channels and of an absence of very cold winter situations.

(4) The fourth stage from 2500 - 3000 y.B.P. until present is characterized by the general dominance of Arctica islandica and a decrease of Abra alba. Simultaneously, the organic carbon content increases. The environmental conditions are inferred to be approximately similar to the present with frequent intervals of oxygen deficiency and in addition probably also occasionally cold winters.

Table 5-2: Recent and fossil distribution patterns of molluscs in Kiel Bight

Species	Present day ecological characteristics	Occurrence in sediment column
Arctica islandica	Boreal species, tolerant of O_2-deficiency and salinity fluctuations	Throughout the profile but increasing since 5000 y.B.P., maximum after 2000 y.B.P., at decreased sedimentation rates. Not occurring in channel center and in coarse-grained northern slope sediments.
Macoma baltica	Mainly sandy bottom, shallower than 17 m	Abundant in early periods after transgression 6500 y.B.P.
Abra alba	Characteristic species in Kiel Bight, 11-33 m water depth, but mainly in muddy sediment between 18 m and 28 m. Little tolerance of oxygen deficiency and extremely cold winter seasons	Common soon after transgression, but drastically reduced in southern slope sediments since ca. 5000 - 3000 y.B.P. to present.
Aloidis gibba	In the range of the thermo-halocline, 15-22 m water depth. Preferably in oxygen-rich bottom water	Common when water depth increased during transgression, strongly reduced since 4700 y.B.P.
Nassarius reticula-tus	In moderately warm water, rare in the Baltic Sea	Time-transgressive on southern slope sediments, during their accretion between 7500 and 5400 y.B.P. No younger occurrence.

5.5.2   Bioturbation

Introduction

The impact of bioturbation on the sediment provides environmental signals for the geological record to a degree that is related to the state of knowledge on the environmental relationships of biogenic structures. In basins where the benthic fauna is restricted by environmental factors, the relative amount of bioturbated sediment as indicated by the preservation of primary sedimentary structures ('degree of bioturbation') can be useful for the interpretation of the environmental record (HERTWECK 1972, HOWARD and FREY 1975).

Commonly in Kiel Bight, the sediments in the deeper sand and muddy sand zones are completely bioturbated throughout the profile. This is plausible if one considers the small accumulation rate in these areas (see 5.3.3). In mud sediments, however, the situation is different. Erosion events or periods of stagnation may cause breaks in the bioturbation. Analysing the sedimentary structures in the sediment column may thus provide evidence for significant environmental changes or events (see 5.3.6).

Description of Biogenic Structures

To get a basis for better interpretation of biogenic structures laboratory experiments were carried out by DOLD (1980). He analysed the activities of the more important burrowing organisms and defined morphologically the generated structures in artificial sediments for diagnosis in natural sediments. Some of his results applicable for interpretation of biogenic structures found in sediment cores are referred to in the following.

Arctica islandica (syn. = Cyprina islandica). This large pelecypod lives near the surface of mud and sandy mud bottoms and occasionally moves through the sediment. This motion affects the sediment by several physical processes which lead to characteristic structures. The horizontal motion generates a sharp boundary to the sediment below like a plough sole and the organism leaves behind a disturbed shadow zone of well mixed sediment (Fig. 5-63). As the digging depth of the adult animal is relatively large (60-80 mm) in relation to other burrowing animals, this motion can have the apparent effect of erosion, by cutting nonhorizontal (biogenic) structures of the host sediment. The plough sole is mostly enhanced by a thin sediment layer consisting of well-sorted grains. The disturbance of the fabric during the movement of the clam may lead to a liquefaction of the mud in the immediate wake allowing a gravitative grain size differentiation. In sandy mud, therefore, a thin sand layer forms with upwards grading into silt and/or clay. Under conditions of steady sedimentation and continuous population this lamination will be vertically repeated, and thus an anorganic lamination structure can be imitated (Fig. 5-64b). Possibly many of the laminated layers in Eckernförder Bucht sediments in the sandy mud zone formerly attributed to storm layers by WERNER (1968), actually are plough sole marks.

A diagnostic criterion for distinguishing plough sole marks from primary laminae is their slight concave-upward shape and the short extent transverse to the direction of motion. The whole track thus has an elongated shape (Fig. 5-63).

WERNER and WETZEL (1982) have shown to what extent the preservation of biogenic structures depends on the relative depth of the burrows. The studies of the burrowing fauna by DOLD (1980) and the analyses of sedimentary structures in sediment cores (WERNER 1968, WINN 1974, WINN et al. 1982) evidence that in Kiel Bight Arctica islandica is hardly surpassed by other animals with respect to borrowing depth. Some polychaetes which may burrow deeper than 8 cm, construct only vertical tubes which will not be competitive to the horizontally moving bivalves. Only Pectinaria koreni may be an occasional competitor. According to the ethological classification of SEILACHER (1953), such plough sole traces belong to the palichnological category of Repichnia.

Little is known about the intensity of locomotion of A. islandica under in situ conditions. The abundance of their traces in many Holocene sediments indicates, however, its geological significance.

A second definite biogenic structure formed by Arctica islandica by the same horizontal motion is due to sediment filling of the furrow left behind by the moving animal. The resulting characteristic finely laminated V- or U-shaped sand body is formed only in areas with at least periodically strong sand supply. In mud sediments with little sand content analogous structures were not yet detected. Palichnologically, this structure is a vertical spreiten burrow similar to the Teichichnus type (SEILACHER 1955) although the mechanism of its generation is completely different.

There are a number of other bivalves which produce also plough sole traces, including Macoma calcarea, Astarte spp., Aloidis gibba, and also the very common Abra alba (DOLD 1980). But all these species are much smaller and shallower digging than A. is-landica and may therefore not be seriously competitive to this bivalve under normal conditions (see compilation of trace producers in Table 5-3).

Pectinaria koreni. The traces of this sediment-feeding polychaete, which lives within a mobile, cone-shaped and sub-vertically oriented tube also have a high potential of preservation, since the animal

- moves horizontally through the sediment, leaving behind a zone of deformation structures (SCHÄFER 1962);

- digs in sediment depths up to 10 cm in Kiel Bight according to DOLD 1980;
- produces grain size differentiations by its digging activity (DOLD 1980);
- normally occurs in dense populations of hundreds to thousand individuals per m^2 (DOLD 1980).

The experiments by SCHÄFER (1962) and DOLD (1980) have revealed the mechanism generating the traces of this animal. Its peristaltic motions necessary for locomotion cause enrichment of coarser grains on the bottom of the trace. In contrast to traces of Arctica and those of other bivalves, they have no bilateral symmetry with concave cross sections. According to DOLD (1980), dense populations of Pectinaria koreni finally produce a continuous plough sole stratum.

In addition to the simple (plough sole) trail, DOLD (1980) describes other characteristic structures formed by the horizontal movement of P. koreni. These structures are caused by the shape of funnels and mounds produced by the feeding activity of the animal. However, most of these traces have little chance of being pre-served in the geological record because they belong to the near-surface layer.

Table 5-3 suggests a higher preservation potential for the traces of P. koreni and A. islandica than for the traces of most of the other organisms.

Tube-dwelling organisms. A further characteristic constituent of biogenic structures found in the sediment cores are tubes of a few millimeters in diameter (Fig. 5-6c). They may be hollow or filled with sediments. Many of them are U-tubes (WERNER 1968). DOLD (1980) describes a number of more common organisms of Kiel Bight which build single tubes or branched tube systems, or which merely leave tube-like traces in the sediment (Table 5-3). Most of the significant tube-dwelling organisms are several millimeters in diameter. Thus, smaller tubes (1-2 mm or < 1 mm) abundant in any estuarine, shallow marine (HERTWECK 1972, HOWARD and FREY 1975) and deep-sea sediments (JÜRGENSEN et al. 1981, WETZEL 1980) were very rarely found in Kiel Bight sediment cores. In Table 5-3, Heteromastus filiformis is the only organism listed which builds such very small tubes. Even deep-burrowed tubes have a chance for preservation only, if their formation is associated with any kind of positive or negative grain size selection. Heteromastus does not seem to fulfil these conditions (HERTWECK 1972). In many cores of mud sediments in Kiel Bight, U-burrows of 3-4 mm diameter are the only tubes found in the deeper horizons (e.g. WERNER 1968).

There are, however, many sections without any tubes. This may indicate that the very frequent but not lined tubes of such polychaetes as Nephtys ciliata and Scoloplos armiger are subjected to extinction, or that large tube builders such as Trochochaeta or Rhodine are not commonly present in the mud basins.

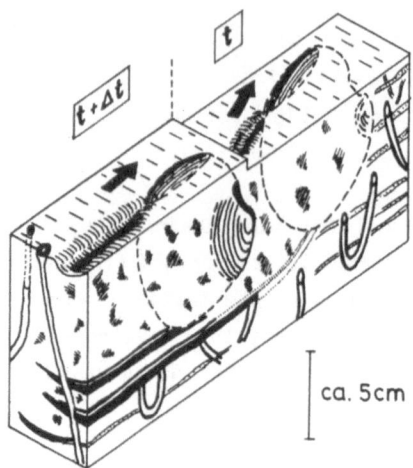

ca. 5cm

Fig. 5-63: Block diagram illustrating bioturbation activity of Arctica islandica. During horizontal motion through the sediment, older structures are eliminated, and plough sole laminae are generated. Situations (t) and (t + Δt) illustrate how a vertical sequence of plough sole lamination is generated. With a complete ploughing of the sediment, tube burrows will be preserved only when penetrating deeper than the Arctica ploughing depth.

Fig.5-64a: Photograph of box sample showing sandy backfill structure of Arctica islandica. Eckernförder Bucht, 23 m water depth.

b: X-ray radiograph (negative) of sandy mud sediment showing bioturbation structure characterized by plough sole traces (arrows) and U-tubes with line walls (partly cut by plough soles). Southern Eckernförder Bucht, 24 m water depth. Core no. GIK 14 304, 142-152 cm below top.

c: X-ray radiograph (negative) of clayey mud sediment with plough sole laminae (arrows) and walled tubes. Northern Kiel Bight, 25 m water depth. Core no. GIK 14 311-2, 186-201 cm below top.

d: X-ray radiograph (negative) of muddy sand sediment showing deformational structure with V-tube sections. Middle of Kiel Bight, 22 m water depth. Core no. GIK 10 006, 27-39 cm below top.

e: X-ray radiograph (negative) of sandy mud sediment showing alternation of laminated and bioturbated portions, consisting partly of plough sole traces and tube burrows. Middle of Vejsnaes Rinne, 25 m water depth. Core no. GIK 11 344, 220-228 cm below top.

f: X-ray radiograph (negative) of sandy mud sediment showing laminated core section of lower Holocene. Same core as 6-64e, 328-335 cm below top. Length of bars on all figures: 5 cm.

257

Abb. 5-64:

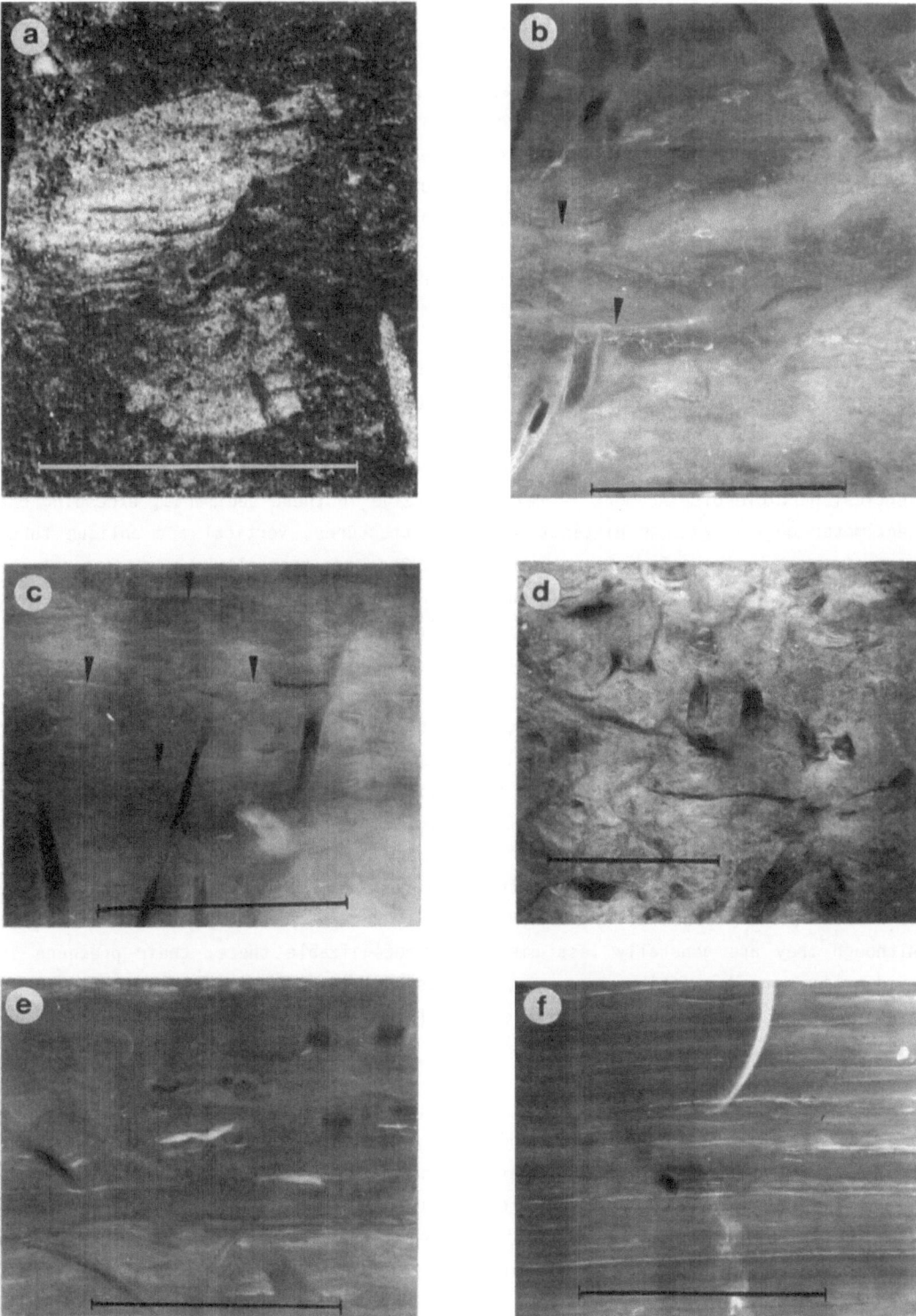

Holocene Ichnofacies Distribution

As discussed above, the potential producers of biogenic structures have different chances to produce preservable traces in the sedimentary column. The most effective 'filters', which the traces have to pass for preservation, are given by the relative burrowing depth (WERNER and WETZEL 1982) and by compaction or collapsing which may already occur in the uppermost decimeters in very soft mud sediments (HERTWECK 1972). Some burrow types may be preserved, but occur too rarely to become a characteristic feature of the structural inventory (Table 5-3).

In studying many cores from different morphological units of Kiel Bight, the following ichnofacies types could be discerned.

Deformational-structure-tube facies. This facies is typical for the totally burrowed sandy bottoms at 18 - 21 m water depth, below the zone continuously reworked by waves. The highly variable structures may be best characterized as 'chaotic'. If one reduces the turbulent fabric to 'eddies' with definable diameters (WERNER 1968), the scale of vertical particle displacement is relatively large in these sediments, exceeding one centimeter on average. As distinct biogenic structures, vertical and oblique tubes occur in addition to deformational structures. They are typically one-ended tubes and may belong to several groups of polychaetes (Table 5-3). Their tubes can be preserved due to their large vertical extension, piercing through the upper, densely populated 'mixed layer'. An example of this structure is shown in Fig. 5-64c.

Plough-sole-tube facies. This ichnofacies type is characterized by abundant sandy or silty laminae interpreted as plough sole traces such as generated by <u>Arctica islandica.</u> The short, concave cross sections of these structures found in sediment cores as in the example of Fig. 5-64b are characteristic. Single tubes or those of the lined U-type are common accessorial elements which occur with varying frequency down to being completely absent sometimes. Plough sole structures can also be identified in completely sand-free muds as occurring in the northern mud basin or in the fjords of Kiel Bight. Although they are generally less distinctly recognizable there, their presence is evident by many transitions and by the occasional occurrence of sandy plough sole traces within the otherwise sand-free muds (Fig. 5-64d). The tube frequencies are significantly lower in the older parts of the Holocene column.

Laminated-deformational facies. In the deeper parts of the mud basins, particularly in the more sheltered areas like Eckernförder Bucht, laminated primary structures consisting of alternating clay, silt or sand layers are more or less distinctly developed. These sedimentary structures may be modified by bioturbation. The structures caused by bioturbation are mainly of the deformational type and only little distinct traces such as tubes can be observed. Such kinds of bioturbation phenomena with particle displacement

depth (mm)	species	biogenic structures	sediment	O₂	pop. dens.	horiz. extens.	orien-tation	preserv. potential
0-20	*Abra alba*	(1) C: sediment streaks; dF: homogenized lenses	m	-	c	20-30	h	low
	Astarte elliptica	(1) R: plough-sole marks; dF: homogenized lenses	mx, ms	-	f	30-40	h	moderate
	Macoma calcarea	(1) C: sediment streaks; dF: oblique pellet streaks	sm, ms	-	f	40-60	h/v	moderate
	Polydora ciliata	(2) D: U-tube, wall lined with coarser grains	s, ms	-	r	3	v	moderate
20-50	*Euchone papillosa*	(2) D: vertical mud tube	m, sm	+	c	2	v	low
	Terebellides stroemi	(2) P: thick-lined mud tube, partly horizontally	sm, m	-	a	6-10	h/v	moderate
	Pherusa plumosa	(2) D: meandering; dF: filled feeding hole; Funnel and scour	mx, sm	-	f	50-100	h/v	moderate
50-80	*Halicryptus spinulosus,*	(3) D: wide U tube	m	+	f	5-7	v	moderate
	Arctica islandica	(1) R: (a) laminated plough-sole mark (b) stratified backfill structure	ms, sm, m	+	c	100	h/v	high
	Pectinaria koreni	(2) P: (a) lamin. plough-sole, stratified funnel; dF: backfills	ms, sm, m	-	a	100	h/v	high
	Scoloplos armiger	(2) P: noncharacteristic, non-lined cylindrical burrows	s - m	+	a	3-4	v/h	low
	Nephtys ciliata	(2) P: branched, little characteristic, non-lined cylindrical burrows	s - m	-	a	3-5	v/h	low
> 80	*Trochochaeta multisetosa*	(2) D: U-tubes, anastomosing, lined	sm, m	+	f	4-6	v	high
	Rhodine gracilior	(2) D: tubes lined with org.matter partly horizontal	mx, sm	-	c	2-3	v/h	high (?)
	Heteromastus filiformis	(2) D: thinly-lined, branched tubes	ms,sm,mx	(+)	c	1	v/h	moderate

Table 5-3: Burrow characteristics and geological significance of sediment-dwelling fauna in muddy sediments of Kiel Bight. **Depth:** main maximum living depth in sediment. Numbers behind **species** names: (1) - Bivalvia, (2) - Polychaeta, (3) - Priapulid worm. Types of **biogenic structures** (lebensspuren) according to SEILACHER (1953): C - cubichnia, D - domichnia, R - repichnia, P - pascichnia; dF - deformation structures. **Sediment:** m - mud, sm - sandy mud, ms - muddy sand, s - sand, mx - mixed sediment. O₂: resistance to oxygen deficiency, + signifies tolerance. **Population density:** categories rare to abundant according to the maximum individual numbers in their preferred biotopes: r - $< 10 \, m^{-2}$, f - 10-$100 \, m^{-2}$, c - 100-$500 \, m^{-2}$, a - $> 500 \, m^{-2}$. **Horizontal extensions** of the structures (mm) **Orientation** of biogenic structures: h= horizontal, v - vertical. **Preservation potential:** indicates possibility of geological preservation. Data mainly based on DOLD (1980).

at small scale only has been named "cryptobioturbation" by HOWARD and FREY (1975). Typically, however, completely bioturbated sections alternate with preserved lamination as shown in Fig. 5-64e. In places, shells are concentrated in layers. In Eckernförder Bucht, these layers consist predominantly of Abra alba shells. Shells of Arctica as well as their traces are missing in these horizons. Calcareous foraminifera can also be concentrated in layers (WERNER 1968). This holds true also for cores where otherwise foraminifera shells are normally dissolved. Possibly, the preservation in these cases is due to rapid burying by storm deposition (see 5.3.6, and WEFER 1976a).

In the lower parts of the Holocene column, lamination is generally better developed than in the upper parts. An example from the Vejsnaes Rinne is given in Fig. 5-64f. Also the abundance of tube burrows decreases with core depth as in the plough-sole-tube facies. This is in agreement with the results of the shell stratigraphy presented in section 5.5.1 where Arctica has been found to be a dominant bivalve species in the more recent sections only. But also the other organisms producing traces of high preservation potential (Table 5-3) apparently had low abundances in the older periods of the Holocene transgression.

Density of tube burrows. It may be surprising that polychaete tubes are not as frequently found in Kiel Bight sediments as might be expected from their high population density (KÜHLMORGEN-HILLE 1963, ARNTZ 1971). In the deeper parts of the Holocene sediments dominance of primary lamination may indicate an actually sparse population of tube producing organisms. However, in intensely bioturbated sediments the effect of 'preservation filters' may be the cause for the low abundance of tubes.

Before assessing destructional effects it has to be considered which factors influence the density of tube burrows.

Assuming a certain living population density of a specific tubebuilding organism (T_L) remaining constant in the time interval considered, the density of fossil tube burrows (T_F) as determined in horizontal sections depends on its length (L), on the turn-over time of the population (R), and on the sedimentation rate (S) as follows:

$$T_F = \frac{L}{S \cdot R} \cdot T_L \qquad\qquad (7)$$

The equation implies the supposition that a density of fossil tube burrows has reached its final value not until the sediment accumulation has attained at least a height which equals the vertical length of the tube burrow. The equation satisfies the limiting conditions that

$$\lim_{s \to 0} T_F = \infty \qquad \text{and} \qquad \lim_{S \to L/R} T_F = T_L$$

It becomes evident by these relationships that the sedimentation rate yet significantly influences the burrow density in the sediment column, although one might have considered it as negligibly small in relation to the short life spans of the organisms.

For mud sediments in Kiel Bight, typical ranges of values of the terms in equation (7) are L = 6 to 12 cm, S = 0.05 to 0.2 cm y^{-1} and R between 2 and 5 years (according to WARWICK 1980). These values yield as minimum $T_F \approx 6 \cdot T_L$ and as maximum $T_F \approx 120 \cdot T_L$. For the typical wall-lined U-tubes found in the mud basins, we assume L = 10 cm, S = 0.1 cm $\cdot y^{-1}$ and R = 3 as realistic values, yielding $T_F \approx 30 \cdot T_L$.

With regard to the observation of bioturbation structures in sediment cores made usually by X-ray radiograph sections having a horizontal cross section of 10 cm², this result means that for the observation of one tube in average within a vertical section of 10 cm length, a population density of 33 living animals per m² will be required. Thus the study of a certain number of parallel radiograph sections is necessary to get reliable results on the abundance of tube builders. As a consequence, the relatively low tube densities observed in the laminated lower sections of the Holocene do not necessarily correspond to low living population densities.

5.5.3   Summary

Of the various constitutents in Holocene sediments bearing environmental signals, macrofaunal shell distribution and biogenic structures have been viewed more closely. Although shell contents are generally severely affected by carbonate dissolution, in channel sediments sufficient material was found and four stages of environmental conditions could be established. Furthermore the channel areas have the advantage that the depth zone below the pycnocline is represented in the geological record.

The four stages of faunal development have been dated by ^{14}C determinations. The most recent stage has been attributed to an environment with intermittent periods of oxygen deficiency, but with continuous bottom life.

The high diversity of bottom dwellers producing biogenic structures contrasts remarkably with that reflected by bioturbation structures strictly controlled by a hierarchical order of the preservation potential of different traces. The environmental record of the benthic fauna thus largely depends on a few specific ichnofaunal elements with a high preservation potential. Deep digging, plough sole generating organisms like Arctica islandica and Pectinaria koreni are ranging by far at the top of the hierarchy. The generation of plough sole traces simulating lamination also results in extinction of all other burrows built before.

It appears that only a small part of the different tube burrow types produced by polychaetes is preserved in the sediments, including wall-lined U-tubes. Their abundance observed in the sedimentary record also varies strongly being generally lower in the older sections.

In the mud sediments of Kiel Bight, several ichnofacies types can be distinguished which partly are observed to vary in the vertical sequence as well. In contrast to the upper parts, the lower section generally is characterized by a partial preservation of laminated primary bedding structures which can be explained by a well-developed stratification of the water body causing temporal oxygen deficiency.

Chapter 6.   BIOGENIC CARBONATES IN TEMPERATE AND SUBTROPICAL ENVIRONMENTS:
              PRODUCTION AND ACCUMULATION, SATURATION STATE AND
              STABLE ISOTOPE COMPOSITION

              G. WEFER, W. BALZER, B. v. BODUNGEN and E. SUESS

ABSTRACT

The interactions between water and sediment with regard to the carbonate cycle have been investigated in a multi-disciplinary program comparing the temperate Kiel Bight of the Western Baltic and the subtropical Harrington Sound, Bermuda Islands. Emphasis has been given to:

1. the estimation of carbonate production by organisms and accumulation rates in the sediment as well as to the erosion of fossil carbonates;
2. the determination of the saturation state of the water with respect to different biogenic carbonate mineral phases, and
3. the changes occurring during transition from a live to a dead species assemblage.

Finally, the important carbonate-producing organism groups have been analyzed with respect to their stable isotope compositions.

6.1   INTRODUCTION

Carbonate sediments are distributed worldwide in shallow-water and deep-sea sediments. In general, rapid shallow-water carbonate sedimentation is restricted to tropical and subtropical climates, because some of the major carbonate producers, such as hermatypic corals and calcareous green algae, are limited to warmer climates. Temperate and polar regions are known for detrital sedimentation from terrigenous sources and carbonate sedimentation is considered unimportant there; nevertheless, these environments are inhabited by carbonate-secreting organisms, and the fate of their calcareous skeletons needs to be better understood as well.

A considerable amount of data is available regarding the distribution and composition of calcium carbonate in shallow water of the different environments of the world (for a review see BATHURST 1971, and MILLIMAN 1974). However, relatively little is known quantitatively about the processes of carbonate production, dissolution and accumulation as well as break-down and transportation of carbonate particles.

To further our understanding of the pathways taken by $CaCO_3$ from initial precipitation by an organism to final burial in the sediment, comparative studies have been carried out over the last decade within a multi-disciplinary research program: "Interaction Sea-Seabottom" at the University of Kiel. The major objectives of these studies with regard to the carbonate cycle were:

1. To determine carbonate production by the important carbonate-producing organisms and to compare these rates to the carbonate accumulation rates in the sediment;
2. To determine the relationship between the live and dead assemblages, because very little information is available as to what proportion of the live assemblage enters the sediment and therefore to what degree the species composition in the sediment

is biased (for a discussion see DOUGLAS et al. 1980). This information is important for quantitative paleo-ecological analyses which depend upon species abundance;

3. To determine the carbonate saturation state with respect to calcite, aragonite and different biogenic magnesian calcites. The proportion of the carbonate production which is buried in the sediments depends mainly on the saturation state of the water column and of the sediment/water interface. Further, knowledge of the environmental conditions for carbonate dissolution is important for the uptake of increasing amounts of atmospheric CO_2 by the oceans (see, for example ANDERSEN and MALAHOFF 1977);

4. To determine the variations of the stable isotope ratios in carbonate shells which can be used (1) for age and growth rate determinations, (2) for recordings of seasonal fluctuations and (3) for paleo-ecological parametrization such as yearly temperature and salinity ranges. Of special interest is the clarification of two basic questions: (1) Do organisms precipitate carbonate in thermodynamic equilibrium? and (2) What causes the observed carbon isotope variation within shell carbonates? (for a review see DUPLESSY 1978).

The areas of investigation were the undersaturated temperate brackish waters of the Western Baltic and the supersaturated subtropical full marine waters of the Harrington Sound, Bermuda. These two areas have about the same water depth; both develop stratified water columns during the late summer but show distinct differences in salinity, temperature ranges and magnitudes of biological production. For both areas no major in- or outflow of sediment is expected. In Harrington Sound the narrow inlet and wind-protected watermass keep sediment exchange to a minimum (NEUMANN 1965). In Kiel Bight (Western Baltic), fluvial input from streams of the surrounding catchment and an exchange by currents through the Danish Straits are not thought to be of importance (SEIBOLD et al. 1971).

This article presents a synthesis of our comparative studies on shallow-water carbonates from these areas and is divided into four parts, each dealing with one specific aspect of the carbonate cycle. The first part deals with calcium carbonate production by different organisms. The second part concerns the carbonate saturation state with regard to the different biogenic carbonate minerals (calcite, aragonite and magnesian calcites). The third part addresses transportation and accumulation of carbonate material, and the last part the distribution of stable isotopes in the shell carbonate of the major shallow-water organisms.

6.2   AREAS OF INVESTIGATION

The areas for our comparative studies were the Kiel Bight (Western Baltic Sea) and the Harrington Sound, Bermuda. The specific site at Kiel Bight, the "Hausgarten" of Sonderforschungsbereich 95, is a rectangle with an area of 0.6 km^2 extending from 5 m depth (turbulent zone) down to the margin of a shallow basin about 26 to 28 m in depth (Fig. 6-1). This area is fairly typical for the Western Baltic.

It is hyposaline, with surface salinities from 12 - 19 °/.. S and deep-water salinities between 17 and 22 °/.. S, and during summer and fall there is a distinct pycnocline developed at 10 - 15 m water depth. Sediment influx from river discharge is negligibly small. From the surface down to 13 m depth, there is a glacial platform covered with a thin layer of coarse-grained residual sediment including boulders. The deeper part of the area is a rather flat, low-silled basin floor with muddy sediments, and high sedimentation rates are typical (up to 1.4 $mm \cdot y^{-1}$, ERLENKEUSER et al. 1974, SUESS and ERLENKEUSER 1975). Details are given by LUTZE (1974) and WEFER et al. (1974).

The Harrington Sound is thought to be a water-filled interdunal depression that may have been further deepened by subsequent collapse of solution caves. The Harrington Sound covers an area of about 5 km^2, a quarter of which is occupied by deep basins of up to 25 m; the remainder is shallower than 20 m (Fig. 6-2). Originally, the Sound was a freshwater lake (ERLENKEUSER et al. 1981); today the water is fully marine. The tidal range is between 10 and 24 cm. Because only one narrow channel connects the Sound and the North Lagoon, a small portion of the surface water is exchanged during each tidal cycle (MORRIS et al. 1977). The vertical structure of the Sound waters is similar to that of a monomictic lake. During summer the 17 m thick surface layer warms from 16°C to about 29°C, and below 17 m a pycnocline develops. The stagnant deeper water shows oxygen depletion during late summer and fall. Due to cooling of the surface waters and stronger winds during fall and winter, the entire water column turns over and is thoroughly mixed until the surface water warms again in spring of the following year. A comprehensive study of the sediments of Harrington Sound was conducted by NEUMANN (1965), further details are in WEFER et al. (1981b), HEMPEL and WEFER (1982), HEINRICH (1982), and the reports of the "Bermuda Inshore Waters Investigation" (BIWI) (MORRIS et al. 1977, BARNES and v. BODUNGEN 1978, v. BODUNGEN et al. 1982).

The fully marine, considerably warmer water column of Harrington Sound strongly contrasts to the waters of the Kiel Bight area. The yearly range in temperature and light intensity is smaller than in the Western Baltic. A further contrast lies in the sediment material. In Kiel Bight almost all the sediment is derived from the glacial till, whereas in Harrington Sound the carbonate sediment is almost exclusively produced by benthic organisms or by the erosion of lithified carbonate sands. There

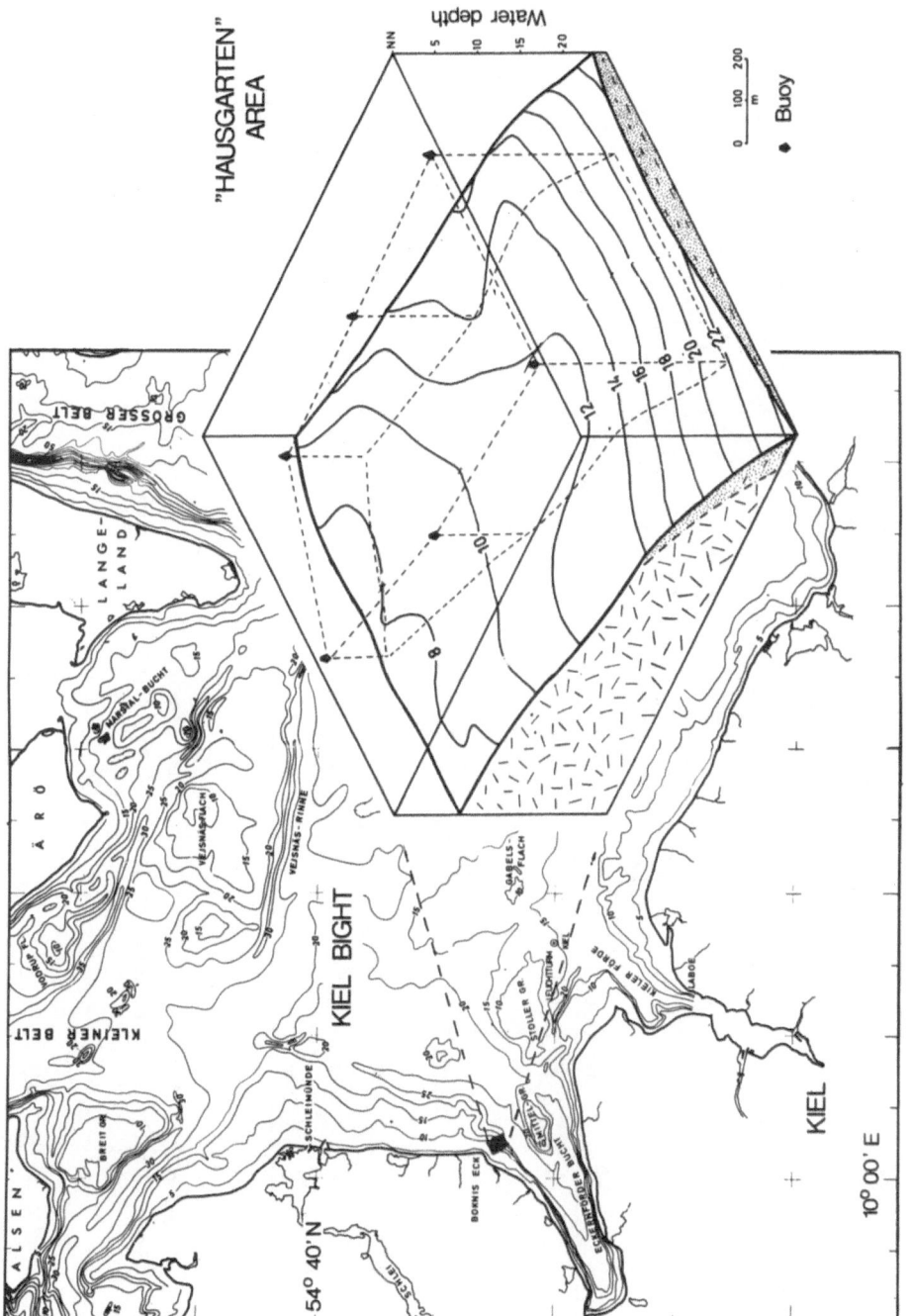

Fig. 6-1: "Hausgarten", the area of investigation of the SFB 95 of Kiel University, located in Kiel Bight (Western Baltic Sea) (modified after WEFER et al. 1974).

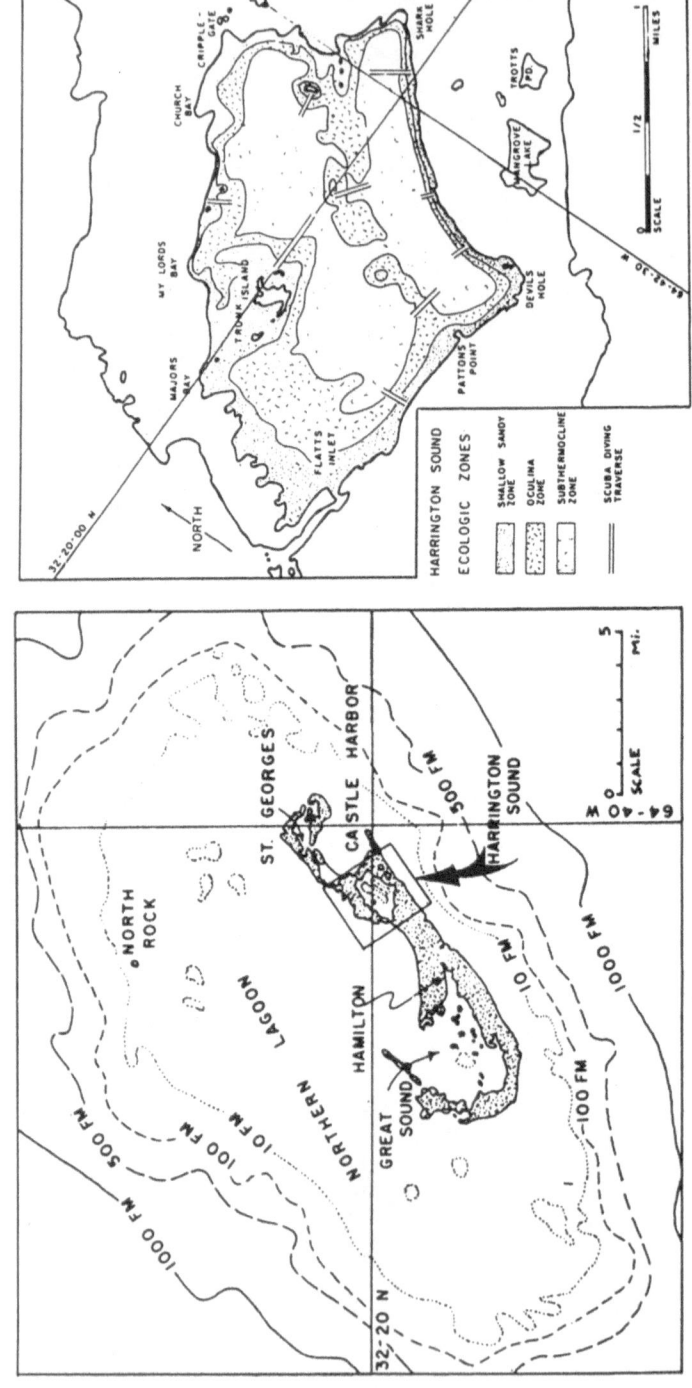

Fig. 6-2: Index map of Bermuda Islands (left side) showing location of Harrington Sound and the ecologic zonation (right side) (after NEUMANN 1965).

are hydrographic similarities between Harrington Sound and Kiel Bight in as far as the development of a pycnocline during late summer enhances oxygen depletion in the deeper basins of both areas.

6.3   CARBONATE PRODUCTION

Two separate approaches can be taken to estimate carbonate production rates. One procedure consists of estimating the rate at which calcium-carbonate production by benthic and planktic communities depletes the overlying water column in $CaCO_3$-related dissolved constituents, i.e. Ca^{2+}, CO_3^{2-}-carbonate alkalinity (for detailed discussion see SMITH 1978). The second procedure consists of estimating the standing stock and turnover rates of the calcifying organisms of the community and is used primarily in the following calculations. For estimating the standing stock of the important carbonate producing organisms defined bottom areas were sampled. The size of the area depends on the size, standing stock and patchiness of the specific organisms and varies between 100 cm^2 (foraminifers) and hundreds of m^2 (echinoderms). For estimating turnover rates different methods were used. In the case of foraminifera, the number of tests produced per year was calculated from the standing stock, the frequency of reproduction and the proportion of individuals that reproduce (WEFER 1976a, 1976b, WEFER and LUTZE 1978, LUTZE and WEFER 1980). For calcareous algae, Alizarin Red-S stain was used as a time marker to allow the measurement of incremental growth (WEFER 1980). Absolute $CaCO_3$ deposition rates in algae and echinoderms were determined using ^{45}Ca as a tracer (BÖHM 1978, NAUEN and BÖHM 1979). For the calculation of growth rates of gastropods, bivalves, corals and larger tropical foraminifera the annual periodicity in the isotopic ratio $^{18}O/^{16}O$, which reflects seasonal temperature and/or salinity fluctuations, was used to establish a time frame (for examples see ERLENKEUSER and WEFER 1982).

6.3.1   Kiel Bight

In Kiel Bight gastropods, bivalves, bryozoans, ostracods, echinoderms and benthic foraminifera are common carbonate producers. A further carbonate supply is from physical breakdown of cliffs and erosion of submarine platforms composed of glacial marl (see chapter 5, this volume). Other carbonate-producing organisms are rare or live only in areas with higher salinity like the coralline algae Phymatolithon studied by BÖHM et al. (1978).

6.3.1.1 Sediment Dwellers

In the study area, carbonate production rates are only available for benthic foraminifera. The carbonate production by molluscs, ostracods and echinoderms has to be estimated from standing stock data reported in the literature.

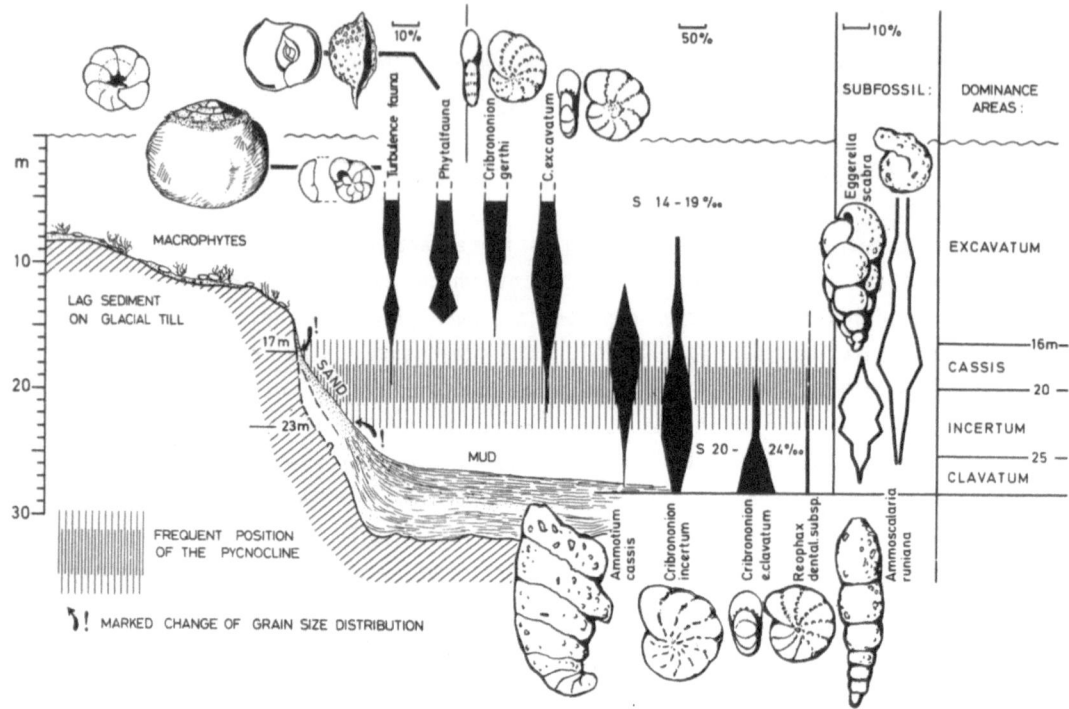

Fig. 6-3: Schematic presentation of a sediment echograph from the "Hausgarten" area.
The distribution of the most important benthic foraminifera is shown by vertical bars. The thickness of the bars corresponds to the average percentages
of each species (modified after LUTZE 1974).

Foraminifera - The distribution of benthic foraminifera in the "Hausgarten" area was
determined by LUTZE (1974) using material from closely sampled traverses obtained at
different seasons (Fig. 6-3). Dominance areas of the four main species (Elphidium
incertum, E. excavatum excavatum, E. e. clavatum and Ammotium cassis) proved to be
constant in time and space, thus encouraging detailed productivity studies with
permanent stations in the centers of the respective populations.

These studies showed that the mean production of foraminiferal carbonate by sediment
dwellers (Elphidium excavatum clavatum, E. incertum, E. e. excavatum) is low on the
erosional platform, varying between 10 and 35 mg m^{-2} y^{-1} (Fig. 6-4) (WEFER 1976a,
WEFER and LUTZE 1978). Production increases slowly with depth where turbulence is
usually lower. Production is much higher in the deeper part of the area, with values
of up to 3,100 mg m^{-2} y^{-1}. Only two species are major contributors of carbonate here:
E. e. clavatum and E. incertum.

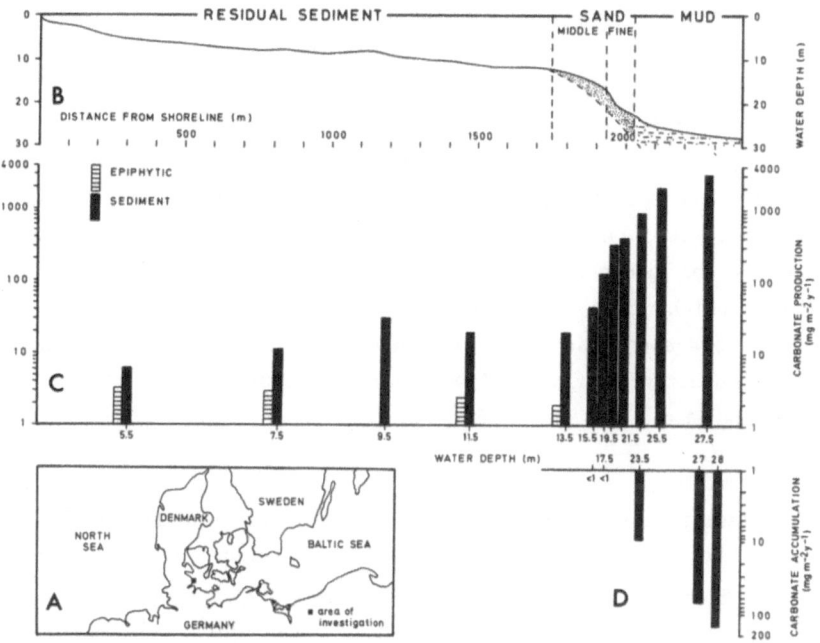

Fig. 6-4: Foraminiferal carbonate production and accumulation in Western Baltic Sea. A = Area of investigation; B = profile showing erosional platform with residual sediment and edge of a small basin with sand and mud; C = carbonate production and D = carbonate accumulation for different water depths (after WEFER and LUTZE 1978).

Echinoderms - Among the echinoderms the common starfish Asterias rubens is very abundant in Kiel Bight. NAUEN (1978a) estimated a mean biomass of 11.7 g m^{-2} wet weight. From a general description of the growth of A. rubens by NAUEN (1978a) it is clear that growth proceeds in three distinct stages: larval stage, waiting stage and growing stage. Using the ^{45}Ca-tracer technique, NAUEN and BÖHM (1979) found a daily growth rate at the waiting stage of 7.56 x 10^{-4} mg CaCO$_3$ mg skeleton^{-1} d^{-1}, which corresponds to 0.09 % d^{-1}. Mean growth at the growing stage as calculated from size frequency distribution and von Bertalanffy growth analysis amounts to 0.6 mm d^{-1} (NAUEN 1978b), which corresponds to 1.1 % d^{-1}. Growth at the growing stage thus exceeds growth at the waiting stage by a factor of 12 (NAUEN and BÖHM 1979). Using a relationship between mean biomass, production and mortality, NAUEN (1978c) estimates the following annual biomass production values for four sediment types: 110 g wet weight m^{-2} y^{-1} on lag sediment, 107 g m^{-2} y^{-1} on sand, 112 g m^{-2} y^{-1} on muddy sand/ sandy mud and 20 g m^{-2} y^{-1} on mud. This corresponds to an annual production of 5 g CaCO$_3$ m^{-2} y^{-1} on lag sediment, 4.8 g m^{-2} y^{-1} on sand, 5 g m^{-2} y^{-1} on muddy sand/ sandy mud and 0.9 g m^{-2} y^{-1} on mud, respectively. For this calculation, the wet weight/skeletal weight relationship of NAUEN and BÖHM (1979) is used.

Ostracodes - ROSENFELD (1976) analyzed sediment and algae samples for the seasonal distribution of ostracode species taken by LUTZE (1974) and WEFER (1976a) at 10 stations in the "Hausgarten" area of Sonderforschungsbereich 95. Most of the ostracode species were present in all of the samples throughout the year; four species were found to be seasonal. Variations in population densities were attributed to changes in length and timing of the reproductive cycles of the different species. The mean production of ostracode carbonate is estimated using standing stock data from ROSENFELD (1976, Fig. 31), an average generation of one per year (ROSENFELD 1979, Table 3) and a measured average weight of a single ostracode shell of 8.5 mg. Carbonate production is low in the area deeper than 17 m water depth; it varies between 17 and less than 1 mg m^{-2} y^{-1}. Production is much higher in the shallower part of the area, with rates of up to 134 mg m^{-2} y^{-1} (Table 6-1).

Molluscs - The molluscs from Kiel Bight have been investigated under different aspects by different investigators. SAMTLEBEN (1973) deciphered the factors responsible for the changes between living populations, dead communities and assemblages of valves from the size-frequency distributions of Mytilus edulis. Further, SAMTLEBEN (1977) used the size frequency distributions in populations of Mytilus edulis to clarify the relationship between the age of individuals and growth increments of their valves. Fig. 6-5 shows the average and maximum length growth of a population from Stoller Grund in Kiel Bight. After about three years the average length of the valves was about 60 mm, while single valves measured more than 80 mm in length. Further, SAMTLEBEN found that growth rates are highest in the summer of the second year, with later growth depending mainly on size of individuals, not on age. After reaching a typical size, individuals can live several years without further length growth. However, valve weight increases steadily and this increase can be used for age grouping of fully grown specimens. Scanning electron micrographs of the ultrastructure of the nacre showed periodic growth allowing age determinations on valves of Mytilus edulis.

MEYER (1975) studied the colonization of different organisms on artificial hard substrates (fouling) to analyze their species succession and to determine production rates both for carbonate and organic matter. Simplified, colonization started in early summer with balanids covering the substrate entirely within about 8 weeks. In July, Mytilus started to settle on the balanids and in the middle of August the whole substrate was covered with a thick carpet of bivalves as illustrated in Fig. 6-6.

From the fouling experiments, MEYER (1975) estimated theoretical/potential carbonate production rates. The production by Mytilus was 35,000 g $CaCO_3$ $m^{-2}y^{-1}$ and by Balanus 2,600 g $m^{-2}y^{-1}$. These figures are in the range of production rates from tropical areas and about 1,000 times higher than average production rates for other organisms from

Table 6-1: Carbonate production by different groups of sediment-dwelling organisms and by glacial till erosion according to water depth in Eckernförder Bucht (Kiel Bight).

Water Depth (m)	Sediment Type[1]	Carbonate Production (mg m^{-2} y^{-1})				
		Benthic Foraminifera[2]	Asteroid Asterias rubens[3]	Ostracodes[4]	Molluscs[5]	Glacial Till[6] Erosion
5 - 6	Lag sediment	6		71		)
7 - 8	Lag Sediment	12	5,000	86		) Average
9 - 10	Lag sediment	36		132		) 108,000
11 - 12	Lag sediment	23				) mg m^{-2} y^{-1}
13 - 14	Sand	31		101		
15 - 16	Sand	44	4,800	60		
17 - 18	Sand	133		17		
19 - 20	Sand	325	5,000	26	~20,000-50,000 up to	
21 - 22	Sand	408		11	400,000 for single	
23 - 24	Mud	903		< 1	Cyprina islandica	
25 - 26	Mud	2,066			patches	
27 - 28	Mud	3,117	900			

1 after WEFER and TAUCHGRUPPE (1974)
2 after WEFER and LUTZE (1978)
3 calculated after NAUEN (1978a, b, c) and NAUEN and BÖHM (1979)
4 calculated after ROSENFELD (1976, 1979)
5 after ARNTZ (1971) and SARNTHEIN (1973)
6 after HEALY and WEFER (1980)

the same area (see Table 6-1). This shows the exceptional position of this biotope in comparison to sediment substrates.

ARNTZ et al. (1976) studied the zonation of molluscs and mollusc shells in the Kiel Bight Channel System; FRITZSCHE (1975) studied a death assemblage of shells on the channel slope of Vejsnäs Rinne, and EXON (1972) and WYNN (1974) the present and postglacial sedimentation in the Flensburger Förde and Great Belt regions.

In all, 30 species of bivalves and 19 of gastropods were sampled by ARNTZ et al. (1976) as living specimens. According to its life span, the bivalve Artica islandica was dominant in the deep and Astarte species on the upper part of the channel slope. Macoma baltica was dominant in a third, more shallow zone, which was actually outside the channel system. Abra alba was the most persistent species in the channels, being present in 86 % of all samples. With the exception of Hydrobia, gastropods were infrequent and, where present, were almost never dominant. On the whole, the composition of the dead specimens corresponded to that of living ones, to the zonation of dominant species and to the distribution and elevation pattern of the maxima of species.

First estimates of shell production are derived from the living to dead ratio of shell samples - notwithstanding the varying amounts of carbonate dissolution. For instance, the production of Astarte species was some 13 times smaller than that of Abra alba and 7 times smaller than that of Artica islandica. Further, a strong change from living to dead shell dominance occurred below the pycnocline at 20 to 24 m water depth.

Over a period of about two years, submarine sedimentary substrates were exposed to benthic colonization to study both the effects of hydrographic and biotic factors as well as those of substrate type. Fig. 6-7 shows an outline of the construction of the submerged platforms taken from the technical description of this experiment (SARNTHEIN and RICHTER 1974). The results regarding the colonization of the meiofauna were reported by SCHEIBEL (1974). The benthic foraminifera were investigated by WEFER and RICHTER (1976). Data on the colonization of molluscs are available from RICHTER (1975) and RICHTER and SARNTHEIN (1977). These experiments were continued through a study of macrobenthic colonization and succession by RUMOHR (1980) and ARNTZ and RUMOHR (1982).

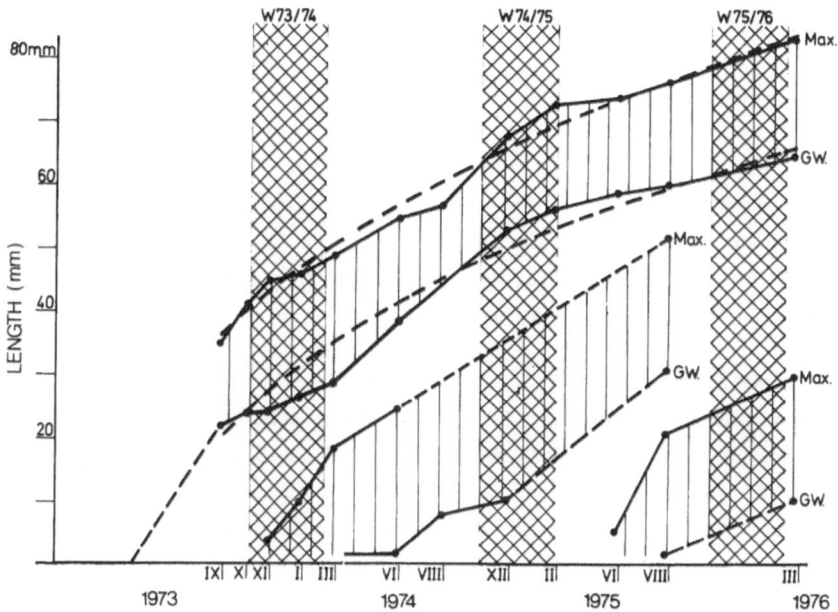

Fig. 6-5: Length growth of <u>Mytilus</u> <u>edulis</u> from Stoller Grund in Kiel Bight between summer 1973 and spring 1976. Max = maximum growth, GW = average growth. Periods between November and February are marked by cross-hatching (after SAMTLEBEN 1977).

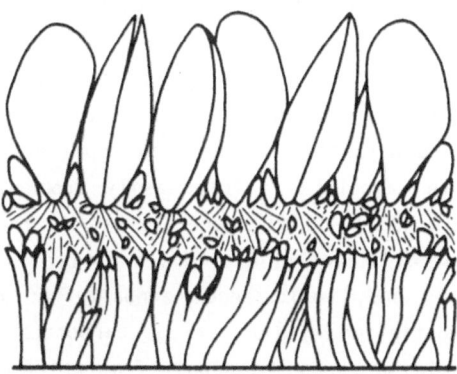

Fig. 6-6: Schematic view of the layered fouling community attached to artificial hard substrate in Kiel Bight. On top the bivalve <u>Mytilus</u> <u>edulis</u>, in the middle small <u>Mytilus</u> embedded in a network of byssus, and on the bottom balanids (after MEYER 1975).

6.3.1.2 Epiphytic Carbonate

In the "Hausgarten" area algae are limited to the erosional platform, since they are found only on boulders. Benthic foraminifera, bryozoans, juvenile bivalves and gastropods as well as calcareous worms live on different species of benthic macroalgae. Common epiphytic foraminifera are Elphidium excavatum excavatum, E. gerthi and Opthalmina kilianensis (WEFER and LUTZE 1978) and bryozoan species living on algae are Callopora craticula, C. aurita, Electra pilosa, Cribrilina punctata and Crisia eburnea (H. RISTEDT, Bonn, pers. comm.).

Further juvenile Mytilus and juvenile Asterias are found in high numbers after reproduction events. In spring (April) carbonate contents are between 2 and 7 % of algae dry weight, during summer between 5 and 25 % and during winter (November) between 4 and 7 %. For the "Hausgarten" area an average algae biomass of 263 g m^{-2} was reported (Black und Kaminski.unpubl.) Using these figures and assuming a turnover rate of one year, then the annual total carbonate production of epiphytic organisms varies between 500 and 6,250 mg CaCO$_3$ m^{-2} y^{-1}. These rates are similar in magnitude to those calculated for the sediment-dwelling organisms (Table 6-1).

6.3.1.3 Carbonate Supply from Glacial Till Erosion

Additional supply of carbonates for recent sedimentation is derived from: (1) adjacent cliff erosion and (2) submarine abrasion. Fluvial input can be neglected. The contribution from currents entering through the Little and Great Belts is not known with certainty but is thought to be insignificant (SEIBOLD et al. 1971). Therefore, cliff erosion and submarine abrasion are the main suppliers. Carbonate supply rates from cliff erosion can be calculated fairly easily from historical maps and mean CaCO$_3$ contents (KANNENBERG 1951). Cliff retreat rates in the area are ~ 0.29 m per year. Allowing an average height for cliffs of 7 m in the "Hausgarten" area, about 2 m^3, corresponding to 4.4 tons, would erode annually on a one meter section of shoreline. This volume multiplied by the average carbonate content of 17.5 % (RUCK 1970) of glacial till yields a carbonate supply from cliff erosion of 0.77 tons per m shoreline per year.

Carbonate supply derived from the submarine platform is much more difficult to esti-mate. WEFER et al. (1976) reported abrasion rates of up to 2.1 - 2.4 cm y^{-1} along a submarine profile at water depths between 1.7 - 3 m. At 10 m of water depth, however, the rates were somewhat less than 0.16 y^{-1}. HEALY and WEFER (1980) presented four models for the supply of cliff detritus vs. submarine abrasion. Depending on the model, supply of sediment from a 1 m wide section across the submarine platform varies between 0.45 and 6.7 m^3 y^{-1}. Accepting the lowest figure, annual abrasion from the submarine platform yields about 0.5 m^3 of sediment, corre-

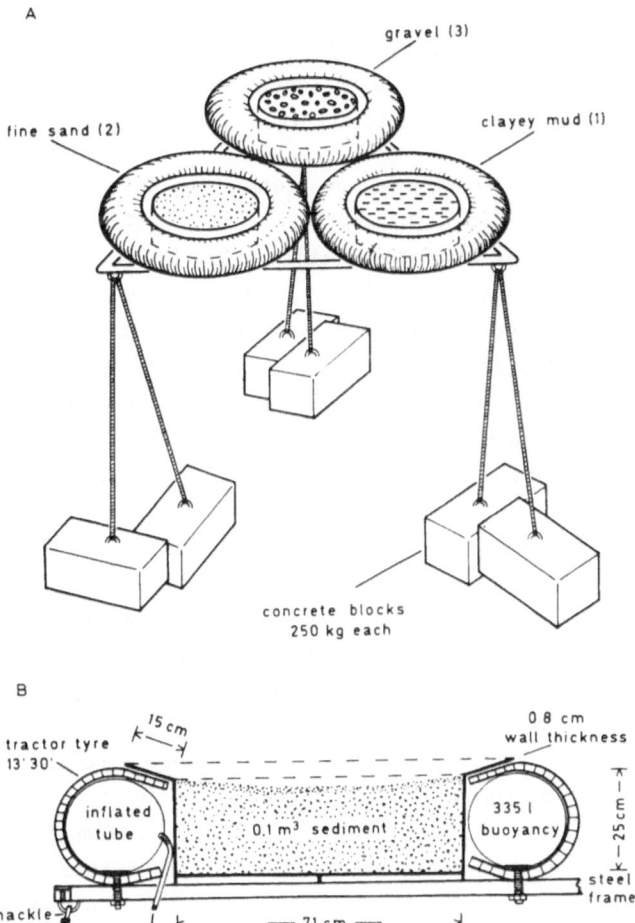

Fig. 6-7: Outline of construction plan of "submerged" platforms for benthic colonization. Tires lend sediment containers streamlined outer shape; buoyancy of
inflated inner tubes gives platform stability and elasticity. (A) General
view (not to scale); (B) cross-section sketch (to scale) (after SARNTHEIN
and RICHTER 1974).

sponding to 1.1 tons and 0.19 tons of CaCO$_3$ over a 1 m wide section across the platform. In the "Hausgarten" area the abrasion platform is about 1750 m wide (Fig. 6-4B), and using an average carbonate content of 17.5 % for the glacial till, this then yields an annual average carbonate supply of 108 g m^{-2}.

Of special importance for the abrasion of sediment on the submarine platform is the boring clam Barnea candida (FLEMMING and WEFER 1973, ARNTZ and RUMOHR 1973, RICHTER and RUMOHR 1976). In shallow waters population densities can be as high as 800 individuals m^{-2}. Under experimental conditions, using a submerged substrate platform in 19 m water depth (for a description see SARNTHEIN and RICHTER 1974), RICHTER and RUMOHR (1976) determined a mean shell growth of 21 mm y^{-1} and calculated an abrasion of 2.1 mm y^{-1} from a population of 162 individuals m^{-2}. Because of far denser populations in shallower waters, RICHTER and RUMOHR (1976) assumed that the abrasion potential might be 3 - 6 times higher than at 19 m water depth.

6.3.2   Harrington Sound, Bermuda

A striking characteristic of Harrington Sound is the marked depth zonation of the major benthic organisms. According to NEUMANN (1965) these zones are: rocky zone, shallow sandy zone, Oculina zone and sub-thermocline zone (Fig. 6-2).

Rocky zone - The short cliffs and collapse blocks that circle the Sound provide a nearly continuous band of rocky substrate to which a variety of carbonate-producing organisms is attached. But also, many kinds of boring organisms live within the rock. NEUMANN (1966) ascribed the origin of extensive notches that deeply undercut the cliffs of Harrington Sound to the erosive action of rock-boring and grazing organisms. Large patches of the exposed rock are colored bright orange by the boring sponge Cliona lampa. Experiments in the laboratory and under natural conditions show that Cliona lampa is capable of removing as much as 6 to 7 kg of material from 1 m^2 of carbonate substrate in 100 days (NEUMANN 1966). This compares to an erosion rate of 1 m of calcarenite per 70 years.

On the cliff faces below tide level, the rocky substrate is covered by the calcareous algae Padina and Amphiroa. On the fronds of Padina plants WEFER (1980) measured an average linear growth of 0.81 mm d^{-1}. On the cliffs and submarine rocks of Harrington Sound, at a depth of 1 - 3 m, the standing stock is up to 350 g dry weight m^{-2}. Assuming a mean carbonate content of 38 % of dry weight (BÜHM and SCHRAMM 1977), and a possible turnover of up to 8 times per year, the annual carbonate production could be as high as 1,060 g m^{-2} y^{-1} (the average is 240 g m^{-2} y^{-1}) (WEFER 1980). The red alga Amphiroa fragilissima also abounds on rocky substrates and may be a major carbonate producer in Harrington Sound.

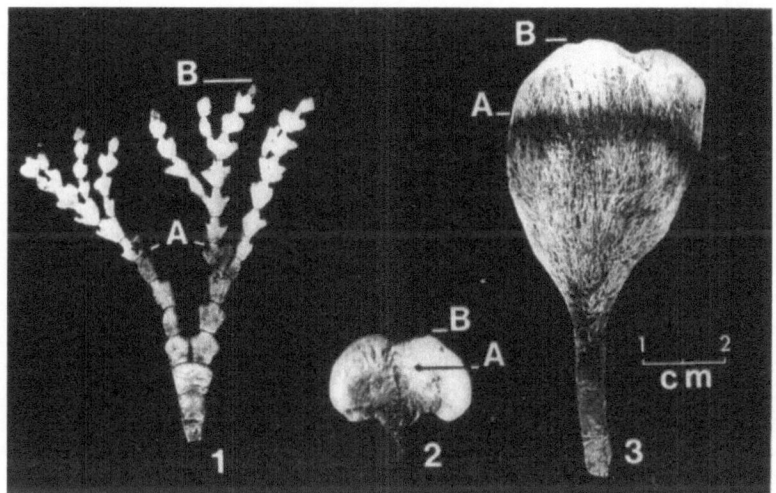

Fig. 6-8: Benthic calcareous algae: Halimeda incrassata (1), Padina sanctae-crucis (2) and Penicillus capitatus (3). Photographs of Alizarin-stained plants 7 days (Halimeda), 11 days (Padina) and 15 days (Penicillus) after the staining event. A = top of the Alizarin line; A-B = increment between staining and sampling (after WEFER 1980).

Shallow sandy zone - The sandy floor from the surface to about 10 m water depth exhibits a more variable distribution of organisms than the other zones (NEUMANN 1965). In the shallow part of the sandy zone, calcareous algae such as Penicillus and Halimeda are especially abundant. WEFER (1980) measured their growth rates using Alizarin Red-S as a time marker (see Fig. 6-8). The measured growth rates suggest that the algae renew their standing stock approximately once every month (Halimeda) or once every one and a half months (Penicillus) during their growing season which lasts only about seven months. For Halimeda, WEFER (1980) calculated a carbonate production of 50 g m^{-2} y^{-1} which is based on a turnover rate of 7 times per year and an average standing stock of 6.7 g $CaCO_3$ m^{-2} which is the mean value found at 31 stations arbitrarily selected in 1 - 4 m of water depth. For Penicillus, the carbonate production is 30 g m^{-2} y^{-1} based on a turnover rate of 4 times per year and an estimated average standing stock of 5.6 g $CaCO_3$ m^{-2} from the same 31 stations where the Halimeda estimate was obtained.

Another important algal carbonate contributor is Halimeda monile, which is only common in areas at or shallower than 3 m and may contribute as much carbonate as H. incrassata. Other green algae like Udotea, Cymopolia and Rhipocephalus are mostly solitary, rare or absent in Harrington Sound and are not important as carbonate ducers there. WEFER (1980) estimates that the total carbonate production by algae in the sandy area of Harrington Sound is nearly 100 g $CaCO_3$ m^{-2} y^{-1} and on the rocky substrate nearly 500 g $CaCO_3$ m^{-2} y^{-1}, which is about half of the carbonate deposition in the shallow water area of the Sound.

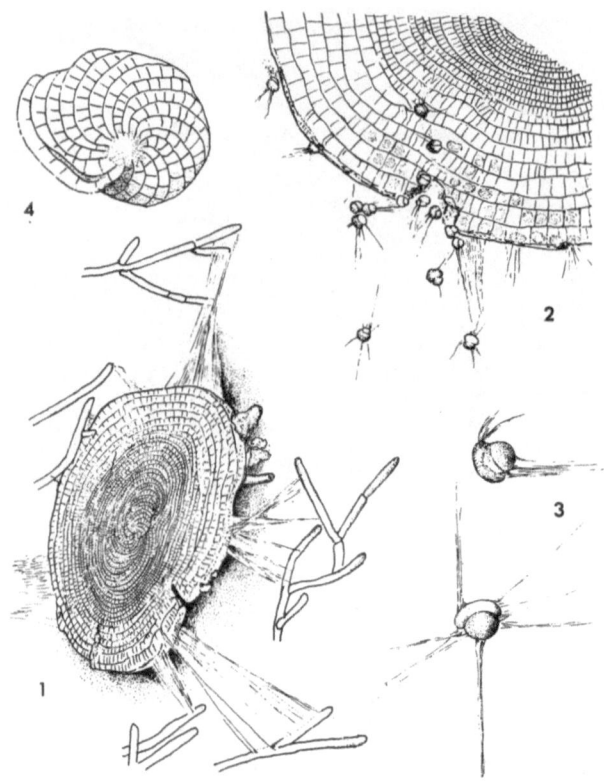

Fig. 6-9: Benthic foraminifera <u>Cyclorbiculina</u> <u>compressa</u>: (1) Adult specimen with
reproduction chambers attached to the alga <u>Cladophora</u> by means of pseudopodial
network (diameter 6.5 mm). (2) Portion of the same specimen with embryos
departing from the parental test. (3) Embryos (proloculus diameter 135 μm).
(4) Juvenile individual, approximately 1.5 months old (test diameter 550 μm)
(after LUTZE and WEFER 1980).

A further important producer of carbonate in Harrington Sound is the larger foraminifera
<u>Cyclorbiculina</u> <u>compressa</u> (Fig. 6-9) living attached to algae and on the sediment in
8 to 12 m of water depth. LUTZE and WEFER (1980) observed asexual reproduction in the
field and in the laboratory and assumed a turnover rate of one per year, which is in
accordance with estimates from stable isotope data (WEFER and BERGER 1980, WEFER
et al. 1981a). Using this turnover rate and an average adult population density of
7,500 m^{-2} LUTZE and WEFER (1980) estimated that the carbonate production should
average around 60 g m^{-2} y^{-1}.

Oculina zone - The transition from sand to mud is very abruptly marked at about 10 m of water depth by the mud-living coral Oculina (NEUMANN 1965). The coral supports itself on the soft mud by developing a brush-like growth on its own debris or on bivalve shells like Arca zebra. The mean standing stock of two important carbonate producers, Arca zebra and Oculina, is 2,500 g $CaCO_3$ m^{-2} and 1,000 g $CaCO_3$ m^{-2} respectively. ERLENKEUSER and WEFER (1982) deciphered the shell growth history of two Arca zebra specimens from Harrington Sound by means of oxygen isotope analyses. Both individuals, one 74 mm long and from 3 m of water depth and the other 66 mm long and from 16.5 m of water depth, showed an age of about 10 years. By using the mean standing stock data and a mean of 10 years for the Arca zebra population, then the carbonate production would be 100 g $CaCO_3$ m^{-2} y^{-1}. Using the annual periodicity in the $\delta^{18}O$ distribution, WEFER (1985) found a growth rate of 2.5 cm y^{-1} for the uppermost part of the actively growing branches of Oculina valenciensis. Because Oculina bushes are about 30 cm high, an average age of about 10 years is assumed for the Oculina colonies in Harrington Sound. Using this mean age and a standing stock of 2,500 g $CaCO_3$ m^{-2}, then the carbonate production by Oculina would be 250 g $CaCO_3$ m^{-2} y^{-1}. Many other calcareous organisms, but none as common as Oculina and Arca, live attached to the corals, on the mud or buried in the sediment (NEUMANN 1965, SCHWEIMANNS 1979); but their carbonate production is insignificant.

Sub-thermocline zone - The Oculina zone ends at about 18 m of water depth which coincides with the thermocline in late summer. This depth also marks a sharp transition between the mixed, warmer and oxygen-rich waters and the cooler, oxygen-poor waters below (NEUMANN 1965). In this zone calcareous organisms are not very common. Only the bivalves Gouldia cernia (Transenella conradina in NEUMANN 1965) and Macoma tenta are found in higher numbers. Their valves commonly comprise the entire coarse fraction of the sediment (NEUMANN 1965, SCHWEIMANNS 1979).

6.4   CARBONATE SATURATION STATE

The amount of calcium carbonate produced and the fraction buried in the sediment varies greatly between different regions. The relationship between production and accumulation has implications for the interpretation of fossil assemblages. It is by no means certain that a fossil record is always representative and is a reliable indicator of the paleoenvironment. The proportion of the living community which enters the sediment record depends mainly on the carbonate saturation state of the water column, especially at the sediment-water interface, the sedimentation rate, and the mineralogy of the carbonate particles produced.

Another aspect of marine dissolution of biogenic carbonates is the effect which fossil fuel CO_2 exerts on the long-term preservation of calcium carbonate in the shallow parts of the oceans. The models presently used to estimate global budgets assume certain degrees of $CaCO_3$ saturation depending on latitude and water depth, but measurements of the saturation state are rare. We determined the saturation state of the main body of water in both areas of investigation with respect to different naturally occurring carbonate minerals and particularly the saturation state and changes thereof at the water-sediment interface. Most of the fate of biogenic carbonates is determined at this crucial depositional site.

6.4.1   Kiel Bight and Adjacent Western Baltic

During a seasonal cycle the well mixed water column of the winter months is followed by development of a strong thermocline during spring. Consequently, the oxygen supply cannot keep up with the sub-pycnocline benthic demand of oxygen and severe depletion results every year in the bottom waters. Almost regularly, hydrogen sulfide occurs in the deep basins and channels. Due to the restricted vertical exchange, the metabolic products of bacterial O_2 consumption, namely CO_2, accumulate in the sub-pycnocline region forcing the seawater to become less saturated with respect to calcium carbonate. Such seasonal patterns of undersaturation, even with respect to the least soluble calcite, prevail in the bottom waters over large portions of the year (WEFER 1976a). In general, and contrary to the open ocean, the brackish waters of the Western Baltic favor dissolution of carbonates because of their reduced ionic strengths and lower concentrations of calcium and carbonate ions.

Saturometry applied to samples extracted from a long-term bell jar experiment in Kiel Bight revealed two distinct periods of calcium carbonate saturation behavior following the development of stagnation:
1. Rapid input of metabolic CO_2 at suboxic conditions leads to strong undersaturation with respect to calcite;
2. During anoxic conditions, undersaturation of the enclosed water turned to supersaturation partially by dissolution of carbonatic phases more soluble than calcite and partially by the neutralizing action of sulfide on excess hydrogen ions forcing the system towards higher concentrations of CO_3^{2-} and HS^-/H_2S (SUESS 1976b, BALZER 1980b).

Reflecting this intense undersaturation, the sedimentary carbonate contents decreased from about 2 % dry weight in the top 2 cm to a fifth of that in the layers below (BALZER 1978). The accumulation of foraminiferal tests in deeper layers which in recent times are supersaturated with respect to calcite is only less than 4 % of the total input (WEFER and LUTZE 1978) because of having to pass through the highly undersaturated suboxic zone prior to final burial.

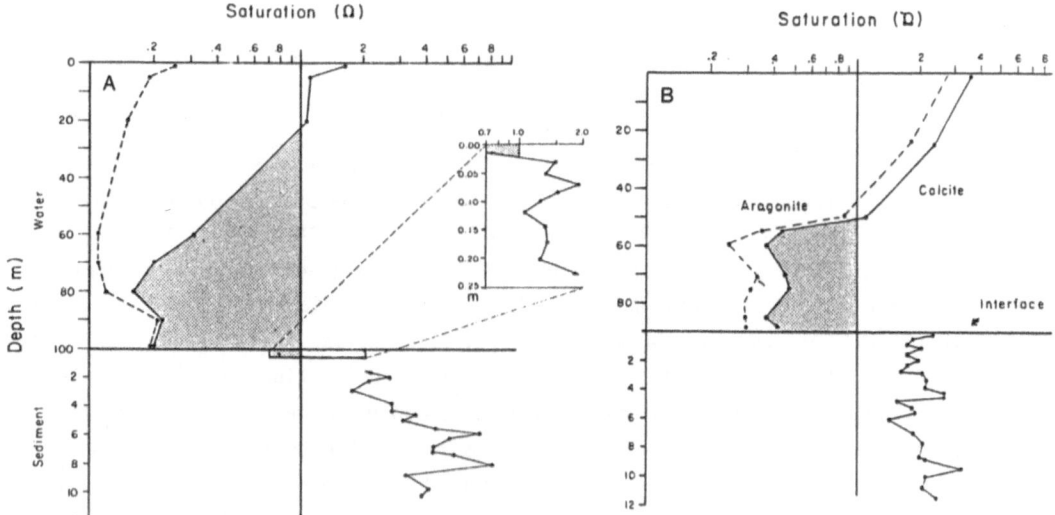

Fig.6-10: Carbonate saturation with respect to calcite and aragonite in the water
column and interstitial waters of Gdansk Bay (A) and Bornholm Basin (B);
the saturation state was determined by saturometry (after SUESS et al.
unpubl.).

Such a vertical sequence of supersaturation in the euphotic zone, undersaturation in
the sub-pycnocline water column and topmost sedimentary layer followed again by
supersaturation in deeper layers is typical for the Baltic Sea and illustrated in
Fig. 6-10 from SUESS et al. (unpublished).

Furthermore, the salinity effect on the overall saturation state of the Baltic waters
is also evident from Fig. 6-10. The Bay of Gdansk waters (Fig. 6-10A), ranging in
salinity from 7 - 12 °/.., are only barely supersaturated with respect to the least
soluble calcium carbonate phase (calcite) and are undersaturated throughout the water
column (~ 80 m thick) with respect to all other carbonate phases, although only
aragonite is shown in Fig. 6-10A. In contrast the Bornholm Basin waters (Fig. 6-10B),
with salinities between 7 and 16 °/.., show a much more substantial portion of the
water column to be supersaturated with respect to both calcite and aragonite. The
undersaturated portion of the water column (shaded area) is diminished to only ~ 40 m
in thickness.

The more crucial site in regard to calcium carbonate preservation and dissolution is
the sediment-water interface. Here, due to the increased production of metabolic CO_2,
undersaturation penetrates into the sediment column. Such a saturation pattern is
shown for the Bay of Gdansk where high sampling rates permit a much more detailed
view of the sediment-water interface (inset Fig. 6-10A). The calcite saturation is

only 20 % in the bottom water, ~ 50 % at the interface and 100 % at 3 cm of sediment depth. Below this, it increases steadily, although with large fluctuations, to 200 % at 1 m and ~ 700 % at 10 m sediment depth.

Any sedimentary carbonate particles, whether biogenic, terrigenous or glacial detrital, residing within these undersaturated waters for an extended period of time suffer severe dissolution. The saturation measurements indicate that even calcite, the least soluble phase, is affected as well. Such dissolution processes are evident from investigations by scanning electron microscopy of early diagenetic, ultrastructural alterations of mollusc and benthic foraminifera tests (LEWY 1975, GROBE and FÜTTERER 1981). Considering the distribution and preservation of molluscs in the sediments, LEWY (1975) proposed that carbonate dissolution occurs mainly on the sea bottom, while extraction of organic matter from within the skeletons proceeds within the sediment. GROBE and FÜTTERER (1981) specified several carbonate destroying processes by ultrastructural patterns of the shell surface and established three zones, each showing different mechanisms of shell fragmentation. From shallow to deeper waters these are: (1) a zone of abrasion in the area of lag sediment where mechanical destruction is dominant; (2) a zone of disintegration in the sandy area where the oxidation of the organic matter between the carbonate crystals prevails, and (3) a zone of corrosion in the muddy area where dissolution of the tests is the dominating shell destruction process (Fig. 6-11). Obviously, the third process is amply substantiated by the saturation patterns and behavior of Kiel Bight waters and those from adjacent regions of the Western Baltic.

6.4.2   Harrington Sound

In Harrington Sound, similarly to the conditions in Kiel Bight, a strong pycnocline regularly develops in late spring leading to oxygen depletion in the near bottom region (Fig. 6-12) and a sharp increase of P_{CO_2} from bacterial degradation of organic carbon. The increase of titration alkalinity clearly indicates that dissolution of carbonates must have taken place since input of CO_2 has no effect on alkalinity by itself and hydrogen sulfide was too low to contribute significantly (BALZER and WEFER 1981). To clarify which of the biogenic carbonates was dissolved, the saturation state of the Harrington Sound with respect to different naturally occurring carbonate minerals was determined by two independent methods: (1) by calculation from the difference in alkalinity prior to and after equilibration with the different solids using the method of BALZER (1980b), and (2) by an in situ dissolution/precipitation experiment, whereby carbonate minerals are placed on mooring lines at depths of 15-24 m for three weeks.

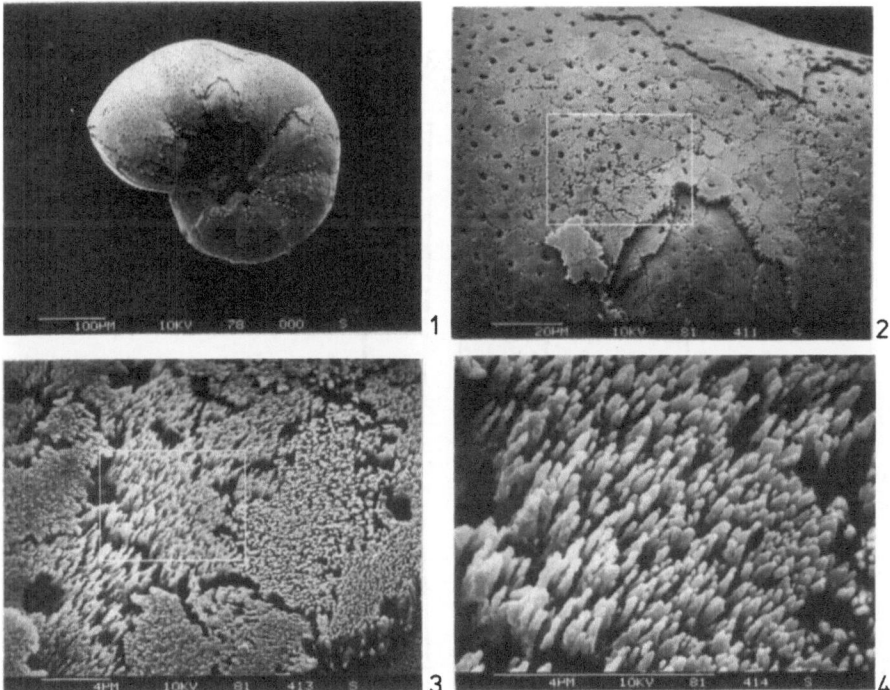

Fig.6-11: Scanning electron micrographs of the benthic foraminifer Elphidium incertum,
sampled in 27 m water depth in Kiel Bight: (1) overall view of the specimen
studied; (2) removal of individual layers with inset seen in (3); (3) and (4)
typical dissolution patterns seen at different magnifications (from GROBE
1981. Table 2, Figs. 5 - 8).

With the saturometry study, equilibration was accomplished by using aragonite from
ground coral Porites porites, magnesium calcite (12 - 12.4 mol% $MgCO_3$) from the sea
urchin Clypeaster rosaecus, and calcite from the Atlantic oyster Ostrea edulis, and
marble as reference material. Under the environmental conditions shown in Fig. 6-12,
all carbonates showed supersaturation of varying degrees in the super-thermocline
zone; the biogenic materials, however, all crossed at 100 % saturation level within a
narrow depth range within the sub-thermocline water body and actually reached
undersaturation. Due either to magnesium calcite impurities in the aragonitic coral
and the calcitic oyster or to magnesium calcite overgrowth, the saturation states of
these two biogenic materials probably also correspond to that of magnesium calcite.
Nevertheless, it was clearly shown that the sub-thermocline waters are acidic enough
to corrode high-magnesium calcites in the 8 - 12 mol% $MgCO_3$ range.

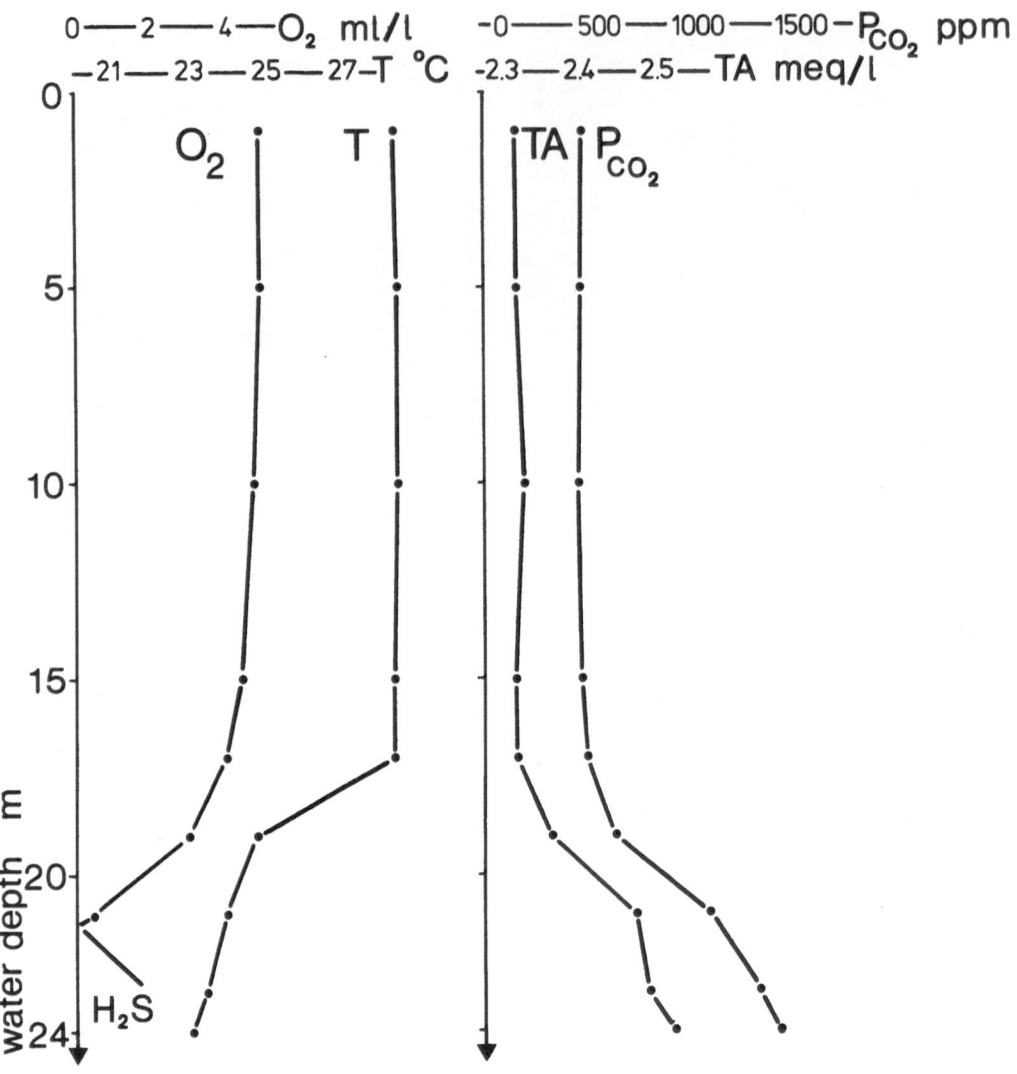

Fig.6-12: Hydrographic parameters (oxygen, temperature, titration alkalinity and partial pressure of CO$_2$) in the water column of Devil's Hole, Harrington Sound, Bermuda (September 1978) (after BALZER and WEFER 1981).

In a second experiment at similar environmental conditions (WEFER and BALZER, unpublished), saturometry and long-term in situ equilibration of carbonates were compared with each other in order to test the reliability of the saturometry technique for possible non-equilibration (BALZER and WEFER 1981). In addition to marble calcite and magnesium calcite from Clypeaster rosaecus, ground aragonite from the snail Strombus gigas and magnesium calcite (17.5 mol% MgCO$_3$) from the benthic red alga Amphiroa fragilissima living in shallow waters of Harrington Sound were used in the saturometry experiments.

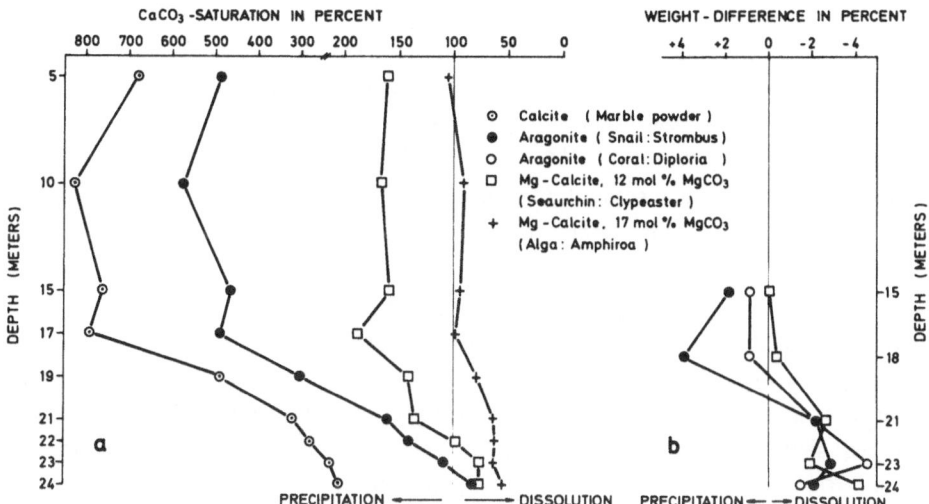

Fig.6-13: Carbonate saturation with respect to different carbonate minerals in the Devil's Hole water column (August/September 1979). The saturation state was determined by saturometry (a) and in situ deployment of carbonates (b) (after WEFER and BALZER unpubl.).

For the in situ experiments, aragonite from Strombus gigas and from the coral Diploria strigosa, and magnesium calcite (12 - 12.4 mol% MgCO3), from the sea urchin Clypeaster rosaecus were exposed to Harrington Sound waters.

The carbonate material was sieved into two size fractions: 63 - 250 μm and 250 - 500 μm, then between 200 and 300 mg were placed into 44 μm mesh sacks. According to the method of METZLER et al. (1982), the sacks were then placed inside plastic spheres, practice golf balls with numerous large holes, allowing water to flow through. Sets of samples were placed at 15, 18, 21, 23 and 24 m water depths on a mooring line in Devil's Hole, Harrington Sound for three weeks. Upon recovery, the nylon sacks including the carbonate material were oxidized with hypochloric acid (one part acid to 5 parts H_2O buffered to pH = 9) to remove organic material attached to the nylon sacks. Subsequently, the samples were rinsed five times in buffered de-ionized water (pH = 10) and dried at 60°C. After weighing, the difference before and after deployment of the carbonates was calculated.

The saturation states for the carbonates employed (as corrected for in situ conditions) are shown in Fig. 6-13. The marble calcite exhibited the largest degree of supersaturation over the entire water column, but saturation was lowered significantly in the sub-thermocline region. Starting from a low degree of supersaturation in the deepest sample, aragonite followed the same trend as calcite

and reached slight undersaturation in the deepest sample, while the sea urchin
sample showed slight supersaturation in surface water and was already corroded a few
meters above the bottom. The high-magnesium alga Amphiroa sample never showed any
supersaturation above the thermocline and was very sensitive to the increased
P_{CO_2}-levels at depth. It may be inferred from this behavior that Amphiroa would never
find its way into the sediment, since even low levels of excess CO_2 normally
prevail at the sediment-water interface which will dissolve this delicate
carbonate framework.

Of ecological significance is the fact that slightly increased levels of atmospheric
CO_2 would shift Amphiroa into undersaturation even in surface waters when they become
equilibrated with the atmosphere. The in situ dissolution/precipitation experiment
showed the same sequence of stability as the results from saturometry for Strombus
gigas and for the sea urchin Clypeaster, but both minerals already reached undersaturation at 20 - 21 m of depth. The most probable reason for this difference in the
transgression depth from super- to undersaturation is that the in situ method gives
an integrated signal of the environmental conditions during the three weeks of
deployment, while saturometry only yields results for the day of sampling. From the
data compiled by v. BODUNGEN et al. (1982), it can be seen that the daily samplings
for saturometry were on the weak side of corrosive intensity in the subthermocline
waters during the three-week period when compared to the integrated in situ experiments.
In addition, however, it cannot be ruled out that combined uncertainties in water
depth measurements for water sampling (saturometry), which was measured from the
sea surface, and for the mooring deployment (in situ experiment), which was measured
from the sea floor, may have caused an error of 1 to 1.5 m. This has significant
consequences for placing the saturation levels because of the strong stratification.

All these different experiments show that the sub-thermocline waters of Harrington
Sound are capable of dissolving high-magnesium calcites which are produced in shallow
areas and transported to the deeper parts of the Sound. Their absence in the sediments
(NEUMANN 1965) points to removal by dissolution. This explanation is corroborated
by an independent experiment. Sediments from the three zones: shallow-water region,
Oculina zone, and sub-thermocline area, were subjected to the same saturometry
procedure as the minerals mentioned above. They show a parallel pattern of saturation
states but with increasing stability from the shallow to the deeper sediments.
Conceding that saturometry applied to different sediment assemblages cannot yield
data with respect to a defined mineral composition, the increasing stability of
mineral assemblages with depth, however, shows that shallow sediments contain easily
soluble high-magnesium calcites which are absent in the sub-thermocline region.
From the saturation experiments there is also strong evidence that aragonite can
be corroded by the near-bottom water during summer stratification.

6.5 CARBONATE ACCUMULATION AND TRANSPORTATION

The amount of carbonate accumulation per area per year can be calculated by multiplying the rate of bulk sediment accumulation by the percent of $CaCO_3$ in the sediment. Accumulation rates were determined by ^{14}C- and ^{210}Pb-measurements or by lithostratigraphic correlations. The carbonate content was determined with a combustion technique (550°C). The individual content of carbonate-producing taxa was determined by counting the number of shells per wet volume of core sediment. For the weight of $CaCO_3$ per volume, this number has to be multiplied by the average weight of the shells.

6.5.1 Kiel Bight

6.5.1.1 Accumulation Rates and Carbonate Contents

At the deepest part of the area near the "Hausgarten", the sedimentation rate determined by ^{14}C measurements was 1.4 mm y^{-1} (ERLENKEUSER et al. 1974, SUESS and ERLENKEUSER 1975). WEFER and LUTZE (1978) calculated accumulation rates from intermediate depths by correlating distinct foraminiferal carbonate maxima in cores from 21.5 and 23.5 m of basin depth with the same maxima in the core analyzed by ERLENKEUSER et al. (1974). The values indicate that the sedimentation rate decreases from 1.4 mm y^{-1} at the deepest part of the basin to 0 at 13.5 m; above this depth no sediment accumulates. Further supporting evidence for a more or less linear increase in sedimentation rate with depth is provided by reflection seismic data (HINZ et al. 1971). The carbonate contents of the sediment are low in the Kiel Bight area, varying between 0 and about 9 % (WEFER et al. 1978a, see also Chapter 4 this volume).

The carbonate fraction of the sediment containing identifiable biogenic skeletal debris consists of:
(1) molluscan shells, mainly bivalves Arctica islandica, Macoma baltica, Abra alba, Astarte spp. and the gastropod Hydrobia;
(2) benthic foraminifera (Elphidium spp.), and
(3) carbonate detritus derived from glacial till erosion, e.g. planktonic foraminifera of Cretaceous and Tertiary age.

Remains of the common starfish Asterias rubens, producing up to 5 g $CaCO_3$ m^{-2} y^{-1}, are almost absent in the sediment. It is likely that the starfish disintegrates easily and dissolves because of the loose formation of the skeletal elements within the animal and the high solubility of magnesian calcites.

Table 6-2 Relation of production to accumulation of foraminiferal carbonate (after WEFER and LUTZE 1978).

	Water depth (m)					
	5-14	15-16	17-18	21-22	23-24	27-28
Carbonate production (mg m^{-2} y^{-1})	10-36	44	133	408	903	3.12
Carbonate accumulation (mg m^{-2} y^{-1})	0	0.19	0.27	3.96	10.5	72.4-132.4
Accumulation/production (in %)	0	0.4	0.2	1.0	1.2	2.3- 4.3

The molluscan shells found in the sediment are aragonitic in mineralogy and therefore are less resistant to dissolution than calcitic shells are. In the undersaturated water column, especially during late summer, and in the supersaturated interstitial waters, only aragonitic infaunal bivalves are preserved which die within the sediment or which are buried rapidly after death to below the undersaturated zone and thereby escape dissolution (LEWY 1975).

WEFER and LUTZE (1978) estimated the mean foraminiferal carbonate accumulation over the last 200 years. It reaches a maximum of 132 mg m^{-2} y^{-1} at 28 m water depth (Fig. 6-4D, Table 6-2). The amount of carbonate decreases progressively toward shallower waters and at 16 m it is < 1 mg m^{-2} y^{-1}. From the average accumulation during the last 200 years and present day production, they further calculated that the percentage of carbonate remaining at 28 m is about 4 %, at 27 m 2 %, at 23.5 and 21.5 m only 1 % of the amount actually produced (Table 6-2). At depths of 18 and 16 m < 0.5 % of the present day production has accumulated over the last 200 years.

6.5.1.2 Relationship Between Live and Dead Assemblage

The molluscan dead species assemblages generally reflect the living ones (ARNTZ et al. 1976). On the whole, they correspond closely to each other in their species composition, dominant species zonation and the distribution and depth pattern of the maxima of species. Downslope transport of shells is inferred by ARNTZ et al. (1976) from a higher abundance of dead-shell species in the deeper part of the area. As measured by the lateral displacement of the mollusc maximum belts, the transport amounts to 1 - 3 m in vertical distance, occasionally up to 7 m at current exposed slopes. These displacements correspond to 30 - 75 m horizontal transport. Besides currents, ARNTZ et al. (1976) suggest extreme wave action as a possible cause for the displacement. The living and dead foraminiferal faunas are also composed of the same

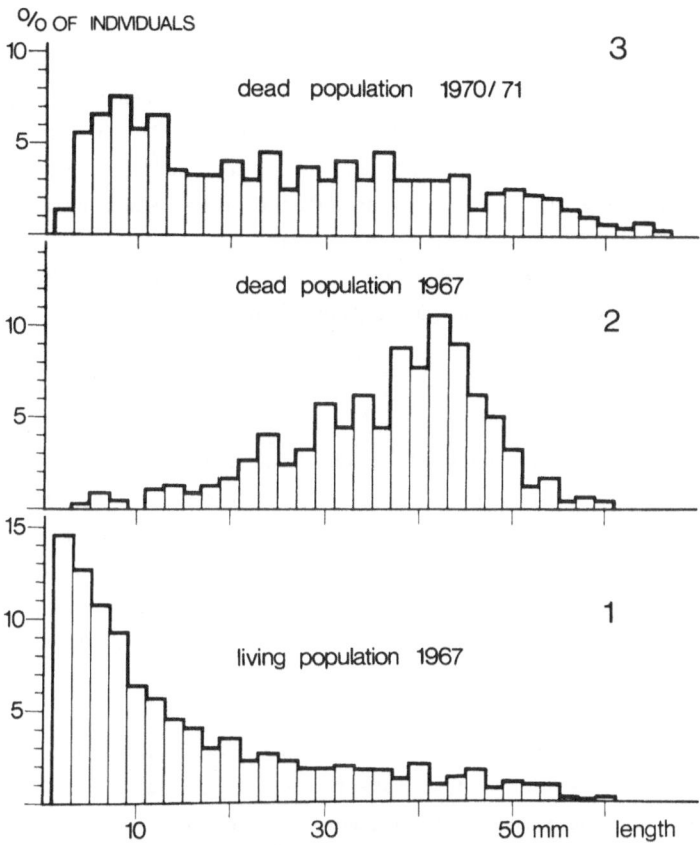

Fig.6-14: Size-frequency distribution of a <u>Mytilus edulis</u> community from Kiel-Holtenau (Kiel Harbour): (1) living population in 1967, (2) autochthonous dead community population in 1967, and (3) autochthonous dead community population in 1970/ 71 after the death of the entire population (after SAMTLEBEN 1973).

species in similar proportions (WEFER and LUTZE 1978). This means that selective dissolution, which generally is important in sedimentation of planktonic foraminifera (BERGER 1971), cannot play much of a role in affecting benthic faunal assemblages. Therefore, it appears that in this shallow-water environment the fossil record of molluscs and benthic calcareous foraminifera should be a reliable indicator of the composition of the original population and thus should allow a realistic assessment of the paleoenvironment. But for benthic foraminifera it cannot be assumed that the proportion of calcareous species in the total population (including arenaceous species) will reflect the original balance, as is obvious from the very low percentage of preserved calcareous tests.

In a comparison between size-frequency distributions of living Mytilus-populations (as an example see Fig. 6-14.1) with those of autochthonous dead communities, SAMTLEBEN (1973) found that juvenile specimens are under-represented or even missing in the dead communities (see Fig. 6-14.2) probably due to the action of predators. Only in the case of a catastrophic death and subsequent burial of the entire population does the size-frequency distribution of an autochthonous dead population correspond to a living population (Fig. 6-14.3). Further, SAMTLEBEN (1973) showed that the size-frequency distribution of an assemblage of dead individuals is no definite indication as to whether the assemblage is of autochthonous or of allochthonous nature.

6.5.2   Harrington Sound

Because of the enclosed waterbody and limited exchange with the sea as well as the supersaturation of the water with respect to carbonate minerals, almost all carbonate produced is accumulated in Harrington Sound. Dissolution of high magnesium calcites is only expected in the area of the sub-thermocline zone. Saturometry measurements are in agreement with X-ray diffraction analysis of NEUMANN (1965), who did not find significant amounts of high magnesium calcites in the deeper regions of the Harrington Sound.

The accumulation rates, which were determined by radiocarbon measurements and ^{210}Pb dating (for a technical description see JENSEN et al. 1977), increase from about 0.1 mm y^{-1} in the shallow sandy zone to more than 1 mm y^{-1} in the subthermocline zone (ERLENKEUSER 1981, ERLENKEUSER et al. 1981).

In Harrington Sound the silt fraction represents a large percentage of the sediment of the deeper zones. NEUMANN (1965) assumed that most of this material is produced originally in the shallow nearshore zones but is winnowed out by turbulence and settles in the protected deeper basins. NEUMANN (1965) could not test his hypothesis,

---

Fig. 6-15: Zonation of the Harrington Sound, Bermuda (A); grain-size distributioin of surface sediments at various stations (B); aragonite, Mg- and Sr-carbonate content of the < 63 μm fraction at above stations (C); corresponding concentrations of mannose (stripped column), fucose (black column) and xylose (white column) given in (D). Fucose and mannose at 10 and 13.5 m depths are indicative of coral (Oculina) and molluscs (Arca) resp. High xylose levels in shallow and again in deep water sediments suggest high algal input. The deep water xylose peak suggests transport of algal sediments (after BÖHM et al. 1981).

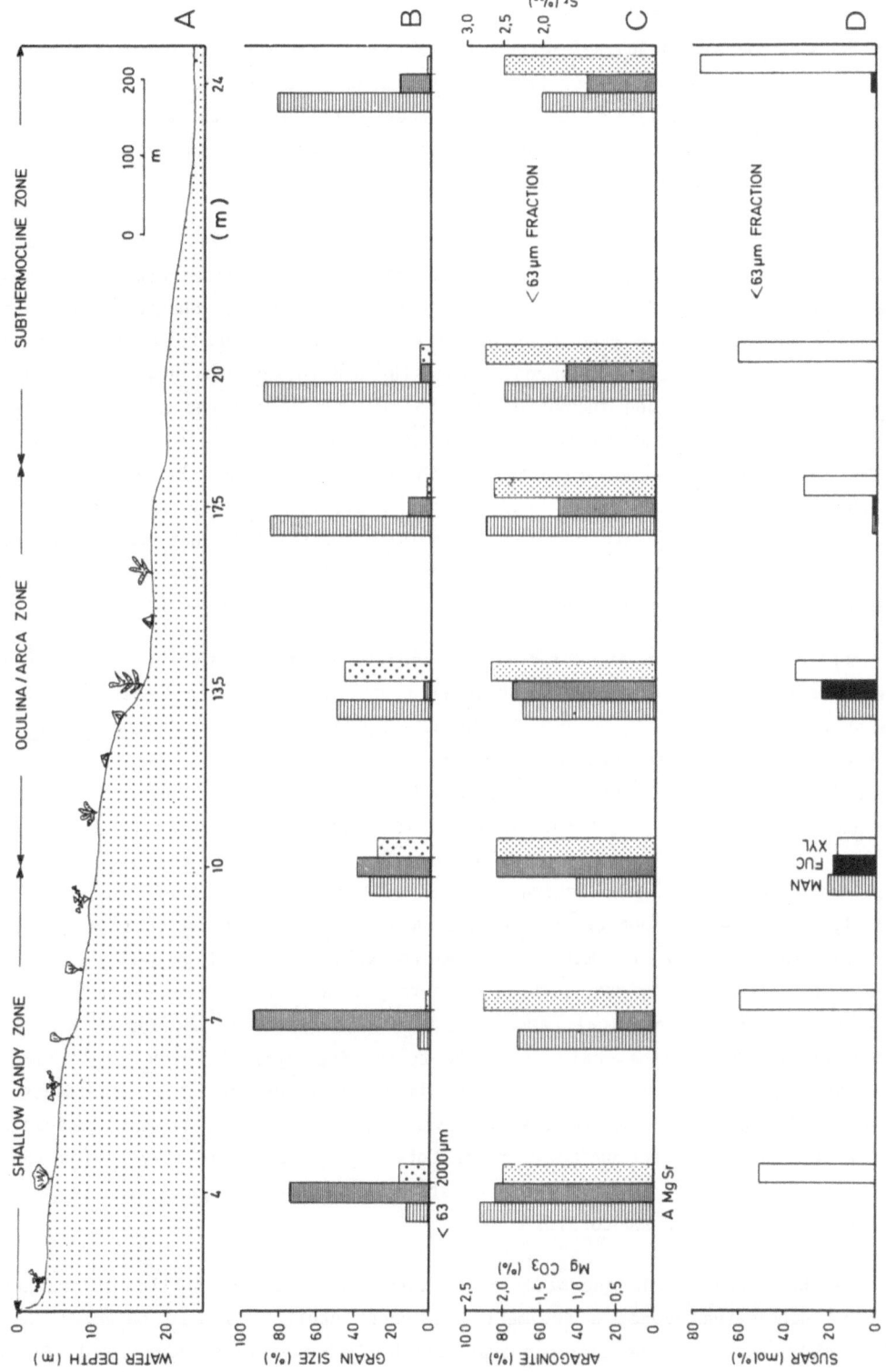

since the identification of the remains of organisms contributing to carbonate sediments by means of scanning electron microscopy is limited to particles of 2 - 20 µm in size (FÜTTERER 1977). Mineralogy and the content of Mg, Sr and trace elements alone are usually insufficient to solve the problem of identification. Therefore, BÖHM et al. (1980) tested the suitability of monosaccharides as markers for taxa identification in carbonate sediments.

The organic matrix of calcareous organisms consists of stable biopolymers such as polysaccharides and glycoproteins which are intimately associated with the carbonate skeleton. Analysis of these hydrolyzed compounds gives rise to characteristic arrays of monosaccharides which provide independent criteria for producer identification. The calcareous green algae Halimeda, Penicillus and Udotea show high xylose and low fucose levels. Xylose and fucose levels are elevated in the red algae Amphiroa but only fucose is prominent in the brown algae Padina. The corals Oculina, Porites, Millepora and Montastrea are relatively rich in fucose and show little or no xylose. In the bivalves Arca, Codakia and in Argopecten, mannose may be characteristic. Analysis of artificial and natural sediments demonstrates that coral and algal aragonite can be distinguished on the basis of the total sugar concentration and respective xylose and fucose levels.

The application of sugar data for such taxonomic identification is demonstrated on seven surface samples from different water depths in Harrington Sound (Fig. 6-15). The shallow water sediment is rich in the 63 - 2,000 µm fraction. In the Oculina-Arca zone at 10 - 18 m, grain size distribution is bimodal with large quantities of < 63 and > 2,000 µm fraction. In the sub-thermocline zone, up to 90 % of the sediment is composed of the < 63 µm fraction. Aragonite is between 60 and 90 % except for the 10 m level where it is 40 %. $MgCO_3$ varies between 0.5 and 2.2 mol%, whereas Sr is between 0.25 and 0.28 % and rather constant. The Sr data dictate that aragonite from corals and algae must be low, not exceeding 30 %. A positive identification is not possible from the inorganic data, but mannose, fucose and xylose levels (Fig. 6-15, lower graph) give additional clues. Xylose is shown to be very high in shallow water and again below 20 m. The occurrence of algal carbonates in shallow water is undisputed from field observations and from the analysis of the coarse fraction. The high xylose level in the sub-thermocline zone suggests that fine fractions of algal sediments are transported and accumulate in deeper water. This conclusion is not contradicted by the low Sr concentration because the total aragonite is relatively low (60 %). Sediment samples from 10 - 18 m depth show fucose and mannose. These sugars are indicative of bivalves and corals.

Besides the winnowing of fine material by turbulence from the shallow to the deeper part of Harrington Sound, a sediment transport "uphill" is also observed. ALHEIT (1980, 1982) estimated the sediment transport by fishes in Harrington Sound. The most

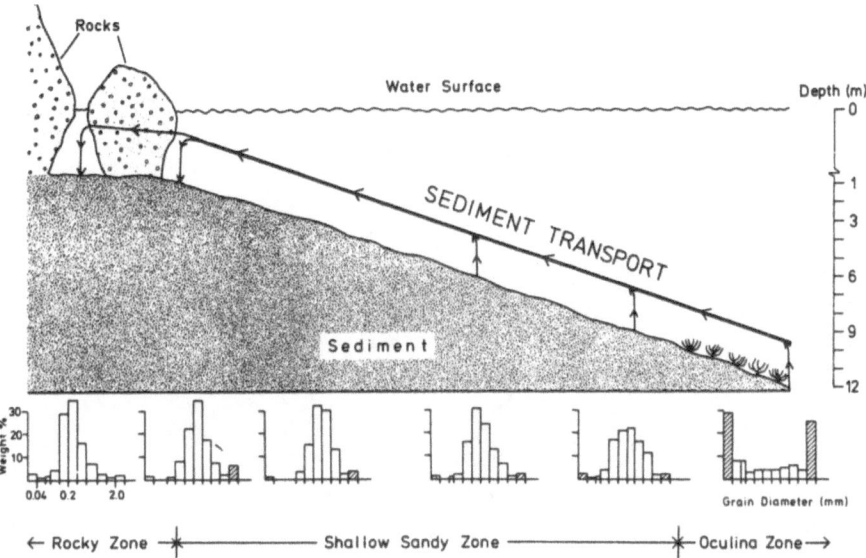

Fig.6-16: Schematic presentation of sediment transport by fishes, mainly grunts and breams, in Harrington Sound, Bermuda (after ALHEIT 1982).

important groups of fishes in this lagoon in terms of biomass, the grunts (Haemulon aurolineatum, H. flavolineatum, H. sciurus) and the bream (Diplodus bermudensis) undertake daily feeding migrations from the shallow rocky zone towards the deeper sand and mud zones. When feeding on zoobenthos they cannot avoid swallowing carbonate sediment particles. These sediment particles pass through the alimentary canal of the fishes and are deposited again, after digestion of the food, as faeces in the shallow zones. Thus, the fishes transport sediment "uphill", as is schematically shown in Fig. 6-16. By recording the fish stock densities, digestion rates and calcium carbonate contents in fish stomachs and guts, it was possible to estimate the amount of sediment transported by this mode. The amounts annually are ~ 4,256 kg calcium carbonate in Harrington Sound.

Based on the estimates and the data discussed here, a preliminary production/ accumulation budget was established by WEFER (1979), which is shown in Fig. 6-17. The total area of Harrington Sound is about 4.8 km^2. The shallow sandy zone occupies about 1.2 km^2, the Oculina zone about 1.4 km^2 and the sub-thermocline zone about 2.2 km^2. The following accumulation rates were estimated from the radiocarbon ages of the sediment and the ^{210}Pb-dating: 0.1 mm y^{-1} in the shallow sandy zone, 0.4 mm y^{-1} in the Oculina zone and about 1 mm y^{-1} in the sub-thermocline zone. Since 1 cm^3 carbonate sand weighs about 1 gr, about 120, 450 and 2,200 tons per year are accumulating in the shallow sandy zone, Oculina zone and sub-thermocline zone, respectively. As

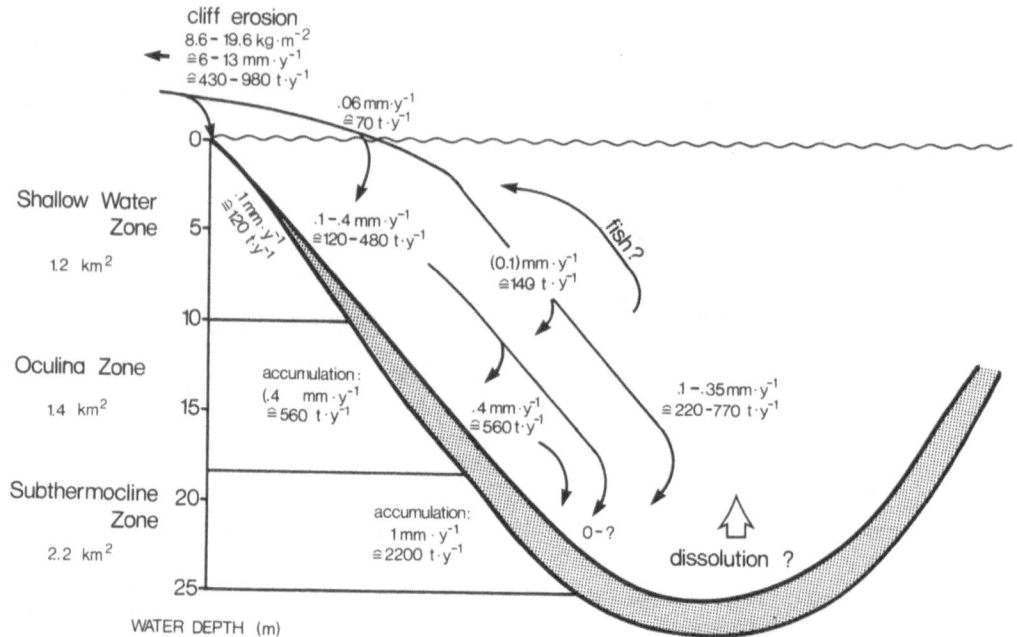

Fig.6-17: General view on the carbonate cycle in Harrington Sound. Depicted are the annual production and accumulation as well as the carbonate supply by cliff erosion in the three ecological zones: sandy shallow zone, Oculina zone and subthermocline zone (modified after WEFER 1979).

production rates, 120 to 480 tons per year were estimated for the shallow sandy zone and about 560 tons per year for the Oculina zone. In the sub-thermocline zone recent carbonate production is negligibly low.

A further important contribution to the sediment budget is that of material eroded from the cliffs surrounding the lagoon. As these carbonates are free of radiocarbon, the relative proportion of fossil carbonate can be calculated by radiocarbon dating (ERLENKEUSER 1981). In Devil's Hole, the ^{14}C-age of the surface material is about 4,000 years, corresponding to a 40 % fraction of fossil carbonate (Fig. 6-18); ^{210}Pb-dating indicates a sedimentation rate slightly greater than 1 mm y^{-1} (ERLENKEUSER 1981). It is estimated that 70, 140 and 220 to 770 tons per year are transported to the sandy shallow zone, Oculina zone and sub-thermocline zone, respectively. This amount of 430 to 980 tons of fossil carbonate eroded corresponds to an annual cliff retreat of 6 to 13 mm for the 14 km long and, on average, 3 to 4 m high cliff surrounding Harrington Sound.

For balancing production and accumulation, significant amounts of carbonate have to be transported into deeper areas of the Sound (Fig. 6-17). It seems that in the shallow sandy zone the surplus amounts to several 100 tons per year. In the Oculina zone, the surplus is lower and a part is transported into the sub-thermocline zone, a

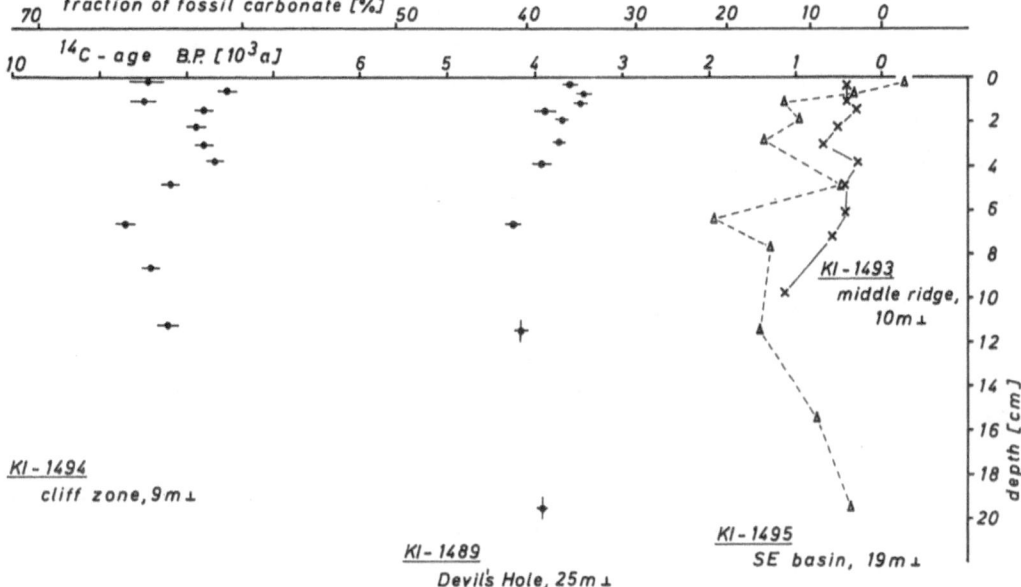

Fig.6-18: ^{14}C age of the bulk carbonate vs. sample depth for four cores from the Harrington Sound, Bermuda (after ERLENKEUSER 1981).

smaller part by fishes into the shallow sandy zone. In the sub-thermocline zone, high magnesium calcites are dissolved.

6.6   DISTRIBUTION OF STABLE ISOTOPES IN THE SHELL CARBONATE OF SHALLOW-WATER ORGANISMS

The knowledge of the isotopic composition of calcareous shallow-water organisms is of interest in both biological and geological studies. As calcareous shallow-water organisms have left a significant paleontological record and have been responsible for the build-up of enormous masses of limestone in the past, their remains are used extensively in biostratigraphy and paleoenvironmental reconstructions.

The most widely used signal in calcareous shells, especially in formaminiferal shells, is the oxygen isotopic composition. The basis of this method is isotopic fractionation which is the different thermodynamic equilibration of calcite and seawater with respect to their $^{18}O/^{16}O$ ratios and further that this difference decreases with increasing temperature. A relationship between temperature of carbonate precipitation and oxygen isotopic composition of the shell material is given by the empirical equation of EPSTEIN et al. (1953), modified by EPSTEIN and MAYEDA (1953)

$$T = 16.5 - 4.3 \; (\delta - A) + 0.14 \; (\delta - A)^2$$

where
$$\delta = \left(\frac{^{18}O/^{16}O \text{ of } CO_2 \text{ gas from sample}}{^{18}O/^{16}O \text{ of } CO_2 \text{ gas from PDB-1 standard}} - 1 \right) \times 1000$$

and
$$A = \left(\frac{^{18}O/^{16}O \text{ of } CO_2 \text{ equilibrated with } H_2O}{^{18}O/^{16}O \text{ of } CO_2 \text{ from PDB-1 standard}} - 1 \right) \times 1000$$

Therefore, δ is defined as the per mill deviation of the sample $^{18}O/^{16}O$ ratio from that of the PDB-1 standard. For information on the theoretical basis of isotope fractionation, standards and different $\delta^{18}O$-temperature relationships see FAURE (1977), HOEFS (1980), and RYE and SOMMER (1980).

Along with the oxygen isotopes, the composition of the stable carbon isotopes is determined, but these have received much less attention than the oxygen isotope composition. For oxygen isotope variations, the main cause appears to be temperature (with the average level determined by the water composition superimposed on a more or less constant vital effect), but no such straightforward explanation has been identified for carbon isotope variations.

The Bermudan waters are especially suitable for isotopic studies, since the seasonal salinity range is only about 0.7 °/··, corresponding to a $\delta^{18}O$ variation of less than 0.2 °/··. Therefore, variations in the stable oxygen isotopes within the shell material can be attributed to either the temperature or vital effects of the organisms. In Kiel Bight, the interpretation of the isotope data is complicated by strong seasonal variations in both temperature and isotopic composition of the seawater. The isotopic composition of calcareous shells can therefore be caused by temperature variations, changes in the isotopic composition of the water, or by non-equilibrium carbonate precipitation.

Our studies were intended to clarify some basic questions:
(1) Are the seasonal temperature ranges and life histories recorded within the shell material by fluctuations of oxygen and carbon isotopes?
(2) Does precipitation occur in equilibrium or not?
(3) Are the equilibrium or disequilibrium constants or variables the same within the same taxonomic unit? and
(4) Are there differences between the various taxa of carbonate-producing organisms in regard to these parameters?

This information is essential for any kind of ecologic or paleoecologic assessment and to determine the utility of stable isotopes in calcareous shells.

From the Bermudan localities, especially from Harrington Sound, the following organisms were studied: large foraminifera (WEFER and BERGER 1980, WEFER et al. 1981a), gastropods (WEFER and KILLINGLEY 1980), bivalves (ERLENKEUSER and WEFER 1982), and calcareous algae (WEFER and BERGER 1981). The large benthic foraminifera Cyclorbiculina compressa from Harrington Sound contains a record of seasonal temperature range and life history stages in its calcareous shell. The estimate for the life span of one year is in accord with the field and laboratory observations of LUTZE and WEFER (1980). These findings imply that the growth rate and, hence, the carbonate production of large foraminifera can be determined even in fossil specimens, if they are well preserved.

Other large benthic foraminifera, almost all common genera belonging to the suborders Rotaliina and Miliolina, from different tropical and subtropical localities have been analyzed by WEFER et al. (1981a). Life spans range between one season to more than two years. Almost all specimens studied showed the expected variations in $\delta^{18}O$ with respect to ambient conditions of seawater isotopic composition and seasonal temperature fluctuations. Taking into account the high $MgCO_3$ content of the shell for the calculation of equilibrium precipitation (see TARUTANI et al. 1969), all specimens seemed to be depleted in ^{18}O independent of their taxonomic position. Specimens from the same species showed about the same level and range of $\delta^{18}O$ values. The miliolid species Marginopora vertebralis, Cyclorbiculina compressa, Archaias angulatus, Peneroplis proteus, and Praesorites orbitolitoides cf. monensis commonly show carbon isotope values up to 2.5 °/·· lighter than expected, independent of sample locality. The rotaliid species Heterostegina depressa, Operculina sp., and Calcarina spengleri show carbon isotope values more than 2 °/·· lighter than expected and also independent of sample locality. All miliolid species analyzed show a tendency, with age, towards lighter-than-equilibrium $\delta^{13}C$ values. The rotaliid species showed the reverse, that is, shells tend toward increased $\delta^{13}C$ values with age.

Within the aragonitic shells of the strombid snails Strombus gigas (Fig. 6-19) and S. costatus, the isotopic ratios of $^{18}O/^{16}O$ and $^{13}C/^{12}C$ show an annual periodicity (WEFER and KILLINGLEY 1980). Strombus gigas exhibits a constant displacement of 0.5 °/·· from oxygen isotopes in accordance with established fractionation relationships (e.g. after EPSTEIN et al. 1953). With this tool, the growth in snails during different stages can be determined on one and the same individual. A S. gigas specimen 24.8 cm long was 7 years old, a S. costatus specimen 19.3 cm long was 5 years, and a S. costatus specimen 7.8 cm long was 2 years of age. From the annual range in oxygen isotopes, WEFER and KILLINGLEY (1980) assumed that growth is restricted to warmer periods of the year, when temperatures were above 19°C. Further, they found that the abundance of ^{18}O and ^{13}C tends to be positively correlated in adult S. gigas and S. costatus specimens, while it was inversely related in a juvenile S. costatus individual.

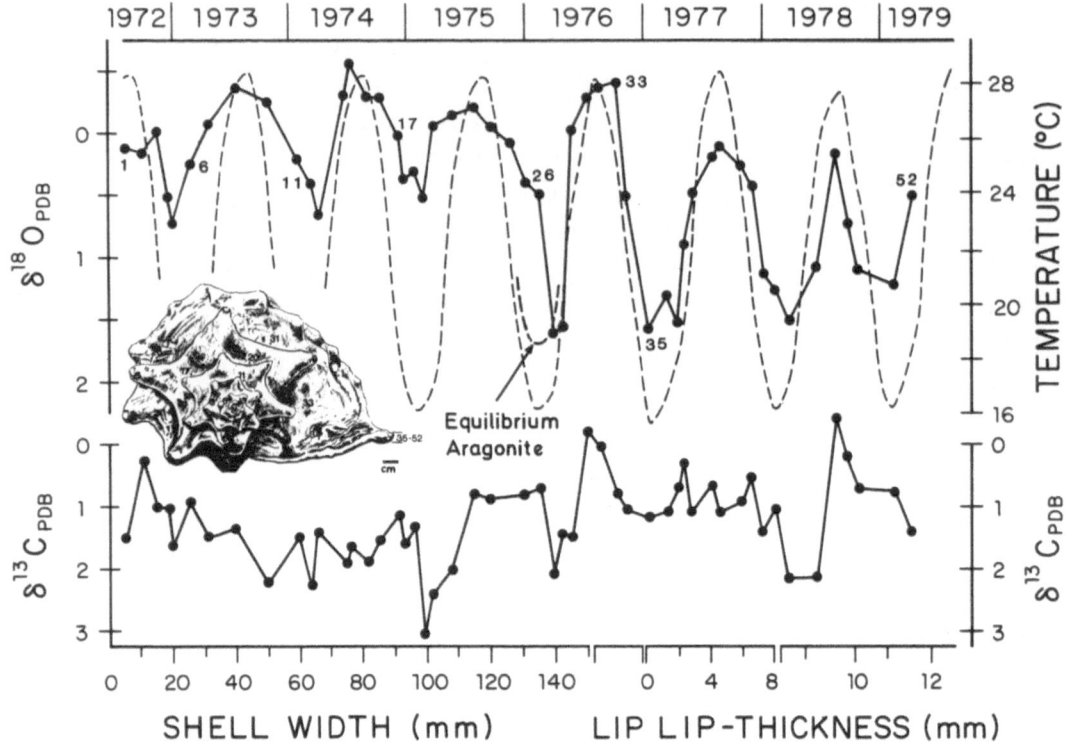

Fig.6-19: Upper: comparison of $\delta^{18}O$ variations in <u>Strombus</u> <u>gigas</u> (solid line) with seasonal changes in surface water temperatures, at 10 m (dashed line) in Northern Lagoon, Bermuda Islands, between August 1972 and September 1979. Temperature and oxygen isotope scales are related by the equation of EPSTEIN et al. (1953).
Lower: $\delta^{13}C$ variations in <u>Strombus</u> <u>gigas</u> with growth. Samples removed from whorl (denoted by shell width) and then across the top surface of the lip, followed by sampling of the lip edge in the direction of lip-thickening. Note scale changes. Numbers at sample points refer to numbers in figure (after WEFER and KILLINGLEY 1980).

Reliable growth data for two common bivalves from Harrington Sound, <u>Arca</u> <u>zebra</u> and <u>Macrocallista</u> <u>maculata</u>, were determined with the aid of the variations of oxygen isotope ratios in the shell carbonate (ERLENKEUSER and WEFER 1982). The two specimens of <u>Arca</u> <u>zebra</u>, one 74 mm long from 3 m of water depth and the other 66 mm long from 16.5 m, were both about 10 years old, and a shell of <u>Macrocallista</u> <u>maculata</u> had grown 71 mm in 11 years. Both species appear to form their shell carbonate in equilibrium with ambient seawater. As in the case of <u>Strombus</u>, shell growth was faster during the warm season and, for the shallower-living specimens, appeared to have halted at temperatures below 19-20°C.

Since very little information was available on the stable isotope composition of benthic calcareous algae in the literature, WEFER and BERGER (1981) analyzed the common genera <u>Halimeda</u>, <u>Penicillus</u>, <u>Acetabularia</u>, <u>Padina</u>, and <u>Amphiroa</u> of the Bermudan environment to clarify some basic questions:

(1) Granted that benthic algae deposit their skeletons at disequilibrium (KEITH and WEBER 1965), just how far from equilibrium does precipitation occur?
(2) Is disequilibrium constant or variable within the same species?
(3) Are there differences between the various kinds of benthic algae regarding disequilibrium precipitation?

WEFER and BERGER (1981) grouped the algae analyzed (5 genera, 3 phyla) into four types according to the $\delta^{18}O$ and $\delta^{13}C$ composition: a "heavy" and a "light" set for oxygen, and a "large variability" and a "small variability" set for carbon. The sets were not congruent. In one alga, different parts of the same skeleton belonged to different oxygen sets. They concluded that oxygen and carbon isotopes in calcareous algae reflect metabolic processes to a very large degree so that environmental signals are subdued, and that no simple explanation applicable to all taxa is obvious for the observed oxygen and carbon variations.

6.7   SUMMARY

In Kiel Bight, carbonate production by organisms is in the range of grams $CaCO_3$ m^{-2} y^{-1} (Table 6-1) and depends on the water depth and the carbonate-producing species. In the lag sediment and sand areas the asteroid <u>Asterias rubens</u> is the most important carbonate producer, while molluscs and benthic foraminifera are dominant in the deeper parts of the area. Similar amounts are produced by epiphytic organisms living on algae in the area of the erosional platform. About 50 to 100 times more carbonate is supplied from the surrounding cliffs and the abrasion platform by glacial till erosion than by biological production.

In Harrington Sound, carbonate production by organisms is two orders of magnitude higher than in Kiel Bight. The production by single species is in the range of several hundreds of grams of $CaCO_3$ m^{-2} y^{-1} and may be as high as 1 kg m^{-2} y^{-1}. Similar amounts are contributed by the erosion of cliffs and submarine outcrops. These figures are about two orders of magnitude smaller than production rates known from other tropical areas. STEARN et al. (1977) reported a gross production rate for the reef surface (excluding loose sediment) of a fringing reef on the west coast of Barbados of 9 kg m^{-2} y^{-1} at which single coral species produced between 10 and 40 kg $CaCO_3$ per m^2 coral surface and year.

The biogenic production rates from Kiel Bight and the Harrington Sound are within the range of the data on benthic carbonate production found worldwide between 25° and

54°N latitude (data are summarized by SARNTHEIN 1973). Other areas where higher values were reported are reefs, sea grass beds and tops of subtropical banks, which yield generally 70 to 1,000 g $CaCO_3$ m^{-2} y^{-1}.

Both areas studied show similar vertical sections of the saturation state of the water with respect to carbonate minerals. Supersaturation or only slight undersaturation is found in the uppermost water column or within the sediment. Undersaturation or less supersaturation was measured in the sub-pycnocline waters, where low oxygen and high CO_2-contents prevailed. However, totally different was the degree of saturation in Kiel Bight. Here also the least soluble calcitic minerals dissolved in the water column, on the sediment surface and in the uppermost sediment surface. In Harrington Sound, undersaturation was observed only for high magnesium calcites showing the highest solubility of the common biogenic carbonate minerals.

As a result of the differences in the carbonate saturation between the two areas, different amounts of carbonates produced are buried in the sediment. Only shells of infaunal organisms or tests which enter the sediment via bioturbation or high sedimentation, are preserved. In contrast to the situation in Kiel Bight, in Harrington Sound almost all carbonate produced is preserved. Only the high magnesium calcites, precipitated by red algae in the shallower part of the Sound and transported into the deeper regions, are being dissolved during the fall, when the CO_2-content increases from enhanced oxidation of organic matter.

Due to the supersaturation of the water with respect to carbonate minerals and the protected waterbody which prevents major sediment transport, in Harrington Sound the dead-shell species assemblage generally corresponds to the live one. Although in Kiel Bight most of the shells produced are dissolved before they enter the sediment, living and dead molluscs and calcareous foraminiferal faunas are composed of the same species in similar proportions. It appears that in this highly undersaturated area the fossil record is a reliable indicator of the original composition and thus should allow a realistic assessment of the paleoenvironment.

The stable isotopes of the dominant groups of calcareous organisms showed that the shells contain information about seasonal temperature ranges and life-history stages. These are recorded as fluctuations of oxygen and carbon isotopic values within the shell material. Almost all specimens showed the expected variations in $\delta^{18}O$ with respect to ambient conditions of water isotopic composition and seasonal temperature fluctuations. Most species incorporate ^{18}O close to equilibrium. Some are depleted (algae, larger foraminifera, corals), while others are enriched in ^{18}O. Almost all specimens analyzed are depleted in ^{13}C with respect to equilibrium precipitation. The $\delta^{13}C$ differences are up to 10 %o in magnitude within one single organism, and in general show numerous different trends.

LITERATURE

AERTEBERG NIELSEN, G., T.S. JACOBSEN, E. GARGAS & E. BUCH 1981: The Belt Project: Evaluation of the physical, chemical and biological measurements.- Nat. Agency Environm. Protection, Denmark, 122 pp.

AIGNER, T. & H.E. REINECK 1982: Proximality trends in modern storm sands from the Helgoland Bight (North Sea) and their implications for basin analysis.- Senckenbergiana marit. 14 (516): 183-215.

ALHEIT, J. 1978: Distribution of the polychaete genus Nephtys: A stratified random sampling survey.- Kieler Meeresforsch., Sdh. 4: 61-67.

ALHEIT, J. 1979: Die Stellung der Fische im Ökosystem einer subtropischen Lagune Bermudas.- Ph.D. Thesis, Univ. Kiel, 136 pp.

ALHEIT, J. 1982: Sediment transport by fishes in Harrington Sound, Bermuda. Proc. IV. Coral reef Symp., Manila (1981),2: 545-552.

ALLEN, J.R.L. 1980: Large transverse bedforms and the character of boundary layers in shallow water environments.- Sedimentology. 27: 317-323.

ALLEN, P.A. 1981: Wave-generated structures in the Devonian lacustrine sediments of south-east Shetland and ancient wave conditions.- Sedimentology 28: 369-379..

ALLER, R.C. 1978: Experimental studies for changes produced by deposit feeders on pore water, sediment, and overlying water chemistry.- Am. J. Sci. 278(9): 1185-1234.

ALLER, R.C. 1980a: Diagenetic processes near the sediment-water interface of Long Island Sound. I. Decomposition and nutrient element geochemistry (S, N, P).- Advances in Geophys. 22: 237-350.

ALLER, R.C. 1980b: Quantifying solute distributions in the bioturbated zone of marine sediments by defining an average microenvironment.- Geochim. Cosmochim. Acta 44(12): 1955-1965.

ALLER, R.C. 1980c: Diagenetic processes near the sediment-water interface of Long Island Sound II. Fe and Mn.- Advances in Geophys. 22: 351-415.

ALLER, R.C. & L.K. BENNINGER 1981: Spatial and temporal patterns of dissolved ammonium, manganese and silica fluxes from bottom sediments of Long Island Sound, USA.- J. Mar. Res. 39: 295-314.

ANDERSEN, N.R. & A. MALAHOFF 1977: The fate of fossil fuel CO_2 in the oceans.- Plenum Press, New York, 749 pp.

ARNTZ, W.E. 1971: Biomasse und Produktion des Makrobenthos in den tieferen Teilen der Kieler Bucht im Jahre 1968.- Kieler Meeresforsch. 27: 36-72.

ARNTZ, W.E. 1978a: The upper part of the benthic food web: the role of macrobenthos in the Western Baltic.- Rapp. Proc. Verb. 173: 85-100.

ARNTZ, W.E. 1978b: Zielsetzung und Probleme struktureller Benthosuntersuchungen in der marinen Ökosystemforschung.- Verh. Ges. Ökol., Kiel 1977: 35-51.

ARNTZ, W.E. 1980: Predation by demersal fish and its impact on the dynamics of macrobenthos.- In: TENORE, K.R. and B.C. COULL (eds.): Marine benthic dynamics.- Univ. South Carolina Press, Columbia: 121-149.

ARNTZ, W.E. 1981: Zonation and dynamics of macrobenthos biomass in an area stressed by oxygen deficiency.- In: G. BARRETT and R. ROSENBERG (eds.): Stress effects on natural ecosystems.- Wiley & Sons, New York: 215-225.

ARNTZ, W.E. & D. BRUNSWIG, 1975: An approach to estimating the production of macrobenthos and demersal fish in a Western Baltic Abra alba community.- Merentutkimuslait. Julk./Havsforskningsinst. Skr. 239: 195-205.

ARNTZ, W.E. & D. BRUNSWIG 1976: Studies on structure and dynamics of Macrobenthos in the Western Baltic carried out by the Joint Research Programme "Interaction Sea-Seabottom" (SFB 95-Kiel).- Proc. Xth European Symp. on Marine Biology 2: 17-42.

ARNTZ, W.E., D. BRUNSWIG & M. SARNTHEIN 1976: Zonierung von Mollusken und Schill im Rinnensystem der Kieler Bucht (Westliche Ostsee).- Senckenbergiana marit. 8: 189-269.

ARNTZ, W.E. & H. RUMOHR 1973: Bohrmuscheln (Barnea candida (L.) and Zirfaea crispata (L.)) in der Kieler Bucht.- Kieler Meeresforsch. 29: 141-143.

ARNTZ, W.E. & H. RUMOHR 1982: An experimental study of macrobenthic colonisation and succession, and the importance of seasonal variation in temperate latitudes.- J. Exp. Mar. Biol. Ecol. 64: 17-45.

BAAS-BECKING, L.G.M., E.J.F. WOOD & I.R. KAPLAN 1957: Biological processes in the estuarine environment. X. The place of the estuarine environment within the aqueous milieu.- Proc. K. Ned. Akad. Wet., Sekt. B 59: 109-123.

BABENERD, B. 1980: Untersuchungen zur Produktionsbiologie des Planktons in der Kieler Bucht (mit einer Auswertung der monatlichen Terminfahrten aus den Jahren 1957-1975).- Ph.D. Thesis, Univ. Kiel, 226 pp.

BAGNOLD, R.A. 1946: Motions of waves in shallow water; interaction between waves and sand bottoms. Proc. Royal Soc. London, Ser. A 187: 1-16.

BAKKEN, L.R. & R.A. OLSEN 1983: Buoyant densities and dry-matter contents of microorganisms: conversion of a measured biovolume into biomass.- Appl. Environ. Microbiol. 45: 1188-1195.

BALTIC PILOT: The Hydrographer of the Navy, Baltic Pilot Vol. 1; Kattegat to Baltic Sea.- 9. ed. London 1974.

BALZER, W. 1978: Untersuchungen über Abbau organischer Materie und Nährstofffreisetzung am Boden der Kieler Bucht beim Übergang vom oxischen zum anoxischen Milieu.- Ph.D. Thesis, Univ. Kiel and Rep. SFB 95, 36, 149 pp.

BALZER, W. 1980a: Calcium carbonate saturometry by alkalinity difference measurement.- Oceanol. Acta 3(2): 237-243.

BALZER, W. 1980b: Redox dependant processes in the transition from oxic to anoxic conditions: An in situ study concerning remineralization, nutrient release and heavy metal solubilization.- In: H.J. FREELAND, D.M. FARMER 6 C.D. LEVINGS (eds.): Fjord Oceanography.- Nato Conf.

Ser. IV, 4: 659-665.

BALZER, W. 1982: On the distribution of iron and manganese at the sediment/water interface: thermodynamic versus kinetic control.- Geochim. Cosmochim. Acta 46: 1153-1161.

BALZER, W. 1984: Organic matter degradation and biogenic element cycling in a nearshore sediment (Kiel Bight).- Limnol. Oceanogr. 29: 1231-1246.

BALZER, W., K. GRASSHOFF, P. DIECKMANN, H. HAARDT & U. PETERSOHN 1983: Redox-turnover at the sediment/water interface studied in a large bell jar system.- Oceanol. Acta 6: 337-349.

BALZER, W., F. POLLEHNE & H. ERLENKEUSER 1986: Cycling of organic carbon in a coastal marine system.- In: P.G. SLY (ed.): Sediments and water interactions.- Springer, N.York: 323-328.

BALZER, W. & G. WEFER 1981: Dissolution of carbonate minerals in a subtropical shallow marine environment.- Marine Chemistry 10: 545-558.

BANSE, K. 1956: Über den Transport von meroplanktischen Larven aus dem Kattegat in die Kieler Bucht.- Ber. dt. wiss. Kommiss. Meeresforsch. 14: 147-164.

BARNER, U. 1964: Untersuchungen an Sedimenten vom Südausgang des Großen Beltes.- Meyniana 15: 1-28.

BARNES, J.A. & B. von BODUNGEN 1978: The Bermuda marine environment. Vol. II.- Bermuda Biol. Station Res., Spec. Publ. 17, 189 pp.

BATHURST, R.G.C. 1971: Carbonate sediments and their diagenesis.- Developments in Sedimentology 12, Elsevier, New York, 620 pp.

BAUERFEIND, S. 1982: Versuche zum Abbau von Plankton- und Detritusmaterial durch natürliche Bakterienpopulationen der Schlei und der Ostsee.- Ph.D. Thesis, Univ. Kiel, 116 pp.

BEHRENDS, R. 1981: Messung bodennaher Turbulenzen mit einem mechanischen Strömungsmesser.- M. Sc. thesis, Univ. Kiel, 98 pp.

BENDER, M.L., G.P. KLINKHAMMER & D.W. SPENCER 1977: Manganese in seawater and the marine manganese balance.- Deep Sea Res. 24: 799-812.

BERGER, W.H. 1971: Sedimentation of planktonic foraminifera.- Mar. Geol. 11: 325-358.

BERNER, R.A. 1971: Principles of chemical sedimentology.- McGraw-Hill, New York, 240 pp.

BERTINE, K.K. & E.D. GOLDBERG 1971: Fossil fuel combustion and the major sedimentary cycle.- Science 173: 233-235.

BLACKADAR, A.K. 1962: The vertical distribution of wind and turbulent exchange in a neutral atmosphere.- J. Geophys. Res. 67(8): 3095-3102.

BOCK, H.H. 1974: Automatische Klassifikation. -Vandenhoek und Ruprecht,Göttingen, 480 pp.

von BODUNGEN, B. 1975: Der Jahresgang der Nährsalze und der Primärproduktion des Planktons in der Kieler Bucht unter Berücksichtigung der Hydrographie.- Ph.D. Thesis, Univ. Kiel, 116 pp.

von BODUNGEN, B., K. von BRÖCKEL, V. SMETACEK & B. ZEITZSCHEL 1975:

Ecological studies on the plankton in Kiel Bight. 1. Phytoplankton.-
Merentutkimuslait. Julk./Havsforskningsinst. Skr. 239: 179-186.

von BODUNGEN, B., K. von BRÖCKEL, V. SMETACEK & B. ZEITZSCHEL 1976a: The
planktontower. I. A structure to study water/sediment interactions in
enclosed water columns.- Mar. Biol. 34: 369-372.

von BODUNGEN, B., K. von BRÖCKEL, V. SMETACEK & B. ZEITZSCHEL 1981: Growth
and sedimentation of the phytoplankton spring bloom in the Bornholm
Sea (Baltic Sea).- Kieler Meeresforsch. Sdh. 5: 49-60.

von BODUNGEN, B., K. GOCKE, V. SMETACEK & B. ZEITZSCHEL 1976b: The
plankton tower. III. The effect of sediment flushing by density
displacement of interstitial water on pelagic primary production and
microbial activity.- Kieler Meeresforsch. Sdh. 3: 87-95.

von BODUNGEN, B., T.D. JICKELS, S.R. SMITH, J.A.B. WARD & G.B. HILLIER
1982: The Bermuda Environment, Vol. III.- Bermuda Biological Station,
Spec. Publ. 18: 123 pp.

BÖHM, E.L. 1978: Application of the ^{45}Ca tracer method for determination
of calcification rates in calcareous algae. Effect of calcium exchange
and differential saturation of algal calcium pools.- Mar. Biol. 47: 9-
14.

BÖHM, E.L., R. DAWSON, G. LIEBEZEIT & G. WEFER 1980: Suitability of
monosaccharides as markers for particle identification in carbonate
sediments.- Sedimentology 27: 167-177..

BÖHM, E.L., W. SCHRAMM & U. RABSCH 1978: Ecological and physiological
aspects of some coralline algae from the Western Baltic. Calcium
uptake and skeleton formation in Phymatolithon calcareum.- Kieler
Meeresforsch. Sdh. 4: 282-288.

BÖLTER, M. 1981: DOC-turnover and microbial biomass production.- Kieler
Meeresforsch. Sdh. 5: 304-310.

BÖLTER, M. 1982a: Submodels of a brackish water environment. I.
Temperature and microbial activity.- Ecological Modelling 17: 311-318.

BÖLTER, M. 1982b: Submodels of a brackish water environment. II.
Remineralisation rates of carbohydrates and oxygen consumption by
pelagic microheterotrophs.- Mar. Ecol. 3(3): 233-241.

BÖLTER, M. & R. DAWSON 1982: Heterotrophic utilisation of biochemical
compounds in Antarctic waters.- Proc. 16th EMBS, Netherlands J. Sea
Res. 16: 315-332.

BÖLTER, M., G. LIEBEZEIT, K. WOLTER & U. PALMGREN 1982: Submodels of a
brackish water environment. III. Microbial biomass production and
related carbon flux.- Mar. Ecol. 3(3): 243-253.

BÖLTER, M., M. MEYER & B. PROBST 1980: A statistical scheme for structural
analysis in marine ecosystems.- Ecological Modelling 9: 143-151.

BOJE, S. 1974: Untersuchungen zum Energie- und Stoffumsatz des
sublitoralen Meeresbodens in der Kieler Bucht.- Ph.D. Thesis, Univ.
Kiel and Rep. SFB 95, 12, 75 pp.

BOJE, S. 1977: Untersuchungen zum Energiefluß in Sedimenten der Kieler
Bucht.- Int. Revue Ges. Hydrobiol. 62: 63-95.

BORDOWSKI, O.K. 1965: Sources of organic matter in marine basins.- Mar.

Geol. 3: 5-31.

BOWDEN, K.F. & M.R. HOWE 1963: Observations of turbulence in a tidal current.- Jour. Fluid Mech. 17: 271-284.

BRECK, W.G. 1974: Redox levels in the sea.- In: E.D. GOLDBERG (ed.): The Sea. Marine Chemistry, Vol. 5, Wiley & Sons, New York: 153-180.

BRESSAU, S. 1957: Abrasion, Transport und Sedimentation in der Beltsee.- Die Küste 6: 64-102.

BREY, T. 1984: Die Makrofauna sandiger Böden der Kieler Bucht in 5 bis 15 m Tiefe.- Ber. Inst. Meereskunde, Kiel 123, 124 pp.

von BRÖCKEL, K. 1975: Der Energiefluß im pelagischen Ökosystems vor Boknis Eck (Westliche Ostsee).- Ph.D. Thesis, Univ. Kiel and Rep. SFB 95, 10, 96 pp.

von BRÖCKEL, K. 1978: An approach to quantify the energy flow through the pelagic part of the shallow water ecosystem off Boknis Eck (Eckernförde Bay).- Kieler Meeresforsch. Sdh. 4: 233-243.

BURNS, R.G. 1980: Microbial adhesion to soil surfaces: consequences for growth and enzyme activities.- In: R.C.W. BERKELEY, J.M. LYNCH, J. MELLING, P.R. RUTTER & B. VINCENT (eds.): Microbial adhesion to surfaces.- Ellis Horwood Limited, Chichester: 249-262.

CAMMEN, L.M. 1982: Effect of particle size on organic matter content and microbial abundance within four marine sediments.- Mar. Ecol. Prog. Ser. 9: 273-280.

CLOUGH, K.S. & J.C. SUTTON 1978: Direct observation of fungal aggregates in sand dune soil.- Can. J. Microbiol. 24: 333-335.

COACHMAN, L.K. & J. WALSH 1981: A diffusion model of cross-shelf exchange of nutrients in the southeastern Bering Sea.- Deep-Sea Res. 28: 819-846.

COOPER, L.H.N. 1937: Oxidation-reduction potential of sea-water.- J. Mar. Biol. Ass. U.K. 22:167-176.

CORPE, W.A. & H. WINTERS 1972: Hydrolytic enzymes of some periphytic marine bacteria.- Can. J. Microbiol. 18: 1483-1490.

DAHM, H.D. 1956: Diatomeenuntersuchungen zur Geschichte der westlichen Ostsee.- Meyniana 5: 7-50.

DALE, N.G. 1974: Bacteria in intertidal sediments: factors related to their distribution.- Limnol. Oceanogr. 19: 509-518.

DALE, T. 1978: Total, chemical and biological oxygen consumption of the sediments in Lindaspollene, Western Norway.- Mar. Biol. 49: 333-341.

DALRYMPLE, R.W., R.J. KNIGHT & J.J. LAMBIASE 1978: Bedforms and their hydraulic stability relationships in a tidal environment, Bay of Fundy, Canada.- Nature 275, No. 5676: 100-104.

DANSKE LODS: Den Danske Lods, Kongl. Sokort-Archiv, Kobenhavn 1983.

DAWSON, R. & G. LIEBEZEIT 1981: The analytical methods for the characterisation of organics in seawater.- In: E.K. DUURSMA & R. DAWSON (eds.): Marine organic chemistry.- Elsevier, Amsterdam: 445-496.

DAWSON, R., G. LIEBEZEIT & B. JOSEFSON 1983: Determination of amino acids and carbohydrates.- In: K. GRASSHOFF, M. EHRHARDT & K. KREMLING (eds.): Methods of seawater analysis.- Verlag Chemie, Weinheim: 319-346.

DeFLAUN, M.F. & L.M. MAYER 1983: Relationships between bacteria and grain surfaces in intertidal sediments.- Limnol. Oceanogr. 28: 873-881.

DEGENS, E.T., M. BEHRENDT, B. GOTTHARDT & E. REPPMANN 1968: Metabolic fractionation of carbon isotopes in marine plankton. II. Data on samples collected off the coast of Peru and Ecuador.- Deep-Sea Res. 15: 11-20.

DHI: Deutsches Hydrographisches Institut, Nautischer Funkdienst III, Hamburg 1979.

DHI: Deutsches Hydrographisches Institut, Ostsee-Handbuch IV. Teil: von Flensburg bis Utklippan und zur polnisch-sowjetischen Grenze, 12. ed., Hamburg 1978.

DHI: Deutsches Hydrographisches Institut, Verzeichnis der Leuchtfeuer und Signalstellen.Teil II. Ostsee, südwestlicher Teil und Gewässer zwischen Ost- und Nordsee, Hamburg 1983.

DIETRICH, G. 1951: Oberflächenströmungen im Kattegat, im Sund und in der Beltsee.- Deutsche Hydrogr. Zeitschr. 4: 129-150.

DINGLER, J.R. 1974: Wave formed ripples in nearshore sands.- Ph.D. Thesis, San Diego, Univ. Calif., 136 pp.

DINGLER, J.R. 1979: The threshold of grain motion under oscillatory flow in a laboratory wave channel.- J. Sed. Petrol. 49: 287-294.

DINGLER, J.R. & D.L. INMAN 1977: Wave-formed ripples in nearshore sands.- Proc. 15th Conf. on Coastal Engin., Honolulu, 2: 2109-2126.

DJAFARI, D. 1976: Mangan-Eisen-Akkumulate in der Kieler Bucht.- Ph.D. Thesis, Univ. Kiel and Rep. SFB 95, 27, 116 pp.

DJAFARI, D. 1977: Mangan-Eisen-Akkumulate in der Kieler Bucht.- Meyniana 29: 1-9.

DÖRJES, J., S. GADOW, H.-E. REINECK & I.B. SINGH 1969: Die Rinnen der Jade (südliche Nordsee). Sedimente und Makrobenthos.- Senckenbergiana marit. 1:5-62.

DOLD, R. 1980: Zur Ökologie, Substratspezifität und Bioturbation von Makrobenthos auf Weichböden der Kieler Bucht.- Ph.D. Thesis, Univ. Kiel, 449 pp.

DOUGLAS, R.G., J. LIESTMAN, C. WALCH, G. BLAKE & M.L. COTTON 1980: The transition from live to sediment assemblages in benthic foraminifera from the southern California borderland.- In: M.E. FIELD et al. (eds.): Quaternary depositional environments of the Pacific Coast. Pacific Section of SEPM: 257-280.

DUGDALE, R.C. & J.J. GOERING 1967: Uptake of new and regenerated forms of nitrogen in primary productivity.- Limnol. Oceanogr. 12: 196-206.

DUPLESSY, J.-C. 1978: Isotope studies.-In: J. GRIBBIN (ed.): Climatic Change.- Cambridge Univ. Press, Cambridge: 46-67.

DURAND, B. (ed.) 1980: Kerogen. Insoluble organic matter from sedimentary rocks.- Editions Techniques, Paris, 519 pp.

EGLINTON, J & M.T.J. MURPHY (eds.) 1969: Organic geochemistry. Methods and results.- Springer, Berlin, 828 pp.

EHRHARDT, M. & A. WENCK 1984: Wind pattern and hydrogen sulphide in shallow waters of the Western Baltic Sea, a cause effect relationship?-Meeresforsch. 30: 101-110.

EINSELE, G., R. OVERBECK, H.U. SCHWARZ & G. UNSÖLD 1974: Mass physical properties, sliding and erodibility of experimentally deposited and differently consolidated clayey muds (approach, equipment and first results).- Sedimentology 21: 339-372.

EINSELE, G. & A. SEILACHER (eds.) 1982: Cyclic and event stratification.- Springer, Berlin, 536 pp.

EINSTEIN, H.A. 1950: The bed load function for sediment transportation in open channel flows.- Technical Bull. 1026, U.S. Dept. of Agriculture, Soil Conservation Service, Washington D.C., 71 pp.

EINSTEIN, H.A. & S.A. EL SAMNI 1949: Hydrodynamic forces on a rough wall.- Reviews of modern Physics 21, 3: 520-524.

ELBRÄCHTER, M. 1971: Untersuchungen über die Populationsdynamik und Ernährungsbiologie von Dinoflagellaten im Freiland und im Labor.- Ph.D. Thesis, Univ. Kiel, 91 pp.

ELLERY, W.N. & M.H. SCHLEYER 1984: Comparison of homogenization and ultrasonication as techniques in extracting attached sedimentary bacteria.- Mar. Ecol. Prog. Ser. 15: 247-250..

ENGVALL, A.-G. 1973: Ammonia release from a reduced sediment.- In: R.O. HALLBERG, L. BAGANDER, A.-G. ENGVALL, M. LINDSTRÖM, S. ODEN and F.A. SCHIPPEL (eds.): The chemical microbiological dynamics of the sediment-water interface.- Contrib. Askö Lab., Univ. Stockholm, Sweden, 2: 1-64.

EPSTEIN, S., R. BUCHSBAUM, H.A. LOWENSTAM & H.C. UREY 1953: Revised carbonate-water isotopic temperature scale.- Geol. Soc. Am. Bull. 64: 1315-1325.

EPSTEIN, S. & T. MAYEDA 1953: Variation of ^{18}O content of waters from natural sources.- Geochim. Cosmochim. Acta 4: 213-224.

ERLENKEUSER, H. 1976: 14-C and 13-C isotope concentration in modern marine mussels from sedimentary habitats.- Naturwiss. 63: 338.

ERLENKEUSER, H. 1979: Environmental effects on radiocarbon in coastal marine sediments.- In: R. BERGER & H.E. SUESS (eds.): Radiocarbon dating.- Proc. Ninth Intern. Conf. Los Angeles and La Jolla, 1976. Univ. of California Press, Berkeley: 453-469.

ERLENKEUSER, H. 1981: Fossil carbonates in the recent sediments of the Harrington Sound, Bermuda.- In: G. WEFER, R. DAWSON & G. HEMPEL (eds.): The Harrington Sound Project of Kiel University.- Bermuda Biological Station, Spec. Publ. 19: 15-22.

ERLENKEUSER, H., R. DAWSON, D. FÜTTERER, H. HEINRICH, G. LIEBEZEIT, D. MEISCHNER, P. MÜLLER & G. WEFER 1981: Environmental changes during the last 9000 years as reflected in a sediment core from Harrington Sound, Bermuda.- In: G. WEFER, R. DAWSON & G. HEMPEL (eds.): The Harrington Sound Project of Kiel University. Bermuda Biological Station, Spec.

Publ. 19: 23-60.

ERLENKEUSER, H., H. METZNER & H. WILLKOMM 1975: University of Kiel Radiocarbon Measurements VIII.- Radiocarbon 17(3): 276-300.

ERLENKEUSER, H. & K. PEDERSTAD 1984: Recent sediment accumulation as depicted by ^{210}Pb-dating.- Norsk geol. tidskrift 64: 135-152.

ERLENKEUSER, H., E. SUESS & H. WILLKOMM 1974: Industrialization affects heavy metal and carbon isotope concentrations in recent Baltic Sea sediments.- Geochim. Cosmochim. Acta 38: 823-842..

ERLENKEUSER, H. & G. WEFER 1982: Seasonal growth of bivalves from Bermuda recorded in their O-18 profiles.- Proc. IV Intern. Coral Reef Symp., Manila, 1981, vol. 2: 643-648.

EXON, N. 1972: Sedimentation in the outer Flensburg Fjord area (Baltic Sea) since the last glaciation.- Meyniana 22: 5-62.

FÄLT, L.M. 1982: Late Quaternary sea-floor deposits off the Swedish West Coast.- Geologiska Institutionen, Univ. Göteborg Publ. A37, 259 pp.

FAHLTEICH, R. 1981: Zur Frühjahrsentwicklung der Copepoden in der Kieler Bucht.- M.S. Thesis, Univ. Kiel, 94 pp.

FALLON, R.D., S.Y. NEWELL & C.S. HOPKINSON 1983: Bacterial production in marine sediments: will cell-specific measures agree with whole-system metabolism?- Mar. Ecol. Prog. Ser. 11: 199-127.

FAUBEL, A. 1982: Determination of individual meiofauna dry weight values in relation to definite size classes.- Cah. Biol. Mar. 23: 339-345.

FAUBEL, A. & L.-A. MEYER-REIL 1983: Measurement of enzymatic activity of meiobenthic organisms: methodology and ecological application.- Cah. Biol. Mar.. 24: 35-49.

FAURE, G. 1977: Principles of isotope geology.- John Wiley & Sons, New York, 464 pp.

FENCHEL, T. 1977: The significance of bacterivorous protozoa in the microbial community of detrital particles.- In: J. CAIRNS (ed.): Aquatic microbial communities.- Garland, New York: 529-544.

FENCHEL, T. & C.B. JÖRGENSEN 1977: Detritus food chains in aquatic ecosystems: The role of bacteria.- Advances in microb. ecol. 1: 1-58.

FERGUSSON, R.L. & P. RUBLEE 1976: Contribution of bacteria to standing crop of coastal plankton.- Limnol. Oceanogr. 21: 141-145.

FLEMMING, B. & G. WEFER 1973: Tauchbeobachtungen an Wellenrippeln und Abrasionserscheinungen in der Westlichen Ostsee südöstlich Bokniseck.- Meyniana 23:9-18.

FLOOD, R.D. 1981: Distribution, morphology, and origin of sedimentary furrows in cohesive sediments, Southhampton Water.- Sedimentology, 28: 511-529.

FLOOD, R.D. 1982: Observations, classification, and dynamics of furrows in cohesive sediments.- Bull. Inst. Geol. Bassin d'Aquitaine, Bordeaux 31: 167-179.

FRITZSCHE, R. 1975: Tauch- und Laborstudien an einem Schillstreifen auf der Rinnenböschung SE Gulstav-Flach (Westliche Ostsee).- M.Sc. Thesis,

Univ. Kiel, 59 pp.

FROELICH, P.N., G.P. KLINKHAMMER, M.L. BENDER, N.A. LUEDTKE, G.R. HEATH, D. CULLEN & P. DAUPHIN 1979: Early oxidation of organic matter in pelagic sediments of the eastern equatorial Atlantic: suboxic diagenesis.- Geochim., Cosmochim. Acta 43(7):1075-1090.

FÜTTERER, D. 1977: Die Feinfraktion (Silt) in marinen Sedimenten des ariden Klimabereichs: Quantitative Analysenmethoden, Herkunft und Verbreitung.- Habilitation, Univ. Kiel, 246 pp.

GADOW, S. & H.E. REINECK 1969: Ablandiger Sandtransport bei Sturmfluten.- Senckenbergiana marit. 1: 63-78.

GAGOSIAN, R.B. & C. LEE 1981: Processes controlling the distribution of biogenic organic compounds in seawater.- In: E.K. DUURSMA & R. DAWSON (eds.): Marine organic chemistry.- Elsevier, Amsterdam, Oceanogr. Ser. 31: 91-123.

GAST, V. 1983: Untersuchungen über die Bedeutung der Bakterien als Nahrungsqelle für das Mikrozooplankton der Schlei und der Ostsee unter besonderer Berücksichtigung der Ciliaten.- Ph.D. Thesis, Univ. Kiel, 159 pp.

GERLACH, S.A. 1971: On the importance of marine meiofauna for benthos communities.- Oecologia, 33: 55-69.

GERLACH, S.A. 1978: Food-chain relationships in subtidal silty sand marine sediments and the role of meiofauna in stimulating bacterial productivity.- Oecologia (Berlin) 33: 55-69..

GEYER, D. 1965: Eigenschwingungen und Erneuerung des Wassers in der Eckernförder Bucht unter besonderer Berücksichtigung der Sturmlage vom 5.-6. Dezember 1961.- Kieler Meeresforsch. 21: 33-54.

GHIORSE, W.C. 1980: Electron microscopic analysis of metal depositing micro-organisms in surface layers of Baltic ferromanganese concretions.- In: P.A. TRUDINGER, M.R. WOLTER & B.J. RALPH (eds.): Ancient and modern environments.- Australian Academy of Science and Springer, New York: 345-354,

GIESEKING, J.E. 1975: Soil components. Vol. I: Organic components.- Springer, Berlin, 534 pp.

GOCKE, K. 1975: Studies on short-term variations of heterotrophic activity in the Kiel Fjord.- Mar. Biol. 33: 49-55.

GOCKE, K. 1977a: Untersuchungen über die heterotrophe Aktivität in der zentralen Ostsee.- Mar. Biol. 40: 87-94.

GOCKE, K. 1977b: Comparison of methods for determining the turnover times of dissolved organic compounds.- Mar. Biol. 42: 131-141.

GOCKE, K., R. DAWSON & G. LIEBEZEIT 1981: Availability of dissolved free glucose to heterotrophic microorganisms.- Mar. Biol. 62: 209-216.

GOLDHABER, M.B., R.C. ALLER, J.K. COCHRAN, J.K. ROSENFELD, C.S. MARTENS & R.A. BERNER 1977: Sulfate reduction, diffusion and bioturbation in Long Island Sound sediments: Report of the FOAM-Group.- Am. J. Sci. 277(3): 193-237.

GOLTERMAN, H.L. 1973: Vertical movement of phosphate in freshwater.- In: E.J. GRIFFITH, A. BEETOR, J.M. SPENCER, and D.T. MITCHELL (eds.):

Environmental Phosphorous Handbook.- J. Wiley Sons, New York: 509-538.

GORDON, C.M. & C.F. DOHNE 1973: Some observations of turbulent flow in a tidal estuary.- J. Geophys. Res. 78(12): 1971-1978.

GRAF, G., W. BENGTSSON, U. DIESNER, R. SCHULZ & H. THEEDE 1982: Benthic response to sedimentation of a spring phytoplankton bloom: process and budget.- Mar. Biol. 67: 201-208.

GRAF, G., W. BENGTSON, A. FAUBEL, L.-A. MEYER-REIL, R. SCHULZ, H. THEEDE & H. THIEL 1984: The importance of the spring phytoplankton bloom for the benthic system of the Kiel Bight.- Rapp. Proc. Verb. ICES C.M. 1982, No. 24, 22 pp.

GRAF, G., R. SCHULZ, R. PEINERT & L.-A. MEYER-REIL 1983: Benthic response to sedimentation events during autumn to spring at a shallow-water station in the Western Kiel Bight. I. Analysis of processes on a community level.- Mar. Biol. 77: 235-246.

von GRAFENSTEIN, U. 1982: Zur Erosionswirkung von Seegang: Beobachtungen an Wellenrippeln in der Kieler Bucht (Westliche Ostsee).- M.Sc. Thesis, Univ. Kiel and Rep. SFB 95, 63, 39 pp.

von GRAFENSTEIN, U. 1984: Zur Aussagekraft von Oszillationsrippeln: Ereignisberzogene, sedimentologische Untersuchungen in Gebieten mit unterschiedlichen Seegangsaspekten in der Nord- und Ostsee.- Ph.D. Thesis, Univ. Kiel and Berichte-Reports, Geolog. Paläontolog. Inst., Univ. Kiel 7, 127 pp.

GRASS, A.J. 1971: Structural features of turbulent flow over smooth and rough boundaries.- Journ. Fluid Mechanics 50: 233-255.

GRIFFITHS, R.P., S.S. HAYASAKA, T.M. McNAMARA & R.Y. MORITA 1978: Relative microbial activity and bacterial concentrations in water and sediment samples taken in the Beaufort Sea.- Can. J. Microbiol. 24: 1217-1226.

GRILL, E.V. & F.A. RICHARDS 1969: Nutrient regeneration from phytoplankton decomposing in seawater.- J. Mar. Res. 22: 51-69.

GROBE, H. 1981: Abrasion and Korrosion benthischer Foraminiferen in der Kieler Bucht, Westliche Ostsee.- M. Sc. Thesis, Univ. Kiel, 52 pp.

GROBE, H. & D. FÜTTERER 1981: Zur Fragmentierung benthischer Foraminiferen in der Kieler Bucht (Westl. Ostsee).- Meyniana 33: 85-96.

GRÜNDEL, E. 1980: Ökosystem Seegraswiese - qualitative und quantitative Unterschungen über Struktur und Funktion einer Zostera-Wiese vor Surendorf (Kieler Bucht, Westliche Ostsee).- Ph.D. Thesis, Univ. Kiel and Rep. SFB 95, 56, 117pp (1982).

HARGRAVE, B.T. 1972: Aerobic decomposition of sediment and detritus as a function of particle surface area and organic content.- Limnol. Oceanogr. 17: 583-596.

HARGRAVE, B.T. 1980: Factors affecting the flux of organic matter to sediments in a marine bay.- In: K.R. TENORE & B.C. COULL (eds.): Marine benthic dynamics.- Univ. South Carolina Press: 243-263.

HARGRAVE, B.T. & G.A. PHILLIPS 1977: Oxygen uptake of microbial communities on solid surfaces.- In: J. CAIRNS (ed.): Aquatic microbial communities.- Garland, New York: 545-587.

HARTMANN, M. 1964: Zur Geochemie von Mangan und Eisen in der Ostsee.-

Meyniana <u>14</u>: 3-20.

HARTMANN, M., P.J. MÜLLER, E. SUESS & H.C. van der WEIJDEN 1976: Chemistry of late quaternary sediments and their interstitial waters from the North West African continental margin.- Meteor-Forsch.-Ergebn. Reihe C Nr. <u>24</u>: 1-67.

HARVEY, J.G.& C.E. VINCENT 1977: Observations of shear in near-bed currents in the southern North Sea.- Est. & Coastal Mar. Sci. <u>5</u>: 715-731.

HATJE, G. 1976: Strömungen in der Vejsnäsrinne 1975/1976.- Rep. SFB 95, Univ. Kiel, <u>24</u>, 15 pp.

HATJE, G. 1977: Strömungen in der Vejsnäsrinne 1976/77.- Rep. SFB 95, Univ. Kiel <u>31</u>, 40 pp.

HATJE, G. & G. SCHALLER 1976: Standardaufbereitungen von Aanderaa-Strömungsmesserdaten auf der NOVA 1200.- Rep. SFB 95, Univ Kiel, <u>23</u>, 44 pp.

HEALY, T. 1980: Submarine terraces and morphology in the Kieler Bucht, and their relation to late Quaternary events.- Boreas <u>10</u>: 209-217.

HEALY, T. & G. WEFER 1980: The efficacy of submarine abrasion vs. cliff retreat as a supplier of marine sediment in the Kieler Bucht, Western Baltic.- Meyniana <u>32</u>: 88-96.

HEALY, T. & F. WERNER (in press): Sediment budget for a semienclosed sea in a near homogeneous lithology, example of Kieler Bucht, Western Baltic.- Senckenbergiana marit. <u>19</u>:

HEATHERSHAW, A.D. 1976: Measurements of turbulence in the Irish Sea benthic boundary layer.- In: I.N. McCAVE (ed.): The benthic boundary layer.- Plenum Publishing Corporation, New York: 11-31.

HEINRICH, A.K. 1962: The life histories of plankton animals and seasonal cycles of plankton communities in the oceans.- J. Cons. Int. Explor. Mer. <u>27</u>: 15-24.

HEINRICH, H. 1982: Die limnisch-marine Biofaziesentwicklung im Harrington Sound, Bermuda während des holozänen Meeresspiegelanstiegs.- Ph.D. Thesis, Univ. Kiel and Rep. SFB 95, <u>67</u>, 76 pp. (1983).

HEM, J.D. 1978: Redox processes at surface manganese oxide and their influence on aqueous metal ions.- Chem. Geol. <u>21</u>: 199-218.

HEMPEL, G. 1975: An interdisciplinary marine project at the University of Kiel "Sonderforschungsbereich 95". - Merentutkimuslait. Julk/ Havsforskningsinst. Skr. <u>239</u>: 162-166.

HEMPEL, G. & G. WEFER 1982: Harrington Sound, Systemstudien an einer subtropischen Lagune.- Naturw. Rdsch. <u>35</u>(5): 198-203.

HENDRIKSON, P. 1975: Auf- und Abbauprozesse partikulärer organischer Substanz anhand von Seston- und Sinkstoffanalysen (7.3.1973-5.4.1974 in der Westlichen Ostsee bei Boknis Eck).- Ph.D. Thesis, Univ. Kiel, 160 pp.

HENDRIKSON, P. 1976: Abbauraten von organischem Kohlenstoff im Seston und in Sinkstoffen der Kieler Bucht.- Kieler Meeresforsch. Sdh. <u>3</u>: 105-119.

HERTWECK, G. 1972: Distribution and environmental significance of lebensspuren and in situ skeletal remains.- Senckenbergiana marit. 4: 125-167.

HESSE, K.J. 1979: Produktions- und Biomassemessungen an pelagischen Heterotrophen unter besonderer Berücksichtigung des Protozooplanktons.- M.Sc. Thesis, Univ. Kiel, 103 pp.

HILLE, P. 1984: Analyse der Einzelkornmechanik und Experimente zum Zusammenhang zwischen auf Sedimentkörner wirkenden fluktuiernden Kräften und Geschwindigkeiten der turbulenten Grenzschichtströmung.- Ph.D. Thesis, Univ. Kiel 1984, 161 pp.

HILLE, P., R. VEHRENKAMP & E.O. SCHULZ-DUBOIS 1983: Measurement of small secondary flow components by photon correlation velocimetry.- Optica Acta 30(2): 139-153.

HINTZ, R.A. 1958: Sedimentpetrographische und diluvialgeologische Untersuchungen im Küstenbereich des Landes Angeln.- Meyniana 6: 116-126.

HINZ, K., F.-C. KÖGLER & E. SEIBOLD 1969: Reflexions-seismische Untersuchungen mit einer pneumatischen Schallquelle und einem Sedimentecholot in der westlichen Ostsee. Teil I. Meßmethoden, Instrumentarium, Interpretation.-Meyniana 19: 91-102.

HINZ, K., F.-C. KÖGLER, J. RICHTER & E. SEIBOLD 1971: Reflexionsseismische Untersuchungen mit einer pneumatischen Schallquelle und einem Sedimentecholot in der Westlichen Ostsee. Teil II. Untersuchungsergebnisse und geologische Deutung.- Meyniana 21: 17-24.

HOEFS, J. 1980: Stable isotope geochemistry.- Springer-Verlag, Berlin 208 pp.

HOFFMANN, C. 1956: Untersuchungen über die Remineralisation des Phosphors im Plankton.- Kieler Meeresforsch. 12: 25-36.

HOPPE, H.-G. 1977: Analysis of actively metabolizing bacterial populations with the autoradiographic method.- In: G. RHEINHEIMER (ed.): Microbial ecology of a brackish water environment.- Springer, Berlin: 179-197.

HOPPE, H.-G. 1981: Vergleichende Untersuchungen über die bakterielle Aktivität und deren Ursachen im Brackwasser der Kieler Förde und einem verunreinigten Fluß.- Habilitationsschrift, Univ. Kiel, 127 pp.

HOWARD, J.D. & R.W. FREY 1975: Estuaries of the Georgia Coast, U.S.A.: Sedimentology and biology. II. Regional animal-sediment characteristics of Georgia estuaries.- Senckenbergiana marit. 7: 33-103.

INMAN, D.L. 1957: Wave-generated ripples in nearshore sands.- U.S. Army Corps Engineers Beach Erosion Board, Techn. Memo 100, 67 pp.

INTERNATIONAL BALTIC SEA FISHERY COMMISSION. Proc. 1st Session Warsaw, October 28-31, 1974, doc. I/S/74/5 Annex II, 1975.

ITURRIAGA, R. 1979: Bacterial activity related to sedimenting particulate matter.- Mar. Biol. 55: 157-169.

JAECKEL, S., jun. 1952: Zur Ökologie der Molluskenfauna in der Westlichen Ostsee.- Schr. naturw. Ver. Schlesw.-Holst. 26: 18-50.

JAGO, C.F. & J.P. BARUSSEAU 1981: Sediment entrainment on a wave-graded shelf, Roussillon, France.- Mar. Geol. 42: 279-299.

JAKOBSEN, T.S. 1979: Recent results on the water exchange of the Baltic.-
ICES. C.M. 1979/C, 16, 25 pp.

JAKOBSEN, T.S. 1980: Sea water exchange of the Baltic. Measurements and
methods.- Ph.D. Thesis, Univ. Copenhagen and The Belt Project,
Copenhagen, 106 pp.

JANSSON, B.-O. 1978: The Baltic- a systems analysis of a semi-enclosed
sea.- In: H. CHARNOCK & G. DEACON (eds.): Advances in Oceanography.-
Plenum Press, New York: 181-183.

JENNE, E.A. 1968: Controls on Mn, Fe, Co, Ni, Cu and Zn concentration in
soils and water: Significant role of hydrous Mn and Fe oxides.- Adv.
Chem. Ser. 73: 337-387.

JENSEN, J.M., W. ENGE, H. ERLENKEUSER & H. WILLKOMM 1977: Age
determination of sediments by PB-210 using a plastic detector
technique.- Nuclear Instruments and Methods 147: 97-99.

JÖRGENSEN, P., H. ERLENKEUSER, H. LANGE, J. NAGY, J. RUMOHR & F. WERNER
1981: Sedimentological and stratigraphical studies of two cores from
the Skagerrak.- Spec. Publs. Int. Ass. Sedimentologists 5: 397-414.

JOHNSON, K.M. & J.McN. SIEBURTH 1977: Dissolved carbohydrates in seawater.
I. A precise spectrophotometric analysis for monosacgarides.- Mar.
Chem. 5: 1-13.

JONES, J.G. 1980: Some differences in the microbiology of profundal and
littoral lake sediments.- J. Gen. Microbiol. 117: 285-292.

JORGENSEN, B.B., N.P. REVSBECH, T.H. BLACKBURN & Y. COHEN 1979: Diurnal
cycle of oxygen and sulfide microgradients and microbial phtosynthesis
in a cyanobacterial mate sediment.- Appl. Environ. Microbiol. 38: 46-
58.

KALLE, L. 1966: The problem of "Gelbstoff" in the sea.- Oceanogr. Mar.
Biol. Ann. Rev. 4: 91-104.

KANNENBERG, E.G. 1951: Die Steilufer der Schleswig-Holsteinischen
Ostseeküste.- Schr. Geogr. Inst. Univ. Kiel 14, 101 pp.

KARG, S. 1979: Vergleichende Produktivitätsuntersuchungen am
Mikrophytobenthos in der Kieler Förde.- M. Sc. Thesis, Univ. Kiel and
Rep. SFB 95, 50, 73 pp.

KATZ, A. & S. BEN-YAAKOV 1980: Diffusion of sea water ions. II. The role
of activity coefficients and ion pairing.- Mar. Chem. 8(4): 263-280.

KEITH, M.L. & J.N. WEBER 1965: Systematic relationships between carbon and
oxygen isotopes in carbonates deposited by modern corals and algae.-
Science 150: 498-501.

KENYON, N.H. 1970: The origin of some transverse sand patches in the
Celtic Sea.- Geol. Mag. 107: 389-394.

KEPKAY, P.E., R.C. COOKE & J.A. NOVITSKY 1979: Microbial autotrophy: a
primary source of organic carbon in marine sediments.- Science 204:
68-69.

KHANDRICHE, A. 1985: Auswirkungen von Oststurmlagen des Winters 1978/79
auf die Sedimentation im Schlickbereich der Eckernförder Bucht.- M.
Sc. thesis, Univ. Kiel, 51 pp., unpubl.

KLUMP, J.V. & C.S. MARTENS 1981: Biogeochemical cycling in an organic rich coastal marine basin. II. Nutrient sediment - water exchange processes.- Geochim. Cosmochim. Acta 45: 101-121.

KNAUER, G.A., J.H. MARTIN & K.W. BRULAND 1979: Fluxes of particulate carbon, nitrogen and phosphorus in the upper water column of the north-east Pacific.- Deep-Sea Res. 26: 97-108.

KNEBEL, H.J., S.W. NEEDELL & C.J. O'HARA 1982: Modern sedimentary environments on the Rhode Island inner shelf off the eastern United States.- Mar. Geol. 49: 241-256.

KNOPPERS, B. 1976: Die Vertikalverteilung motiler Protozoan in Abhängigkeit von Umweltfaktoren und der Tageszeit in einer abgeschlossenen Wassersäule.- M.Sc. Thesis, Univ. Kiel and Rep. SFB 95, 22, 86 pp.

KNOPPERS, B. 1981: Zur Charakterisierung partikulär-organischer Substanz im Meer mit biologischen, chemischen und optischen Methoden.- Ph.D. Thesis, Univ. Kiel and Rep. SFB 95, 66, 47 pp. (1982).

KÖLMEL, R. 1976: Ökosysteme im Wechsel zur Anaerobiose. Zoobenthos und Abbau in zeitweise anoxischen Biotopen der Kieler Bucht.- Ph.D. Thesis, Univ Kiel and Rep. SFB 95 Univ. Kiel 33, 403 pp. (1977).

KÖLMEL, R. 1979: The annual cycle of macrozoobenthos: Its community structures under the influence of oxygen deficiency in the Western Baltic.- In: E. NAYLOR and R.G.HARTNOLL (eds.): Cyclic phenomena in marine plants and animals.- Pergamon Press, Frankfurt: 19-28.

KOLDERUP ROSENVINGE, L. 1909: The marine algae of Denmark, Part I, Introduction, Rhodophycea I.- D. Kgl. Danske Vidensk. Selsk. Skrifter (Naturvid. Mathem. Afd.)7, 150 pp.

KOMAR, P.D. 1974: Oscillatory ripple marks and the evaluation of ancient wave conditions and environments.- J. Sed. Petrol. 44: 169-180.

KOMAR, P.D. & M.C. MILLER 1973: The threshold of sediment movement under oscillatory water waves.- J. Sed. Petrol. 43: 1101-1110.

KRAMBECK, C., H.-J. KRAMBECK & J. OVERBECK 1981: Microcomputer assisted biomass determination of plankton bacteria on scanning electron micrographs.- Appl. Environ. Microbiol. 42: 142-149.

KRAMER, J.R., S.E. HERBES & H.E. ALLEN 1972: Phosphorus: Analysis of water, biomass and sediment.- In: H.E. ALLEN & J.R. KRAMER (eds.): Nutrients in natural wates.- J. Wiley Sons, New York: 51-100.

KRAUSKOPF. K.B. 1956: Factors controlling the concentrations of 13 rare metals in sea water.- Geochim. Cosmochim. Acta 9: 1-32.

KREMLING, K., C. OTTO & H. PETERSEN 1979: Spurenmetall-Untersuchungen in den Förden der Kieler Bucht. Datenbericht von 1977/1978.- Berichte Inst. für Meereskunde, Univ. Kiel, 66, 38 pp.

KREY, J. 1974: Das Plankton.- In: L. MAGAARD & G. RHEINHEIMER (eds.): Meereskunde der Ostsee.- Springer-Verlag, New York: 103-130.

KROM, M.D. & R.A. BERNER 1980: The diffusion coefficients of sulfate, ammonium and phosphate ions in anoxic marine sediments.- Limnol. Oceanogr. 25: 327-337.

KÜHLMORGEN-HILLE, G. 1963: Quantitative Untersuchungen der Bodenfauna in der Kieler Bucht und ihre jahreszeitlichen Veränderungen.- Kieler Meeresforsch. 19: 42-66.

KÜHLMORGEN-HILLE, G. 1965: Qualitative und quantitative Veränderungen der Bodenfauna der Kieler Bucht in den Jahren 1953-1965.- Kieler Meeresforsch. 21: 167-191.

KUIJPERS, A. 1974: Trace elements at the depositional interface and in sediments of the outer parts of the Eckernförder Bucht, Western Baltic.- Meyniana 26: 23-38.

KUIJPERS, A. 1978: Effects of bottom currents in the Northern part of the Sound, mapped with side-scan sonar.- Rep. SFB 95, Univ. Kiel, 39, 18 pp.

KUIJPERS, A. 1980: Sediment patterns and bedforms and their relationship to the flow regime in the Belt Sea and the Sound.- Ph.D. Thesis, Univ. Kiel, 138 pp.

KUIJPERS, A. 1985: Current-induced bedforms in the Danish Straits between Kattegat and Baltic Sea.- Meyniana 37: 97-127.

LANCE, G.N. & W.T. WILLIAMS 1966: A general theory of classification sorting strategies. I. Hierarchical Systems.- Computer Z. 9: 373-380.

LARSSON, U. & A. HAGSTRÖM 1979: Phytoplankton exudate release as an energy souce for the growth of pelagic bacteria.- Mar. Biol. 52: 199-206.

LENZ, J. 1965: Zur Ursache der an die Sprungshicht gebundenen Echostreuschichten in der westlichen Ostsee.- Ber. dt. wiss. Komm. Meeresforsch. 18: 111-161.

LENZ, J. 1974: On the amount and size distribution of suspended organic matter in Kiel Bight.- Ber. dt. wiss. Komm. Meeresforsch. 23: 209-225.

LENZ, J. 1977: On detritus as a food source for pelagic filter feeders.- Mar. Biol. 41: 39-48.

LENZ, J. 1981: Phytoplankton standing stock and primary production in the Western Baltic.- Kieler Meeresforsch. Sdh. 5: 29-40.

LESHT, B.M. 1979: Relationship between sediment resuspension and the statistical frequency distribution of bottom shear stress.- Mar. Geol. 32: M19-M27.

LEWY, Z. 1975: Early diagenesis of calcareous skeletons in the Baltic Sea, Nothern Germany.- Meyniana 27: 29-33.

LI, Y.-H. & S. GREGORY 1974: Diffusion of ions in sea water and in deep-sea sediments.- Geochim. Cosmochim. Acta 38: 703-714.

LIEBEZEIT, G. 1980: Aminosäuren und Zucker im marinen Milieu - neuere analytische Methoden und ihre Anwendung.- Ph.D. Thesis, Univ. Kiel and Rep. SFB 95, 52: 195 pp.

LIEBEZEIT, G., M. BÖLTER, I.F. BROWN & R. DAWSON 1980: Dissolved free amino acids and carbohydrates at pycnocline boundaries in the Sargasso Sea and related microbial activity.- Oceanol. Acta 3: 357-362.

LIEBEZEIT, G. & R. DAWSON 1982: The analysis of natural organic compounds in sea water.- Kontakte (Merck) 2: 19-28.

LINDROTH, P. & K. MOPPER 1979: High performance liquid chromatographic determination of subpicomole amounts of amino-acids by precolumn fluorescence derivatisation with o-phthaldialdehyde.- Anal. Chem. 51: 1667-1674.

LOHMANN, H. 1908: Untersuchung zur Feststellung des vollständigen Gehaltes des Meeres an Plankton.- Wiss. Meeresunters. Abt. Kiel, N.F. 10: 131-370.

LONSDALE, P. 1982: Sediment drifts of the Northeast Atlantic and their relationship to the observed abyssal currents.- Bull. Inst. Geol. Bassin d'Aquitaine, Bordeaux 31: 147-149.

LÜTHJE, H. 1978: The macrobenthos in the red algal zone of Kiel Bay (Western Baltic).- Kieler Meeresfórsch. Sdh. 4: 108-114.

LUTZE, G.F. 1965: Zur Foraminiferen-Fauna der Ostsee. Meyniana 15: 75-142.

LUTZE, G.F. 1974: Foraminiferen der Kieler Bucht (Westl. Ostsee). I. "Hausgartengebiet" des Sonderforschungsbereiches 95.- Meyniana 26: 9-22.

LUTZE, G.F. & G. WEFER 1980: Habitat and asexual reproduction of Cyclorbiculina compresssa (Orbigny), Soritidae.- J. Foram. Res. 10: 251-260.

MANTZ, P.A. 1977: Incipient transport of fine grains and flakes by fluids. Extended Shields diagram.- Proc. ASCE, J. Hydr. Div. 103, no. HY6: 601-615.

MARINELEITUNG 1922: Ostseehandbuch, südlicher Teil, 7. ed. Berlin.

MARTENS, C.S. 1978: Some of the chemical consequences of microbially mediated degradation of organic materials in estuarine sediments.- In: Biogeochemistry of estuarine sediments.- Proc. UNESCO/SCOR Workshop, Melreux, Belgium: 266-278.

MARTENS, P. 1975: Über die Qualität und Quantität der Sekundär- und Tertiärproduzenten in einem marinen Flachwasserökosystem der Westlichen Ostsee.- Ph.D. Thesis, Univ. Kiel, 111 pp.

MARTENS, P. 1976: Die planktische Sekundär- und Tertiärproduzenten im Flachwaserökosystem der Westlichen Ostsee.- Kieler Meeresforsch. Sdh. 3: 60-71.

McCAFFREY, R.J., A.C. MYERS, E. DAVEY, G. MORRISON, M. BENDER, N. LUEDTKE, D. CULLEN, P. FROEHLICH & D. KLINKHAMMER 1980: The relation between pore water chemistry and benthic fluxes of nutrients and manganese in Narragansett Bay, Rhode Island. Limnol. Oceanor. 25(1): 31-44.

McKINNEY, T.F., W.L. STUBBLEFIELD & D.J.P. SWIFT 1974: Large-scale current lineations on the central New Jersey shelf: Investigation by side-scan sonar.- Mar. Geol. 17: 79-102.

McCLEAN, S.R. 1981: The role of non-uniform roughness in the formation of sand ribbons and Langmuir circulations.- Mar. Geol. 42: 49-74.

McCLEAN, S.R. 1983: Turbulence and sediment transport measurements in a North Sea tidal inlet (The Jade).-In: J. SÜNDERMANN and W. LENZ (eds.): North Sea Dynamics.- Springer, Berlin: 436-452.

MELVASALO, T., J. PAWLAK, K. GRASSHOFF, L. THORELL & A. TSIBAN (eds.) 1981: Assessment of the effects of pollution on the natural resources of the Baltic Sea, 1980.- Baltic Sea Environment Proc. 5B, 426 pp.

METZLER, C.V., C.R. WENKAM & W.H. BERGER 1982: Dissolution of calcium carbonate in the Eastern equatorial Pacific. An in situ experiment.- J. Foram. Res. 12: 362-368.

MEYER, H.U. 1976: Zur Ökologie und Produktivität von Bewuchsgemeinschaften auf neubesiedelten Hartsubstraten in der Kieler Bucht.- Ph.D. Thesis, Univ. Kiel, 116 pp.

MEYER-REIL, L.-A. 1978: Uptake of glucose by bacteria in the sediment.- Mar. Biol. 44: 293-298.

MEYER-REIL, L.-A. 1981: Enzymatic decomposition of proteins and carbohydrates in marine sediments: Methodology and field observations during spring.- Kieler Meeresforsch. Sdh. 5: 311-317.

MEYER-REIL, L.-A. 1983: Benthic response to sedimentation events during autumn to spring at a shallow water station in the Western Kiel Bight. II. Analysis of benthic bacterial populations.- Mar. Biol. 77:247-256.

MEYER-REIL, L.-A. 1984a: Bacterial biomass and heterotrophic activity in sediments and overlying waters.- In: J.E. HOBBIE & P.J. le B. WILLIAMS (eds.): Heterotrophic activity in the sea.- Plenum Press, New York, London: 523-546.

MEYER-REIL, L.-A. 1984b: Seasonal variations in bacterial biomass and decomposition of particulate organic material in marine sediments.- Arch. Hydrobiol. Beih. Ergebn. Limnol. 19: 201-206.

MEYER-REIL, L.-A., M. BÖLTER, G. LIEBEZEIT & W. SCHRAMM 1979: Short-term variations in microbiological and chemical parameters.- Mar. Ecol. Progr. Ser. 1: 16.

MEYER-REIL, L.-A., M. BÖLTER, R. DAWSON, G. LIEBEZEIT, H. SZWERINSKI & K. WOLTER 1980: Interrelationships between microbiological and chemical parameters of sandy beach sediments, a summer aspect.- Appl. Environ. Microbiol. 39: 797-802.

MEYER-REIL, L.-A., R. DAWSON, G. LIEBEZEIT & H. TIEDGE 1978: Fluctuations and interactions of bacterial activity in sandy beach sediments and overlying waters.- Mar. Biol. 48: 161-171.

MEYER-REIL. L-A., W. SCHRAMM & G. WEFER 1981: Microbiology of a tropical coral reef system (Mactan; Philippines).- Kieler Meeresforsch., Sdh. 5: 431-432.

MICHAELIS, W., K. MOPPER, C. GARRASI & E.T. DEGENS 1976: Organische Substanzen im Sediment und Wasser der Hamburger Alster.- Mitt. Geol.-Paläont. Inst. Univ. Hamburg, Sonderband Alster: 173-188.

MILLER, M.C. & P.D. KOMAR 1980a: Oscillation sand ripples generated by laboratory apparatus.- J. Sed. Petrol. 50: 173-182.

MILLER, M.C. & P.D. KOMAR 1980b: A field investigation of the relationships between oscillaton ripple spacing and the near-bottom water orbital motions.- J. Sed Petrol. 50: 183-191.

MILLIMAN, J.D. 1974: Marine Carbonates.- Springer-Verlag, Berlin, 395 pp.

MÖLLER, H. 1979: Significance of coelentevates in relation to other plancton organisms.- Berichte dt. wiss. Komm. Meeresforsch. 27: 1-18.

MÖLLER, H. 1980: Population dynamics of Aurelia aurita medusae in Kiel Bight, Germany (FRG).- Mar. Biol. 60: 123-128.

MONIN, A.S. & A.M. YAGLOM 1971: Statistical fluid dynamics: Mechanics of turbulence.- MIT Press, Cambridge, 769 pp.

MONTAGNA, P.A. 1982: Sampling design and enumeration statistics for bacteria extracted from marine sediments.- Appl. Environ. Microbiol. 43: 1366-1372.

MOPPER, K., R. DAWSON, G. LIEBEZEIT & V. ITTEKOT 1980: The monosaccharide spectra of natural waters.- Mar. Chem. 10: 55-66.

MORIARITY, D.J.W. 1980: Measurement of bacterial biomass in sandy sediments.- In: P.A. TRUDINGER, M.R. WALTER & B.J. RALPH (eds.): Biogeochemistry of ancient and modern environments.- Australian Acad. Sci., Canberra, and Springer Verlag, Berlin, 131-139.

MORIARITY, D.J.W. & A.C. HOWARD 1982: Ultrastructure of bacteria and the proportion of gram-negativ bacteria in marine sediments.- Microb. Ecol. 8: 1-14.

MORIARITY, D.J.W. & P.C. POLLARD 1982: Dial variation of bacterial productivity in seagrass (Zostera capricornio) beds measured by rate of thymidine incorporation into DNA.- Mar. Biol. 72: 165-173.

MORRIS, B., J. BARNES, F. BROWN & J. MARKHAM 1977: The Bermuda environment.- Bermuda Biological Station. Spec. Publ. 15, 120 pp.

MORRISON, S.J. & D.C. WHITE 1980: Effects of grazing by estuarine gammaridean amphipods on the microbiota of allochthonous detritus.- Appl. Environ. Microbiol. 40: 659-671.

MORTON, R.A. 1981: Formation of storm deposits by wind-forced currents in the Gulf of Mexico and the North Sea.- In: S.D. NIO, R.T.E. SCHÜTTENHELM & Tj.C.E. van WEERING (eds.): Holocene Marine Sedimentation in the North Sea Basin.- Spec. Publ. Int. Ass. Sedimentol. 5: 387-396.

MÜLLER, G., J. DOMINIK & R. REUTHER 1980: Sedimentary record of environmental polution in the Western Baltic Sea.- Naturwiss. 67: 595-600.

MÜLLER, P.J. & E. SUESS 1979: Productivity, sedimentation rate and sedimentary organic matter in the oceans. I. Organic carbon preservation.- Deep Sea Res. 26: 1347-1362.

MURRAY, J.W. & P.G. BREWER 1977: Mechanisms of removal of manganese, iron and other trace metals from seawater.- In: G.P. GLASBY (ed.): Marine manganese deposits.- Elsevier: 291-326.

NALEWAJKO, C. & D.W. SCHINDLER 1976: Primary production, extracellular release and heterotrophy in two lakes in the ELA, Northwestern Ontario.- J. Fish. Board Can. 23: 219-226.

NAUEN, C.E. 1978a: Populationsdynamik and Ökologie des Seesterns Asterias rubens in der Kieler Bucht. Ph.D. Thesis, Univ. Kiel, 215 pp.

NAUEN, C.E. 1978b: Population dynamics of the sea star Asterias rubens L., an important competitor to demersal fish stocks in Kiel Bay. Rapp. Proc. verb. ICES.

NAUEN, C.E. 1978c: The growth of the sea star Asterias rubens, and its role as benthic predator in Kiel Bay.- Kieler Meeresforsch. Sdh. **4**: 68-81.

NAUEN, C.E.& L. BÖHM 1979. Skeletal growth in the echinoderm Asterias rubens L. (Asteroidae, Echinodermata) estimated by ^{45}Ca labelling.- J. Exp. Mar. Biol. Ecol. **38**: 261-269.

NELLEN, W., M. FIEDLER, A. TEMMING & M. WEIGELT: Eine Analyse der fischereibiologischen und fischereilichen Verhältnisse in einem für die Ölförderung genutzten Offshore-Bereich des deutschen Ostseegebiets.- Ber. Inst. Meereskd. Kiel **136**: 80 pp.

NELSON, C.H. & S.D. NIO 1982: The northeastern Bering shelf: new perspectives of epicontinental shelf processes and depositional products - an introduction.- Geol. en Mijnb., **61**: 2-4.

NEUMANN, A.C. 1965: Processes of recent carbonate sedimentation in Harrington Sound, Bermuda.- Bull. Mar. Sci. **15**: 987-1035.

NEUMANN, A.C. 1966: Observations on coastal erosion in Bermuda and measurements of the boring rate of the sponge Cliona lampa.- Limnol. Oceanogr. **11**: 92-108.

NEWTON, R.S., E. SEIBOLD & F. WERNER 1973: Facies distribution patterns on the spanish Sahara continental shelf mapped with side-scan sonar.- Meteor-Forsch. Erg. C **15**: 55-77.

NEWTON, R.S. & F. WERNER 1972: Transitional-size ripple marks in Kiel Bay (Baltic Sea).- Meyniana **22**: 89-94.

NICHOLS, F.H. 1977: Dynamics and production of Pectinaria koreni (Malmgren) in Kiel Bay, West Germany. In: KEEGAN, B.F., P.O. CEIDIGH & P.J.S. BOADEN (eds.): Biology of benthic organisms.- Proceedings of the 11th European Symposium on Marine Biology, Galway, Ireland, Oct. 1976. Pergamon Press, Oxford, New York: 453-463.

NICKELS, J.S, R.J. BOBBIE, R.S. MARTZ, G.A. SMITH, D.C. WHITE, and N.L. RICHARDS 1981: The fact of silica grain shape, structure, and location on the biomass and community structure of colonizing marine microbiota.- Appl. Environ. Microbiol. **41**: 1262-1268.

NIEMISTÖ, L. & A. VOIPIO 1974: Studies on the recent sediments in the Gotland Deep.- Merentutkimuslait. Julk./Havsforskninginst. Skr. **238**: 13-23.

NÖTHIG, E. 1984: Experimentelle Untersuchungen an natürlichen Planktonpopulationen unter besonderer Berücksichtigung heterotropher Organismen.- M.Sc. Thesis, Univ. Kiel, 105 pp.

ODUM, H.T. 1972: An energy circuit language for ecological and social systems: its physical basis.- In: B.C. PATTEN (ed.): Systems analyses and simulation in ecology. II vol.- Academic Press New York, London: 140-211.

OVERBECK, R. 1979: Erosionsexperimente an künstlich abgelagerten und konsolidierten Kaolin- und Bentonitsedimenten.- Rep. SFB 95, Univ. Kiel, **51**, 135 pp.

PALMGREN, U. 1981: Untersuchungen über die Abhängigkeit der Bakterienpopulation von abiotischen und biotischen Parametern in einem Brackwassergebiet.- Ph.D. Thesis, Univ. Kiel, 102 pp.

PAMATMAT, M.M. 1977: Benthic community metabolism: A review and assessment of present status and outlook.- In: B.C. COULL (ed.): Ecology of marine benthos.- Univ. South Carolina Press, Columbia: 89-111.

PAMATMAT, M.M. 1982: Heat production by sediment: ecological significance.- Science 215: 395-397.

PAMATMAT, M.M., G. GRAF, W. BENGTSSON & C.S. NOVAK 1981: Heat production, ATP concentration and electron transport activity of marine sediments.- Mar. Ecol. Prog. Ser. 4: 135-143.

PEINERT, R. 1981: Die Sedimentation von Plankton und Detritus unter verschiedenen Umweltbedingungen.- M. Sc. Thesis, Univ. Kiel, 70 pp.

PEINERT, R., A. SAURE, P. STEGMANN, C. STIENEN, H. HAARDT & V. SMETACEK 1982: Dynamics of primary production and sedimentation in a coastal ecosystem.- Neth. J. Sea Res. 16: 276-289.

PETERSEN, C.G.J. 1914: Valuation of the sea. II. The animal communities of the sea bottom and their importance for marine zoogeography.- Rep. Dan. Biol. St. 21: 1-68.

PFISTER, G., U. GERDTS, A. LORENZEN & K. SCHÄTZEL 1983: Hardware and software implementation of on-line velocity correlation measurements in oscillatory and turbulent rotational Couette Flow.- Proc. 5th Conf. on Photon Correlation Techniques in Fluid Mechanics, Univ. Kiel: 256-262.

POLLEHNE, F. 1977: Aspekte der räumlichen und zeitlichen Verteilung von Zooplanktonpopulationen in abgeschlossenen Wasserkörpern in der Kieler Bucht.- M.Sc. Thesis, Univ. Kiel, 77 pp.

POLLEHNE, F. 1980: Die Sedimentation organischer Substanz, Remineralisation und Nährsalzrückführung in einem marinen Flachwasserökosystem.- Ph.D. Thesis, Univ. Kiel and Rep. SFB 95, 57, 149 pp.

POMEROY, L.R., E.E. SMITH & C.M. GRANT 1965: The exchange of phosphate between estuarine water and sediments.- Limnol. Oceanogr. 10: 167-172.

PRANGE, W. 1978: Der letzte weichselzeitliche Gletschervorstoß in Schleswig-Holstein - das Gefüge überfahrener Schmelzwassersande und die Entstehung der Morphologie.- Meyniana 30: 61-75.

PRATJE, O. 1939: Die Sedimentation in der südlichen Ostsee.- Annal. Hydrogr. u. marit. Meterol. 67: 209-221.

PRATJE, O. 1948: Die Bodenbedeckung der südlichen und mittleren Ostsee und ihre Bedeutung für die Ausdeutung fossiler Sedimente.- Dtsch. Hydrogr. Zeitschr. 1: 45-61.

RAYMONT, J.E.G. 1981: Plankton and productivity in the oceans. Vol. 1. Phytoplankton.- 2nd Edition, Pergamon Press, 489 pp.

REDFIELD, A.C., B.H. KETCHUM & F.A. RICHARD 1963: The influence of organisms on the composition of sea water.- In: M.N. HILL (ed.): The Sea. vol. 2 ,Wiley Sons ,N.York: 26-77.

REIMERS, C.E. & E. SUESS 1983: The partitioning of organic carbon fluxes and sedimentary organic matter decomposition rates in the ocean.- Mar. Chem. 13: 141-168.

REIMERS, T. 1976: Anoxische Lebensräume: Struktur und Entwicklung der Mikrobiozönose an der Grenzfläche Meer/Meeresboden.- Ph.D. Thesis and

Rep. SFB 95 Univ. Kiel, 20, 134 pp.

REIMERS, T. & R. KÖLMEL 1976: Beiträge des Sediments zum Stoffumsatz in der Kieler Bucht. I. Salzgehaltsschwankungen im oberflächennahen Porenwasser und der Stoffaustausch zwischen Interstitial und Bodenwasser.- Kieler Meeresforsch. Sdh. 3: 96-104.

REMANE, A. 1940: Einführung in die zoologische Ökologie der Nord- und Ostsee.- In: G. GRIMPE & H. WAGLER (eds.): Die Tierwelt der Nord- und Ostsee. Ia.- Leipzig, 238 pp.

REMANE, A. & C. SCHLIEPER 1971: Biology of brackish water.- Wiley & Sons, New York, Toronto, Sydney, 372 pp.

RESIG, J.M. 1965: Lösungserscheinungen an Foraminiferen der Ostseesedimente.- Ph.D. Thesis, Univ. Kiel, 81 pp.

REVSBECH, N.P. & B.B. JÖRGENSEN 1981: Primary production of microalgae in sediments measured by oxygen microprofile, $H^{14}CO$ -fixation, and oxygen exchange methods.- Limnol. Oceanogr. 26: 717-730.

RHEINHEIMER, G. 1977: Regional and seasonal distribution of saprophytic and Coliform bacteria.- In: G. RHEINHEIMER (ed.): Microbial ecology of a brackish water environment.- Springer, Berlin: 121-137.

RHOADS, D.C. 1974: Organism-sediment relations on the muddy sea floor.- Oceanogr. Mar. Biol. Ann. Rev. 12: 263-300.

RICHTER, W. 1975: Besiedlungsexperimente zur benthischen Karbonatproduktion vor Boknis-Eck (Westliche Ostsee).- M.Sc. Thesis, Univ. Kiel, 93 pp.

RICHTER, W. & H. RUMOHR 1976: Untersuchungen an Barnea candida (L.): Ihr Beitrag zur submarinen Geschiebemergelabrasion in der Kieler Bucht.- Kieler Meeresforsch. Sdh. 3: 82-86.

RICHTER, W. & M. SARNTHEIN 1977: Molluscan colonization of different sediments on submerged platforms in the Western Baltic Sea.- In: B.F. KEEGAN, P.O. CEIDIGH & P.J.S. BOADEN (eds.): Biology of Benthic Organisms.- Pergamon Press, Oxford: 531-539.

RILEY, G.A. 1956: Oceanography of Long Island Sound, 1952-1954. IX. Production and utilization of organic matter.- Bull. Bingham Oceanogr. Coll. 15: 325-344.

ROSENFELD, A. 1976: Ostracoden der Ostsee.- Ph.D. Thesis, Univ. Kiel, 108 pp.

ROSENFELD, A. 1979: Seasonal distribution of recent ostracods from Kiel Bay, Western Baltic Sea.- Meyniana 31: 59-82.

ROSENFELD, J.K. 1981: Nitrogen diagenesis in Long Island Sound sediments.- Am. J. Sci. 281: 436-462.

ROWE, G.T., C.H. CLIFFORD, K.L. SMITH & P.L. HAMILTON 1975: Benthic nutrient regeneration and its coupling to primary productivity in coastal waters.- Nature 255: 215-217.

RUBIN, D.M. & D.S. McCULLOUGH 1980: Single and superimpopsed bedforms: a synthesis of San Francisco Bay and flume observations.- Sediment. Geol. 26: 207-231.

RUBLEE, P.A. 1982: Seasonal distribution of bacteria in salt marsh sediments in North Carolina.- Estuarine Coastal Shelf Sci. 15: 67-74.

RUBLEE, P.A. & B.E. DORNSEIF 1978: Direct counts of bacteria in the sediments of a North Carolina salt marsh.- Estuaries 1: 188-191.

RUCK, K.-W. 1970: Baugeologie der Lockergesteine im Nord- und Ostseeraum.- In: Grundbau Taschenbuch, Band I, Ergänzungsband.- W. Ernst & Sohn, Berlin: 161-217.

RUMOHR, H. 1979: Automatic camera observations on common demersal fish in the Western Baltic.- Meeresforsch. 27: 198-202.

RUMOHR, H, 1980: Der "Benthosgarten" in der Kieler Bucht. Experimente zur Bodentierökologie.- Ph.D. Thesis, Univ. Kiel and Rep. SFB 95, 55, 195 pp.

RYE, D.M. & M.A. SOMMER II. 1980: Reconstruction of paleotemperature and paleosalinity regimes with oxygen isotopes.- In: F.G. STEHLE (ed.): Topics of Geobiology. Vol. I. In: D.C. RHOADS & R.L. LUTZ (eds.): Skeletal growth of aquatic organisms.- Plenum Press, New York: 169-202.

SAMTLEBEN, C. 1973: Größenverteilung von Populationen, Totengemeinschaften und Klappengemeinschaften der Muschel Mutilus edulis L..- Meyniana 23: 69-92.

SAMTLEBEN, C. 1977: Klappenwachstum und Entwicklung von Größenverteilungen in Populationen von Mytilus edulis L..- Meyniana 29: 51-69.

SARNTHEIN, M. 1973: Quantitative Daten über benthische Karbonatsedimentation in mittleren Breiten.- Festschrift Heibel, Univ. Innsbruck 86: 267-269.

SARNTHEIN, M. & W. RICHTER 1974: Submarine experiments on benthic colonization of sediments in the Western Baltic Sea. I. Technical layout.- Mar. Biol. 28: 159-164.

SAYLES, F.L. 1979: The composition and diagenesis of interstitial solutions. I. Fluxes across the seawater-sediment interface in the Atlantic Ocean.- Geochim. Cosmochim. Acta 43: 527-545.

SCHÄFER, V. 1979: Wellenmaßanlage vor der Probstei - Meßeinrichtung.- Mitt. Leichtweiss-Inst. f. Wasserbau, Techn. Univ. Braunschweig 65: 299-320.

SCHÄFER, W. 1962: Aktuo-Paläontologie.- Frankfurt a.M., Kramer, 666 pp.

SCHAETZEL, K. 1983: Noise in photon correlation and photon structure functions.- Optica Acta 30: 155-166.

SCHAUER, U. 1982: Zur Bestimmung der Schubspannung am Meeresboden aus der mittleren Strömung.- Berichte Inst. Meereskunde, Univ. Kiel, 105, 88 pp.

SCHEIBEL, W. 1974: Submarine experiments on benthic colonization of sediments in the Western Baltic Sea. II. Meiofauna.- Mar. Biol. 28: 165-168.

SCHEIBEL, W. 1976: Qualtitative Untersuchungen am Meiobenthos eines Profils unterschiedlicher Sedimente in der Westlichen Ostsee.- Helgoländer wiss. Meeresunters. 28: 31-42.

SCHEIBEL, U. & W. NOODT 1975: Population densities and characteristics of meiobenthos in different substrates in the Kiel Bay.-

Merentutkimuslait. Julk/Havsforskningsinst. Skr. <u>239</u>: 173-178.

SCHLICHTING, H. 1965: Grenzschichttheorie.- G. Braun, Karlsruhe, 736 pp.

SCHMIDT, C. 1978: Untersuchungen zum Stoffumsatz und zur Dynamik von Abbauprozessen bei benthischen Makroalgen und Seegras der Kieler Bucht.- Ph.D. Thesis, Univ. Kiel and Rep. SFB 95, <u>41</u>, 96 pp.

SCHMIDT, D. 1981: Isotopenverhältnis des organischen Kohlenstoffs aus marinen Sedimenten.- Staatsexamensarbeit, Univ. Kiel, 125 pp.

SCHROEDER, R.A. 1975: Absence of ß-alanine and γ-aminobutyric acid in cleaned foraminiferal shells.- Earth Planet. Sci. Lett. <u>25</u>: 274-278.

SCHULZ, R. 1983: Die Wirkung von Sedimentationsereignissen auf die benthische Lebensgemeinschaft.- Ph.D. Thesis, Univ. Kiel, 116 pp.

SCHULZ, S. 1969: Das Makrobenthos der südlichen Beltsee (Mecklenburger Bucht und angrenzende Seegebiete).- Beitr. Meereskunde <u>26</u>: 21-46.

SCHULZ, T. & R. SIARA 1976: Probleme der automatischen Isoliniendarstellung.- Interocean: 1103.

SCHULZ-DUBOIS, E.O. (ed.) 1983: Photon correlation techniques in fluid mechanics.- Proc. 5th Intern. Conf., Kiel-Damp 1982, Springer, New York, 399 pp.

SCHULZ-DUBOIS, E.O. & I. REHBERG 1981: Structure function in lieu of correlation function.- Applied Physics <u>24</u>: 323.

SCHÜTZLER, A. & W. ALTHOF 1969: Nautische Grenzen der Ozeane und Meere.- Publ. 8834, DDR Seehydrogr. Dienst, Rostock: 1-75.

SCHWEIMANNS, M. 1979: Die Verteilung von Gastropoden und Bivaviern (Mollusca) im Harrington Sound, Bermuda.- M.Sc. Thesis, Univ. Kiel, 75 pp.

SCHWEIMER, M. 1976: Erosionshäufigkeit in der Westlichen Ostsee als Folge des Seegangs.- M.Sc. Thesis, Univ. Kiel and Rep. SFB 95, <u>21</u>, 59 pp.

SCHWINGHAMER, P. 1981: Characteristic size distributions of integral benthic communities.- Can. J. Fish. Aquat. Sci. <u>38</u>: 1255-1263.

SEIBOLD, E., N. EXON, M. HARTMANN, F.-C. KÖGLER, H. KRUMM, G.F. LUTZE, R.S. NEWTON & F. WERNER 1971: Marine geology of Kiel Bight.- In: G. MÜLLER (ed.): Sedimentology of parts of Central Europe.- Guidebook, 8th Internat. Sed. Congr., Heidelberg: 209-235.

SEIFERT, G. 1954: Das mikroskopische Korngefüge des Geschiebemergels als Abbild der Eisbewegung, zugleich Geschichte des Eisabbaues in Fehmarn, Ost- Wagrien und dem Dänischen Wohld.- Meyniana <u>2</u>: 124-190.

SEILACHER, A. 1953: Studien zur Palichnologie. I. Über die Methoden der Palichnologie.- N.Jb. Geol. Paläont. Abh. <u>96</u>: 421-451.

SEILACHER, A. 1955: Spuren und Fährten im Unterkambrium.- In: O.H. SCHINDEWOLF & A. SEILACHER: Beiträge zur Kenntnis des Kambrium in der Salt Range (Pakistan).- Akad. Wiss. Lit. Mainz, math.- nat. Kl., Abh. <u>10</u>: 11-143.

SEILACHER, A. 1982: General remarks about event deposits.- In: G. EINSELE & A. SEILACHER (eds.): Cyclic and event stratification.- Springer, Berlin: 161-174.

SEIP, K.L., G. LUNDE, S. MELSOM, E. MEHLUM, A. MELHUUS & H.M. SEIP 1979: A mathematical model for the distribution and abundance of benthic algae in a Norwegian fjord.- Ecological Modelling 6: 133-166.

SEKI, H., J. SKELDING & T.R. PARSONS 1968: Observations on the decomposition of a marine sediment.- Limnol. Oceanogr. 13: 440-447.

SHAW, T.I. 1962: Halogens.- In: R.A. LEWIN (ed.): Physiology and biochemistry of algae.- Academic Press, New York: 247-253.

SHIELDS, A. 1936: Anwendung der Ähnlichkeitsmechanik und der Turbulenzforschung auf die Geschiebebewegung.- Mitteilungen Preuss. Versuchsanstalt Wasserbau u. Schiffbau, Berlin 26, 42 pp. (see also translation by OTT, W.P. & J.C. van UCHELEN: U.S. Dept. Agriculture, Soil Conservation Service Coop. Lab., Calif. Inst. Techn.).

SIBERT, J.R. & R.J. NAIMANN 1980: The role or detritus and the nature of estuarine ecosystems.- In: K.R. TENORE & B.C. COULL (eds.): Marine benthic dynamics.- Univ. South Carol. Press: 311-323.

SIMONEIT, B.R.T. 1978: The organic chemistry of marine sediments.- In: J.P. RILEY & R. CHESTER (eds.): Chemical Oceanography, Vol. 7.- Academic Press, London: 233-311.

SKJOLDAL, H.R. & C. LÄNNERGREN 1978: The spring phytoplankton bloom in Landaspollene, a land-locked Norwegian fjord. II. Biomass and activity of net- and nanoplankton.- Mar. Biol. 47: 313-323.

SMETACEK, V. 1975: Die Sukzession des Phytoplanktons in der westlichen Kieler Bucht.- Ph.D. Thesis, Univ.- Kiel and Rep. SFB 95, 29, 151 pp.

SMETACEK, V. 1980a: Zooplankton standing stock, copepod faecal pellets and particulate detritus in Kiel Bight.- Estuarine and coastal Mar. Sci. 2: 477-490.

SMETACEK, V. 1980b: Annual cycle of sedimentation in relation to plankton ecology in Western Kiel Bight.- Ophelia Suppl. 1: 65-76.

SMETACEK, V. 1981: The annual cycle of protozooplankton in Kiel Bight.- Mar. Biol. 63: 1-11.

SMETACEK, V. 1984: The supply of food to the benthos.- In: M.J. FASHAM (ed.): Flows of energy and materials in marine ecosystems: Theory and practice.- Plenum Press, New York: 517-548.

SMETACEK, V. 1985a: The annual cycle of Kiel Bight plankton: A long-term analysis.- Estuaries 8: 145-157.

SMETACEK, V. 1985b: Role of sinking in diatom life-history cycles: ecological, evolutionary and geological significance.- Mar. Biol. 84: 239-251.

SMETACEK, V., B. von BODUNGEN, K. von BRÖCKEL & B. ZEITZSCHEL, 1976: The plankton-tower. II. Release of nutrients from sediments due to changes in the density of bottom water.- Mar. Biol. 34: 373-378.

SMETACEK, V., B. von BODUNGEN, B. KNOOPERS, R. PEINERT, F. POLLEHNE, P. STEGMAN & B. ZEITZSCHEL 1982: Seasonal stages characterizing the annual cycle of an inshore pelagic system.-Rapp. Proc. verb. ICES 183: 126-135.

SMETACEK, V., B. von BODUNGEN, B. KNOOPERS, F. POLLEHNE & B. ZEITZSCHEL

1982: The plankton tower. IV. Interactions between water column and sediment in enclosure experiments in Kiel Bight.- In: G.D. GRICE & M.R. REEVE (eds.): Marine Mesocosms: Biological and chemical research in experimental ecosystems.- Springer, New York: 205-210.

SMETACEK, V., K. von BRÖCKEL, B. ZEITZSCHEL & W. ZENK 1978: Sedimentation of particulate matter during a phytoplankton spring bloom in relation to the hydrographical regime.- mar. Biol. 47: 211-226.

SMETACEK, V. & P. HENDRIKSON 1979: Composition of particulate organic matter in Kiel Bight in relation to phytoplankton succession.- Oceanol. Acta 2: 287-298.

SMITH, J.D. 1977: Modelling of sediment transport on continental shelves.- In: E.D. GOLDBERG (ed.): The Sea, Vol. 6.- Wiley-Interscience, New York: 539-577.

SMITH, J.D. & S.R. McCLEAN 1977: Spatially averaged flow over a wavy surface.- J. Geophys. Res. 82(12): 1735-1746.

SMITH, S.V. 1978: Alkalinity depletion to estimate the calcification of coral reefs in flowing waters.- In: D.R. STODDART & R.E. JOHANNES (eds.): Coral Reefs: Research Methods: 397-404.

SOULSBY, R.L. 1981: Measurements of the Reynolds stress components close to a marine sand bank.- Mar. Geol. 42: 35-47.

SOULSBY, R.L. & K.R. DYER 1981: The form of the near-bed velocity profile in a tidally accelerating flow.- J. Geophys. Res. 86, C9: 8067-8074.

STEARN, C.W., T.P. SCOFFIN & W. MARTINDALE 1977: Calcium carbonate budget of a fringing reef on the west coast of Barbados. Part 1. Zonation and productivity.- Bull. Mar. Sci. 27: 479-510.

STEELE, J.H. 1974: The structure of marine ecosystems.- Harvard Univ. Press, 128 pp.

STEGMANN, P. 1981: Beziehungen zwischen Phytoplankton und Zooplankton in der Kieler Bucht.- M.Sc. Thesis, Univ. Kiel, 87 pp.

STEGMANN, P. & R. PEINERT 1984: Interrelationships between herbivorous zooplankton and phytoplankton and their effect on production and sedimentation of organic matter in Kiel Bight.- Limnologica (Berlin) 15: 487-495.

STEPHAN, H.-J. 1971: Glazialgeologische Untersuchungen im Raum Heiligenhafen (Ostholstein).- Meyniana 21: 67-86.

STERNBERG, R.W. 1968: Friction factors in tidal channels with differing bed roughness.- Mar. Geol. 6: 243-260.

STOFFERS, H. 1976: Untersuchungen zum Isotopenverhältnis des Kohlenstoffs bei Rotalgen aus der Kieler Bucht.-Staatexamsarbeit, Univ. Kiel, 81 pp.

STUMM, W. & J.J. MORGAN 1970: Aquatic chemistry.- Wiley-Interscience, New York, 583 pp.

SUESS, E. 1976a: Nutrients near the depositional interface.- In: I.N. McCAVE (ed.): The benthic boundary layer.- Plenum Press, New York: 57-80.

SUESS, E. 1976b: Porenlösungen mariner Sedimente - ihre chemische Zusammensetzung als Ausdruck frühdiagenetischer Vorgänge.-

Habilitationsschrift, Univ. Kiel, 193 pp.

SUESS, E. 1978: Distinction between natural and anthropogenic materials in sediments.- Proc. UNESCO/SCOR Workshop, Melreux, Belgium, Paris: 224-237.

SUESS, E. 1979: Mineral phases formed in anoxic sediments by microbial decomposition of organic matter.- Geochim. Cosmochim. Acta 43: 339-352.

SUESS, E. 1980: Particulate organic carbon flux in the oceans surface productivity and oxygen utilization.- Nature 288: 260-263.

SUESS, E. & D. DJAFARI 1977: Trace metal distribution in Baltic Sea ferro-manganese concretions: Inferences on accretion rates.- Earth Planet. Sci. Lett. 35: 49-54.

SUESS, E. & H. ERLENKEUSER 1975: History of metal pollution and carbon input in Baltic Sea sediments.- Meyniana 27: 63-75.

SUESS, E. & P.J. MÜLLER 1980: Productivity, sedimentation rate and sedimentary organic matter in the oceans. II. Elemental fractionation.- Colloques Internationaux du C.N.R.S. No 293. Biogéochimie de la matière organique à l'interface eau - sédiment marin.

SUNDBY, B., N. SILVERBERG & R. CHESSELET 1981: Pathways of manganese in an open estuarine system.- Geochim. cosmochim. Acta 45: 293-307.

SVANSSON, A. 1980: Exchange of water and salt in the Baltic and adjacent sea.- Oceanol. Acta 3: 431-440.

SWEENEY, R.E. & I.R. KAPLAN 1980: Natural abundances of ^{15}N as a source indicator for near-shore marine sedimentary and dissolved nitrogen.- Mar. Chem. 9: 81-94.

SZWERINSKI, H. 1981: Investigations on nitrification in the water and the sediment of the Kiel Bight (Baltic Sea).- Kieler Meeresforsch. Sdh. 5: 396-407.

SWIFT, D.J.P., J.W. KOFOED, F.P. SAULSBURY & P. SEARS 1973: Holocene evolution of the shelf surface, central and southern Atlantic shelf of North America.- In: D.J.P. SWIFT, D.B. DUANE & O.H. PILKEY (eds.): Shelf sediment transport: process and pattern.- Dowden, Hutchinson and Ross, Stroudsberg, Pa.: 143-180.

TABAT, W. 1979: Sedimentologische Verteilungsmuster in der Nordsee.- Meyniana 31: 83-124.

TANOUE, E. & N. HANDA 1979: Differential sorption of organic matter by various sized sediment particles in recent sediment from the Bering Sea.- J. Oceanogr. soc. Jap. 35: 199-208.

TARUTANI, T., R.N. CLAYTON & T.K. MAYEDA 1969: The effect of palymorphism and magnesium substitution on oxygen isotope fractionation between calcium carbonate and water.- Geochim. Cosmochim. Acta 33: 987-996.

TAYLOR, P.A. & K.R. DYER 1977: Theoretical models of flow near the bed and their implication for sediment transport. In: E.D. GOLDBERG (ed.): The Sea, Vol. 6.- Wiley-Interscience, New York: 579-601.

TESSENOW, U. 1972: Lösungs-, Diffusions- und Sorptionsprozesse in der Oberschicht von Seesedimenten. I. Ein Langzeitexperiment unter aeroben und anaeroben Bedingungen im Fließgleichgewicht.- Arch. Hydrobiol.

Suppl. <u>38</u>: 353-398.

THEEDE, H. 1981: Studies on the role of benthic animals of the Western Baltic in the flow of energy and organic material.- Kieler Meeresforsch., Sdh. <u>5</u>: 434-444.

THUROW, F. 1970: Über die Fortpflanzung des Dorsches (<u>Gadus morhua</u> L.) in der Kieler Bucht.- Ber. dt. wiss. Komm. Meeresforsch. <u>21</u>: 170-192.

TISSOT, B.P. & B.H. WELTE 1978: Petroleum formation and occurrence.- Springer, Berlin: 538 pp.

TRASK, P.D. 1932: Origin and environment of source sediments of petroleum.- Gulf Publ. Co., Houston, Texas, 1323 pp.

TUREKIAN, K.K. 1977: The fate of metals in the oceans.- Geochim. Cosmochim. Acta <u>41</u>: 1139-1144.

UNSÖLD, G. 1982: Der Transportbeginn rolligen Sohlmaterials in gleichförmigen turbulenten Strömungen: Eine kritische Überprüfung der Shields-Funktion und ihre experimentelle Erweiterung auf feinstkörnige, nichtbindige Sedimente.- Ph.D. Thesis, Univ. Kiel and Rep. SFB 95, <u>70</u>, 141 pp. (1984).

VINCENT, C.E. & J.G. HARVEY 1976: Roughness length in the turbulent Ekman layer above the sea bed.- Mar. Geol. <u>22</u>: M75-M81.

VINOGRADOV, A.P. 1953: The elementary chemical composition of marine organisms.- Memoirs Sears Foundation Mar. Res. II, New Haven, 647 pp.

WALSH, J.J. 1981: Shelf-sea ecosystems.- In: A.R. LONGHURST (ed.): Analysis of marine ecosystems.- Acad. Press: 159-196.

WARWICK, R.M. 1980: Population dynamics and secondary production of benthos.- In: K.R. TENORE & B.C. COULL (eds.): Marine benthic dynamics.- Univ. South Carolina Press: 1-24.

WASSMANN, P. 1983: Sedimentation of organic and inorganic particulate material in Lindaspollene, a stratified, land-locked fjord in western Norway.- Mar. Ecol. Progr. Ser. <u>13</u>: 237-248.

WATTENBERG, H. 1949: Entwurf einer natürlichen Einteilung der Ostsee.- Kieler Meeresforsch. <u>6</u>: 10-15.

WEATHERLY, G.L. 1972: A study of the bottom boundary layer of the Florida Current.- J. Phys. Oceanogr. <u>2</u>: 54-72.

WEFER, G. 1976a: Umwelt, Produktion und Sedimentation benthischer Foraminiferen in der Westlichen Ostsee.- Ph.D. Thesis, Univ. Kiel and Rep. SFB 95, <u>14</u>, 103 pp.

WEFER, G. 1976b: Environmental effects on growth rates of benthic Foraminifera (shallow water, Baltic Sea).- Maritime Sediments, Spec. Publ. <u>1</u>: 39-50.

WEFER, G. 1979: Der Karbonat-Kreislauf in einer subtopischen Lagune.- UMSCHAU in Wissensch. u. Techn. <u>22</u>: 699-705.

WEFER, G. 1980: Carbonate production by algae Halimeda, Penicillus und Padina.- Nature <u>285</u>: 323-324.

WEFER, G. 1985: Die Verteilung stabiler Isotope in Kalkschalern mariner Organismen.- Geologisches Jahrbuch, Reihe A, <u>81</u>: (in press).

WEFER, G. & W.H. BERGER 1980: Stable isotopes in benthic Foraminifera: Seasonal variation in large tropical species.- Science 209: 803-805.

WEFER, G. & W.H. BERGER 1981: Stable isotope composition of benthic calcareous algae.- J. Sed. Petrol. 51: 459-465.

WEFER, G., R. DAWSON & G. HEMPEL 1981b: The Harrington Sound Project: Kiel University.- Bermuda Biological Station, Spec. Publ. 19, 94 pp.

WEFER, G., B. FLEMMING & TAUCHGRUPPE KIEL 1976: Submarine Abrasion des Geschiebemergels vor Boknis Eck (Westl. Ostsee).- Meyniana 28: 87-94.

WEFER, G. & J.S. KILLINGLEY 1980: Growth histories of strombid snails from Bermuda recorded in their O-18 and C-13 profiles.- Mar. Biol. 60: 129-135.

WEFER, G., J.S. KILLINGLEY & G.F. LUTZE 1981a: Stable isotopes in recent larger foraminifera.- Palaeogeogr. Palaeoclimat. Palaeoecol. 33: 253-270.

WEFER, G. & G.F. LUTZE 1976: Benthic Foraminifera biomass production in the Western Baltic.- Kieler Meeresforsch., Sdh. 3: 76-81.

WEFER, G. & G.F. LUTZE 1978: Carbonate production by benthic Foraminifera and accumulation in the Western Baltic.- Limnol. Oceanogr. 23(5): 992-996.

WEFER, G. & W. RICHTER 1976: Colonization of artificial substrates by Foraminifera.- Kieler Meeresforsch. Sdh. 3: 72-75.

WEFER, G., E. SUESS, W. BALZER, G. LIEBEZEIT, P.J. MÜLLER, C.A. UNGERER & W. ZENK 1982: Fluxes of biogenic components from sediment trap deployment in cirumpolar waters of the Drake Passage.- Nature, 299: 145-147.

WEFER, G. & TAUCHGRUPPE KIEL 1974: Topographie und Sedimente im "Hausgarten" des Sonderforschungsbereichs 95 der Universität Kiel (Eckernförder Bucht, Westliche Ostsee).- Meyniana 26: 3-7.

WEFER, G., M. WEBER & H. ERLENKEUSER 1978: Sandablagerungen während der postglazialen Transgression in der Eckerförder Bucht (Westliche Ostsee).- Senckenbergiana marit. 10: 39-61.

WEIDEMANN, H. 1950: Untersuchungen über unperiodische und periodische hydrographische Vorgänge in der Beltsee.- Kieler Meeresforsch. 7: 70-86.

WEIGELT, M. 1985: Untersuchungen zur Situation des Benthos nach einer ausgedehnten Periode vollständigen Sauerstoffschwunds im Bodenwasser der Kieler Bucht.- Berichte Inst. Meereskunde, Univ. Kiel, 138, 122 pp.

WEISE, W. & G. RHEINHEIMER 1978: Scanning electron microscopy and epifluorescence investigation of bacterial colonization of marine sand sediments.- Microbial Ecol. 4: 175-188.

WEISE, W. & G. RHEINHEIMER 1979: Fluoreszenzmikroskopische Untersuchungen über die Bakterienbesiedlung mariner Sandsedimente.- Botanica Marina 22: 99-106.

WERNER, F. 1967: Sedimentation und Abrasion am Mittelgrund (Eckernförder Bucht, Westliche Ostsee).- Meyniana 17: 101-110.

WERNER, F. 1968: Gefügeanalysis feingeschichteter Schlicksedimente der Eckernförder Bucht (Westliche Ostsee).- Meyniana 18: 79-105.

WERNER, F. 1979: Die Sedimentverteilung außerhalb der Riffzone vor der Probstei aufgrund von Sidescan-Sonar-Aufnahmen.- Mitt. Leichtweiß Inst. Techn. Univ. Braunschweig 65: 139-163.

WERNER, F., J. ALTENKIRCH, R.S. NEWTON & E. SEIBOLD 1976: Sediment patterns and their temporal variation on abrasion ridges in a moderate flow regime (Stoller Grund, Baltic Sea).- Meyniana 28: 95-105.

WERNER, F., W.E. ARNTZ & TAUCHGRUPPE KIEL 1974: Sedimentologie und Ökologie eines ruhenden Riesenrippelfeldes.- Meyniana 26: 39-62.

WERNER, F. & R.S. NEWTON 1975: The pattern of large-scale bed forms in the Langeland Belt (Baltic Sea).- Mar. Geol. 19: 25-59.

WERNER, F., G. UNSÖLD, B. KOOPMAN & A. STEFANON 1980: Field observations and flume experiments on the nature of comet marks.- Sedimentary Geology 26: 233-262.

WERNER, F. & A. WETZEL 1982: Interpretation of biogenic structures in oceanic sediments.- Bull. Inst. Geol. Bassin d'Aquitaine, Bordeaux 31: 275-288.

WETZEL, A. 1981: Ökologische und stratigraphische Bedeutung biogener Gefüge in quartären Sedimenten am NW-afrikanischen Kontinentalrand.- "Meteor" Forsch. Ergebn. C 34: 1-47.

WHEELER, B. 1966: Phototactic vertical migration in Exuviella baltica. Bot. Mar. 9: 15-17.

WHELAN, J.K. 1977: Amino acids in a surface sediment core of the Atlantic abyssal plain.- Geochim. Cosmochim. Acta 41: 803-810.

WHITE, S.J. 1970: Plane bed thresholds of fine grained sediments.- Nature 228: 152-153.

WHITICAR, M.J. 1978: Relationships of interstitial gases and fluids during early diagenesis in some marine sediments.- Ph D. thesis, Univ. Kiel and Rep. SFB 95, 35, 152 pp.

WHITICAR, M.J. 1982a: Determination of interstitial gases and fluids in sediment collected with an "in situ" sampler.- Anal. Chem. 54: 1796-1798.

WHITICAR, M.J. 1982b: The presence of methane bubbles in the acoustically turbid sediments of Eckernförder Bay, Baltic Sea.- In: K.A. FANNING & F.T. MANHEIM (eds.): The Dynamic Environment of the Ocean Floor.- Lexington Book, Mass.: 219-235.

WHITICAR, M.J. & F. WERNER 1982: Pockmarks: submarine vents of natural gas or freshwater seeps?- Geol. Mar. Letters 1: 193-199.

WILLIAMS, P.J.LeB. 1981: Incorporation of heterotrophic processes into the classical paradigm of the planktonic food web.- Kieler Meeresforsch., Sdh. 5: 1-28. WILLIAMS, P.J.LeB. 1981: Microbial contribution to overall marine plankton metabolism: direct measurements of respiration.- Oceanol. Acta 4: 359-364.

WINN, K. 1974: Present and postglacial sedimentation in the Great Belt channel (Western Baltic).- Meyniana 26: 63-101.

WINN, K., F. AVERDIECK & F. WERNER 1982: Spät- und postglaziale Entwicklung des Vejsnäs-Gebietes (Westliche Ostsee).- Meyniana 34: 1-29.

WINTERHALTER, K.B. 1966: Iron-manganese concretions from the Gulf of Finland.- Geotekn. Julk. 69: 1-77.

WITTSTOCK, R. 1982: Zu den Ursachen bodennaher Strömungen in der nordöstlichen Kieler Bucht.- Berichte Inst. Meereskunde, Univ. Kiel 107, 105 pp.

WITTSTOCK, R., U. SCHAUER & G. SCHALLER 1978: Strömungen in der Vejsnaes-rinne 1977/78.- Rep. SFB 95, Univ. Kiel 45, 87 pp.

WOLTER, K. 1980: Untersuchungen zur Exsudation organischer Substanz und deren Aufnahme durch natürliche Bakterienpopulationen.- Ph.D. Thesis, Univ. Kiel and Rep. SFB 95, 54, 127 pp.

WOLTER, K. 1982 Bacterial incorporation of organic substances released by natural phytoplankton populations.- Mar. Ecol. Prog. Ser. 7: 287-295.

WORTHMANN, H. 1975: Die Makrobenthos- und Fischbesiedlung in verschiedenen Flachwassergebieten der Kieler Bucht (Westl. Ostsee).- M.Sc. Thesis, Univ. Kiel, 141 pp.

WORTHMANN, H. 1976: Die Molluskenfauna verschiedener Flachwassergebiete der Kieler Bucht, Artenzusammensetzung und Produktivität.- Kieler Meeresforsch. Sdh. 3: 25-36.

WYRTKI, K. 1953: Die Dynamik der Wasserbewegungen im Fehmarnbelt (1).- Kieler Meeresforsch. 9: 155-170.

YALIN, M.S. 1963: An expression for bed-load transportation.- Proc. ASCE, J.Hydr. Div. 89, HY3:

YALIN, M.S. 1972: Mechanics of sediment transport.- Pergamon Press, New York, 290 pp.

YINGST, J.Y. & D.C. RHOADS 1980: The role of bioturbation in the enhancement of bacterial growth rates in marine sediments.- In: K.R. TENORE & B.C. COULL (eds.): Marine benthic dynamics.- Univ. South Carolina Press, Columbia: 407-421.

YURKOVSY, A.K. 1971: Results of fraction investigation of the organic substance in the Baltic Sea.- Proc. Joint. Oceangr. Assembly (Tokyo 1970): 466-467.

ZEITZSCHEL, B. 1965: Zur Sedimentation von Seston, eine produktionsbiologische Untersuchung von Sinkstoffen und Sedimenten der Westlichen und Mittleren Ostsee.- Kieler Meeresforsch. 21: 55-80.

ZEITZSCHEL, B. 1980: Sediment-water interactions in nutrient dynamics.- In: K.R. TENORE & B.C. COULL (eds.): Marine benthic dynamics.- Univ. South Carolina Press, Columbia: 195-218.

ZEITZSCHEL, B., P. DIEKMANN & L. UHLMANN 1978: A new multisample sediment trap.- Mar. Biol. 45: 285-288.

ZIMMERMANN, R. 1977: Estimation of bacterial and biomass by epifluorescence microscopy and scaning electron microscopy.- In: G. RHEINHEIMER (ed.): Microbial ecology of a brackish water environment.- Springer, Berlin: 103-120.

INDEX